The Nile Development Game

Mina Michel Samaan

The Nile Development Game

Tug-of-War or Benefits for All?

Carl-Friedrich-Gauß-Faculty
Technical University of Braunschweig
(der Technischen Universität Carolo-Wilhelmina
zu Braunschweig)

A dissertation submitted in partial fulfillment
of the requirements for the degree of
Doctor of Political Science (Dr. rer. pol.)
(zur Erlangung des Grades eines **Doktors der Staatswissenschaften (Dr. rer. pol.)**
vorgelegte Dissertation)

Submission date: 15.02.2017
Disputation date: **23.08.2017**
Referee 1: Prof. Dr. Ulrich Menzel
Referee 2: Prof. Dr. Anja P. Jakobi

Springer

Mina Michel Samaan
Braunschweig University of Technology
Braunschweig, Niedersachsen, Germany

ISBN 978-3-030-02664-6 ISBN 978-3-030-02665-3 (eBook)
https://doi.org/10.1007/978-3-030-02665-3

Library of Congress Control Number: 2018961430

This Springer imprint is published by the registered company Springer Nature Switzerland AG
The registered company address is: Gewerbestrasse 11, 6330 Cham, Switzerland

Endorsements

In this book, Dr. Mina Samaan examines the development of the Nile cooperation over several decades, presenting the interaction and interdependency of cross-border water supply and potential lines for conflict. Water, despite being central to all countries, is not equally shared and the importance of this book is the detailed analysis on how riparian states shift their share, responsibilities, and burdens over time. Readers of this book will find a dense description of current and past water policies along the Nile, presented by Dr. Mina Samaan as an author with a regional and interdisciplinary expertise.

– Prof. Dr. Anja P. Jakobi, Professor of International Relations, TU-Braunschweig

Everyone immediately understands that cooperation is essential, especially in areas vulnerable to the impacts of climate change and where water is already scarce. Political and armed conflicts are definitely no solutions. Moreover, they further exacerbate the problem in unnecessary ways. We all know that international policies and relations are not easy going matters. What really counts at the end of the day is the willingness and serious, concrete intention to find and implement effective solutions. Here, peaceful cooperation is essential, otherwise there will be no winners, only losers.

Talking about losers would also means talking about ecosystems. Uncontrolled overexploitation of transboundary lakes, rivers, and aquifers can jeopardize ecosystem services and have dire consequences for the reliability and sustainability of water supplies, which could additionally contribute to international tension, if those impacts are felt more keenly on the other side of a border.

The Excellence Center on Sustainable Water Management in Developing Countries (EXCEED-SWINDON) at the Technical University of Braunschweig has been working since 2009 on various fields related to water. The EXCEED-SWINDON consists of a network of 30 full member institutions worldwide, each of which is represented by a renowned scientist whose research focuses on water problems.

This book is the research product of the EXCEED-SWINDON PhD three-year scholarship, funded by the German Academic Exchange Service (DAAD), which was awarded to Mina Michel Samaan based on the decision of an international committee of water experts from ten universities out of five different countries. Being the scientific coordinator of the EXCEED-SWINDON, I got to know Mina during his doctoral study and gained an impressive insight into his target-oriented, structured, and thorough research work.

The book mainly highlights the consequences of missing basin-wide cooperation and the formation of partial coalitions, one against another, while taking the example of the "River Nile Problem" into account. The reader gets an in-depth understanding of complex interactions between the Nile riparian countries in planning and implementing their large-scale developmental schemes.

However, one should bear in mind that it is never too late for making a real concerted effort to identify and to implement trustful common grounds. While this might already be a small step forward for all parties involved, it would definitely mean a big step forward for those relying on water every day.

– Prof. Dr. Andreas Haarstrick, Professor of Bioprocess Engineering,
TU-Braunschweig

Foreword

One important question in the context of climate change is how to deal with the 276 transboundary river basins worldwide. These basins cover nearly half of the earth with around one-third of its population. Obviously, there are examples of transboundary river basins where problems of cooperation among riparian states have been solved, such as the Rhine and Danube. However, there are other constellations where intense and almost irresolvable conflicts have arisen, such as the Jordan and Aral lake basins. This type of constellation is particularly the case in arid or semi-arid regions as in the Near and Middle East, where there is limited rainfall, while artificial irrigation from surface water is the basis for life. The popular slogan of the "coming water wars" emerged in the context of transboundary river basins shared among numerous riparian states using the waters and the fertile sediments transported by the river for different purposes such as drinking, irrigation, fishing, hydropower generation, cooling of power plants, navigation, and sewage disposal. In addition, a difficult problem to resolve in theory as well as in practice is saving losses due to evapotranspiration and underground drainage. Only hydraulic engineering is thought to solve these problems and meet those intricate needs. Nevertheless, it requires advanced technology and special expertise to harness nature and provide sufficient water, energy, and food supplies with minimum social and environmental costs.

The Nile, the largest river in the world, not in terms of runoffs as the Amazon, but in terms of length, is a prominent example for the constellation described, because 11 countries of completely different climate zones belong to its basin. While the upstream riparians on the Blue Nile (e.g., Ethiopia) or the White Nile (e.g., Uganda) have enough water, because they rely on the tropics on rainfed not irrigated agriculture and they have several lakes and rivers besides the Nile, the downstream riparians Egypt and Sudan are almost completely dependent on the Nile. Every intervention at the upper reaches of the Nile (e.g., building hydropower dams and water regulators), any water withdrawal for irrigation or other purposes, or even the disposal of wastewater affects both the quantity (permanently or seasonally) and quality of water in the downstream countries. Moreover, the speed by which a reservoir is filled after a dam is completed entails serious consequences with respect to the amount of runoffs. The water situation for Egypt is further fragile in an extreme way, because unlike other rivers, the Nile by reaching the Sahara is no longer fed by rains or tributaries, whereas it loses water by evaporation in an enormous amount, which is clear in the Nasser Lake behind the Aswan High Dam crossing the border to Sudan.

The weak position of Egypt in geopolitical terms is compensated by the fact that the country is the most powerful in the basin, possessing an air force capable of destroying a dam before its completion by an air raid; at least Egypt can threaten to do so. Further, it can raise historical claims about the water, which is a relevant aspect in international water law, and has a long-established tradition and extensive expertise in water management since ancient times, while the upstream riparians did not have the same hydraulic civilization nor used the Nile water for such a long time as Egypt did.

The author of the study was at the beginning of his research confronted with the following situation: Egypt and Sudan, more or less completely dependent on the Nile water in many respects, are no more able to feed their growing populations by the existing cultivated land in the valley and delta of the Nile. All projects to reclaim new land in the desert by artificial irrigation need additional water, which has to come not only from nonrenewable aquifers but also from the Nile. However, draining off more water is threatened by projects of the upstream countries, in constructing hydropower dams and irrigation schemes to produce sufficient energy and food for domestic as well as export purposes. So a tug-of-war situation comes into existence, in which Egypt is not only lacking the additional water for new irrigation projects but having permanently or at least for few years less water available until the reservoirs behind the new dams are filled. The alternative would be to look for a benefits-for-all approach, whereby the downstream riparians participate in energy and food production with upstream riparians, taking into consideration international division of labor and other forms of compensation for the loss of water available for them.

Actually, Egypt and Ethiopia are the main rivals in the basin because more than 80% of the water arriving in Egypt is delivered by the Blue Nile and other tributaries originating in the Ethiopian Highlands. Since 2011, Ethiopia has launched the construction of the Grand Ethiopian Renaissance Dam (GERD), a huge national project that aims to produce water power necessary for its industrialization. The dam's supplementary schemes are implemented with international support, especially by the World Bank and China. For instance, the Chinese investments in Ethiopia involve high-voltage transmission lines to the national grid and the neighboring countries, in addition to industrial parks and a railway connecting Addis Ababa with Djibouti. This may be understandable given the fact that China is the top export destination and import origin for Ethiopia and that the top exports are agricultural products. Thus, insofar as Ethiopia is backed internationally, it can withstand Egypt's political or even military pressure in case of substantial reduction of the Nile runoff during the filling period of the reservoir behind the GERD after its completion in the near future. Egypt's situation has worsened after Sudan has shifted its position, despite their historical coalition as downstream countries, supporting the GERD for its enormous benefits in terms of inexpensive electricity and water regulation, urgently needed to cope with the loss of its former oilfields after the independence of South Sudan. The attempts to resolve water-related conflicts cooperatively through the Nile Basin Initiative (NBI), operating since 1999, have failed, notably after the individual signature of the Cooperative Framework Agreement (CFA) by most of the upstream countries in 2010 and the consequent decision of Egypt and Sudan to suspend their participation in the NBI activities. The failure of multilateralism has led the Nile riparian states to undertake unilateral actions, searching for external partners to provide the required financial and technical assistance, such as the World Bank, China, and the Gulf States.

The problem sketched above has constituted the research question of this study about the variables influencing the strategies adopted by the riparian countries and the resulted outcomes: sole utilization of the water by one riparian or a group of riparians, cooperation among different riparians based on common interests, and contention due to unilateral actions regardless of adverse impacts on the others. By analyzing the four rounds of the "Nile Game" (colonial round, 1880s–1940s; Cold War round, 1950s–1980s; post-Cold War round, 1990s–2000s; and post-2011 round), the author showed that the game context, not only at the domestic and regional levels but even at the global one, is the major driver for the way in which the riparian countries act upon their shared water resources. Thus, the study arrived at the conclusion that in each round of the game equilibrium came into existence, which was determined by the changing political and environmental contexts. Examples of this are the rivalry between the colonial powers Great Britain and France in controlling the whole basin, the East–West Conflict under which Egypt and Ethiopia were supported by the Soviet Union and the USA alternatively, the decolonization and establishment of independent states in the basin, the Near East Wars including the Arab-Israeli and the Ethiopian-Eritrean conflicts, the end of the Cold War and the emergence of new actors in financing projects, and currently the engagement of China in Africa in light of its Silk Road Initiative. Equilibrium also has been significantly influenced by every revolution, military takeover, and period of decentralization, in addition to extreme events of floods and droughts in every round. These changing contextual variables have been responsible for the current status of equilibrium in which the riparian states are not following the multilateral or neo-institutionalist approach favored by the author (benefits for all), but adopting the unilateral or neo-realist approach (tug-of-war). Such findings contribute to rethinking the role of institutions at the different levels in creating adequate context for the Nile riparian states to shift from unilateralism to multilateralism for the sake of present and future generations of the basin.

"Prof. Dr. Ulrich Menzel" who is the main supervisor of the study and the former acting director of the Institute of Social Sciences at TU-Braunschweig, Germany

Acknowledgment

At the outset, I would like to express my grateful thanks to my main supervisor and to the project that funded this doctoral study. In March 2012, I participated in a workshop at Mansoura University, Egypt, organized by the Excellence Center for Sustainable Water Management in Developing Countries (EXCEED), which is sponsored by the German Academic Exchange Service (DAAD) and headquartered at TU-Braunschweig, Germany. Following my presentation, which focused on national development plans and challenges of water scarcity in Egypt, Prof. Dr. Ulrich Menzel, the then acting director of the Institute of Social Sciences (ISW) at TU-Braunschweig, commented that the problem has a further dimension and that it originates from outside, not inside, Egypt, inspiring me to study the development plans in the entire Nile Basin. I applied thereafter for the EXCEED PhD scholarship funded by the DAAD. A committee of international experts selected my proposal and Prof. Dr. Menzel accepted to supervise my doctoral thesis. In German, the main supervisor is called "Doktorvater," literally "the father of the doctorate," which was what I experienced with Prof. Dr. Menzel in every sense of the word. Since the first day of my scholarship in November 2013, he has made an enormous effort and has given me much of his valuable time to put me on the right track and push me forward along the way. In addition, the support of EXCEED to my work exceeded the limits of the scholarship. Professor Dr. Andreas Haarstrick, the project's scientific coordinator, has provided me with his complete support in all my steps before, during, and after the scholarship period. The EXCEED board, namely Prof. Dr. Mufit Bahadir, Prof. Dr. Norbert Dichtl, Prof. Dr. Ulrich Menzel, and Prof. Dr. Andreas Haarstrick, decided to purchase a specialized screen reader software for visually impaired persons so as to enable me to do my duty efficiently, which was initiated by Mr. Mathias Marx, the then ISW librarian. In this respect, I cannot forget the continuous support of Prof. Dr. Zeinab Aboul-Naga and Prof. Dr. Mufeed Batarseh, who invited me to the Mansoura University workshop, introduced me to the program, and ensured the success of the research.

I would like also to express my deep appreciation to my second supervisor, Prof. Dr. Anja Jakobi, the head of the International Relations Chair at TU-Braunschweig, for her massive support and constructive insights. In addition to the supervisors and the EXCEED-SWINDON scientific coordinator, I gratefully thank the other members of the dissertation committee, Prof. Dr. Nils Bandelow and Prof. Dr. Christian Lessmann. Moreover, I would like to acknowledge all of the chair staff for their useful discussions that helped me to develop the theoretical and methodological foundations of the research, particularly Dr. Bastian Loges, Dr. Michael Fuerstenberg, Dr. Steffen Stuebig, Mr. Soeren Koepke, Dr. Elitsa Katsarova, Ms. Janina Kant, Ms. Katharina Mann, and Ms. Jasmin Haunschild. I do appreciate the help of both Ms. Janina Kant in translating the abstract of the dissertation into German and Prof. Dr. Menzel in editing it. I am also very thankful to Ms. Betina Kolodziej and Ms. Doris Hellmann, the secretaries of the ISW and the EXCEED project, respectively, for facilitating all administrative steps, in addition to Ms. Kathrin Reiff and Ms. Cara Transfeld at the International Office, Ms. Imma Braun and Ms. Cornelia Pioch at the Dean's office, and Ms. Nicole Gaschler and Mr. Axel Zschippang at the University Library. Notably, my research stay in Braunschweig could not have been successful without the special care provided to me by my friends Mr. Munis Kairlas, Mrs. Eman Ghali, Dr. Hany Ghaly, and Dr. Mariam Gaid.

Following the dear Braunschweigers, the first person I would like to give special thanks to is Dr. Ramy Lotfy Hanna at the Institute of Development Studies (IDS), University of Sussex, UK, whose doctoral thesis focuses on transnational land-water investments in the Nile Basin. Ramy has been a dear friend, supportive colleague, and a true brother. His continuous encouragement, helpful discussions, and careful proofreading contributed significantly to the book.

Furthermore, I would like to thank the Stockholm International Water Institute for sponsoring my participation at the 2016 World Water Week. There, I was able to interview top water officials from Egypt, Ethiopia, and Sudan, the then executive director of the Nile Basin Initiative (NBI), and a former president of the Nile Basin Discourse (NBD), in addition to the international experts, Prof. Dr. Owen McIntyre, Prof. Dr. Ashok Swain, Dr. Andres Jaegerskog, and Dr. Ana Eliza Cascao. I am so grateful indeed to all those interviewees for their valuable time and informative talks, which substantially helped me in carrying out in-depth analysis for the ongoing events in the Nile Basin. I am also so thankful to all the governmental and nongovernmental organizations that gave me permission to include their maps in the book, namely the US Bureau of Reclamation (Emily Quinn), the Nile Basin Initiative (Milly Mbuliro), and the FAO Aquastat Database (Virginie Gillet).

I would like also to give big thanks to the editorial staff of Springer Nature, NY (Lorraine Klimowich and Kelly Daugherty), in addition to the discussants who reviewed my different papers and provided me with their substantial comments, namely Prof. Dr. Steven T. Seitz and Prof. Dr. James Stoutenborough (MPSA 2016), Prof. Dr. Ken Conca (ISA 2017), and Prof. Dr. Michael Mosser (ISA 2018). I thank also Prof. Dr. Dirk Siepmann for his useful tips on "publishing in English" at TU-Braunschweig, particularly his recommendation to use ozdic.com (a free collocation dictionary).

I should never forget to thank the academic staff at Mansoura University, my home institution where I received my BSc and MSc and worked as a lecturer assistant for 5 years, namely Prof. Dr. Salah El-Sayed, Prof. Dr. Ahmed Rashed, Prof. Dr. Mohamed Esmat El-Attar, Prof. Dr. Osama Farag, Prof. Dr. Lamis El-Gezawy, Prof. Dr. Sherief Sheta, Eng. Ahmed Nabih, and Dr. Moamen Sudani.

Finally, it gives me the most pleasure conveying my heartfelt thanks to my parents who have been overwhelming with their care and have been the light of my way. In this study, they have been my eyes while building the database of resources, selecting the maps, constructing the charts, and managing the formatting of the book. In addition, my brother (Faddy M. Samaan) has been a continuous source of encouragement and inspiration for me.

Abstract

The Nile is a story full of inherited heterogeneities among its riparian states; a mixture of conflicting and complementary interests, asymmetric interdependence, and a long history of mistrust. The basin comprises water-stressed downstream riparians that have almost no alternative freshwater resources vis-á-vis underdeveloped upstream riparians that are freshwater-rich but poor in their energy and food security. All of the Nile riparians, though in varying degrees, give the Nile a central position in their national development plans to meet their growing needs in various sectors. Because any project constructed at any point on the river causes different consequences to the whole basin, development on the Nile cannot be viewed as an independent decision, but it obviously incorporates an intricate network of actions and reactions. To analyze such complex interactions, this study models development on the Nile as an iterated Prisoners' Dilemma game, in which the riparian states are the players, their possible choices are either to cooperate or to act unilaterally, the water development schemes are the sequential moves, and the generated status quo is the game equilibrium. The research hypothesis is that the equilibrium in the Nile game is driven by the changing political and environmental contexts at the multiple levels through their varying influences on the inherited heterogeneities among the riparians over interests, capabilities, and beliefs. Process tracing is used to analyze four rounds of the game: the colonial round (1880s–1940s), the Cold War round (1950s–1980s), the post-Cold War round (1990s–2000s), and the post-2011 round.

Investigation shows clearly that every change in the game equilibrium is induced by a change in the context. The British occupation of Egypt in 1882 under the pretext of securing the Suez Canal that was dug in the 1860s led to utilizing the Nile waters for the first time ever on a large scale upstream of Egypt. This opened the way for cooperation between Egypt and Sudan through the 1929 Agreement of which they expanded their irrigation schemes and for cooperation between Egypt and Uganda through the 1949–1953 Agreements of which the former regulated the White Nile flow and the latter generated hydropower. In the 1950s, Egypt and Sudan entered into difficult negotiations over the construction of the Aswan High Dam and the Roseires Dam, whereas the military takeover in Sudan in 1958 broke the deadlock with Egypt and enabled the conclusion of the 1959 Agreement. The bipolar rivalry between the USA and the Soviet Union during the Cold War fueled the competition between Egypt and Ethiopia over large-scale national projects, but due to civil wars, regional disputes, and severe famines in Ethiopia, Egypt was able to maintain the upper hand on the Nile. Since the 1961–1964 great rains on the Equatorial Lakes, international donors have sponsored different programs to facilitate basin-wide cooperation. Yet, this path was challenged by the general atmosphere of mutual mistrust and reached a complete deadlock by 2010 through the individual signature of most upstream riparians on the Cooperative Framework Agreement (CFA) and through Egypt's and Sudan's subsequent suspension of their participation in the Nile Basin Initiative (NBI). As a result, multilateralism could not replace unilateralism in the basin, which is directly linked to the availability of alternative resources for the riparians to finance their national projects with either domestic funds or international investments, mainly from China and the Gulf States. Thus, the equilibrium of the Nile game moved from Egypt's historical and natural sole utilization of the Nile waters to bilateral cooperation with Sudan and with Uganda in the colonial round. It then tended toward contention between Egypt and Ethiopia in the Cold War round and toward multilateral cooperation in the post-Cold War round, but it ultimately moved to contention between the downstream and upstream riparians. In the recent post-2011 round, since the independence of South Sudan and the Egyptian Revolution in January 2011, followed by Ethiopia's unilateral commencement of the Grand Ethiopian Renaissance Dam (GERD) in April 2011, the balance of interests, capabilities, and beliefs among the Eastern Nile riparians has profoundly changed. On the one hand, Sudan's immediate need for energy and hard currency led it to support the GERD, given that the latter will allow for importing inexpensive electricity and attracting investments in the agricultural sector. On the other hand, Egypt became fully absorbed in domestic affairs, affecting its capacity to halt projects that may cause significant harm downstream. Consequently, the game equilibrium has been swinging between contention and cooperation over the GERD. Moreover, the shadow of the future reflects serious challenges on the basin, mainly the rapidly growing populations, the increasing water, energy and food needs, and the uncertainties associated with climate change, all of which raise the question about the destiny of the equilibrium in the upcoming decades.

Overall, it can be said that the game has been left to the changing contexts that have pushed equilibrium away from multilateral cooperation with no effective institutions that can protect the riparians from such negative influences. The missing basin-wide cooperation and the formation of partial coalitions one against another have turned the game into a tug-of-war. Nevertheless, the key message of this book is that all sides should be aware that the rope they are tugging is a vulnerable river and that the major punishment will be imposed on the current and next generations that are supposed to find the basin more developed but healthier and more peaceful.

Zusammenfassung (Abstract in German)

Der Nil erzählt eine komplexe Geschichte ererbter Heterogenität seiner Anrainerstaaten. Er umfasst gegensätzliche und komplementäre Interessen, asymmetrische Interdependenzen und eine lange Geschichte des Misstrauens. Das Flussbecken schließt flussabwärts Anrainerstaaten ein, die unter Wassernot leiden und die neben dem Nil keine Alternative haben, Frischwasser zu beziehen. Genauso schließt er flussaufwärts Anrainer ein, die große Frischwasservorkommen besitzen aber nur eine geringe Energie- und Ernährungssicherheit haben. Diese Staaten vereint, dass sie, zur Deckung ihres wachsenden Bedarfs dem Nil in unterschiedlicher Art und Weise eine zentrale Rolle in ihren nationalen Entwicklungsplänen zuweisen. Die daraus resultierenden Projekte zur Entwicklung des Nils können nicht unabhängig und isoliert voneinander verstanden werden, da jedes Projekt unterschiedliche Konsequenzen für das ganze Flussbecken hat und ein kompliziertes Netzwerk von Aktionen und Reaktionen einbindet. Um diese komplexen Interaktionen zu analysieren, werden die Entwicklungen am Nil in der folgenden Studie als ein iteratives Gefangenen-Dilemma modelliert. In diesem Model werden die Anrainerstaaten als Spieler, ihre Entwicklungsprojekte als sequentielle Schritte und der generierte Status quo als Gleichgewicht konzeptualisiert. Die Forschungshypothese ist, dass die Dynamik des Gleichgewichts im Nil-Spiel durch die sich wandelnden politischen und ökologischen Kontexte unterschiedlicher Ebenen und ihre jeweils variierenden Einflüsse auf die länderspezifischen Heterogenität hinsichtlich ihrer Interessen, Fähigkeiten und Überzeugungen erzeugt wird. Für die Analyse dieser Dynamiken wird eine Prozessanalyse von vier Spielrunden unternommen: Die Koloniale Runde (1880s–1940s), die Kalte Krieg- Runde (1950s–1980s), die Post-Kalte Krieg-Runde (1990s–2000s) und die Post-2011-Runde.

Meine Untersuchungen haben ergeben, dass Veränderungen im Gleichgewicht des Spiels durch Veränderungen seines jeweiligen Kontexts hervorgerufen wurden. Die britische Okkupation Ägyptens im Jahr 1882, bedingt durch den Bau des Suezkanals in den 1860er Jahren, eröffnete den Weg zur Kooperation zwischen Ägypten und dem Sudan, der im Jahre 1929 zum Abkommen über den Ausbau der Bewässerungssysteme beider Staaten führte. Die Abkommen zwischen 1949 und 1953, in denen Ägypten den Weißen Nils regulierte und Uganda Wasserkraftwerke baute, bereiteten den Weg für eine Kooperation zwischen Ägypten und Uganda. Die militärische Kontrolle über den Sudan im Jahre 1958 ermöglichte es, dass die abgebrochenen Verhandlungen mit Ägypten wieder aufgenommen wurden, zum Abkommen von 1959 führten. Die Rivalität zwischen den USA und der Sowjetunion befeuerten den Wettbewerb zwischen Ägypten und Äthiopien bezüglich ihrer jeweiligen nationalen Großprojekte. Aufgrund des Bürgerkriegs, der regionalen Konflikte und der Hungersnöte in Äthiopien war Ägypten in der Lage, die Oberhand zu behalten. Seit den großen Regenfällen über den Seen im Gebiet des Äquators zwischen 1961 und 1964 haben internationale Geber diverse Programme gefördert, die wiederum die beckenweite Kooperation gestärkt haben. Dieser Ansatz war jedoch geprägt durch eine Atmosphäre des Misstrauens und erreichte mit der Unterzeichnung des Cooperative Framework Agreement (CFA) durch die Mehrheit der Oberlieger des Nils und durch die Aufkündigung der Beteiligung Ägyptens und des Sudans in der Nile Basin Initiative (NBI) 2010 einen neuen Stillstand. Dies dokumentiert, dass der Multilateralismus nicht in der Lage ist, den Unilateralismus am Nil zu ersetzen. Diese Beobachtung kann dadurch begründet werden, dass alternative Ressourcen zur Finanzierung der nationalen Projekte in Form von innerstaatliche Finanzmitteln und internationalen Investments von Seiten Chinas und der Golfstaaten zur Verfügung standen.

Die Gleichgewichtsdynamik im Nil-Spiel lässt sich folgendermaßen zusammenfassen: Das Gleichgewicht bewegte sich in der Kolonialen Runde von einseitiger ägyptischer Ausbeutung des Nils hin zu bilateraler Kooperation mit dem Sudan und Uganda. Danach tendierte es zu Konflikten zwischen Ägypten und Äthiopien in der Runde des Kalten Krieges, bewegte sich dann Richtung Konflikt zwischen den Unterliegern und Oberliegern. In der jüngsten Post-2011 Runde seit der Unabhängigkeit des Südsudans und der ägyptischen Revolution und dem unilateral in Angriff genommenen Bau des Grand Ethiopian Renaissance Dam (GERD) im April 2011 veränderten sich die Interessen, Fähigkeiten und Überzeugungen der Anrainer des Blauen Nils tiefgreifend. Einerseits unterstützte der Sudan, bedingt durch seinen hohen Bedarf nach Energie und konvertibler Währung das Dammprojekt, da GERD den Import von billiger Energie bedeutet und weitere ausländische Investition im Agrarsektor anzieht. Andererseits wurde Ägypten durch seine innerstaatlichen Konflikte so absorbiert, dass es nicht länger in der Lage war, diejenigen Projekte flussaufwärts gelegenen Projekte zu stoppen, die flussabwärts womöglich negative Konsequenzen zeigen würden. Deshalb pendelt das Gleichgewicht des Spiels im Hinblick auf GERD zwischen Kooperation und Konflikt. Darüber hinaus könnten künftige Entwicklungen zu neuen Herausforderungen für das Nilbecken werden. Zu den Faktoren, die das Schicksal des Gleichgewichts in den kommenden Jahren bestimmen, gehören die Konsequenzen des Klimawandels, die schnell wachsende Bevölkerung und der daraus resultierende Bedarf an Rohstoffen.

Insgesamt kann gefolgert werden, dass das Spiel aufgrund des sich ändernden Kontexts sich wegbewegt von einer multilateralen Kooperation, da es keine effektiven Institutionen gibt, die die Anrainer vor den negativen Auswirkungen schützen. Da es keine, sich über das gesamte Becken des Nils erstreckende Kooperation gibt, haben sich Koalitionen gebildet, die gegeneinander arbeiten und das Spiel in ein

Tauziehen verwandelt. Dennoch lautet die zentrale Aussage dieser Arbeit, dass sich alle Parteien darüber im Klaren sein sollten, dass das Tau, an dem sie ziehen, ein verletzlicher Fluss ist. Die negativen Folgen werden die künftigen Generationen zu ertragen haben, die es verdient hätten, ein besser entwickeltes, gesünderes und friedlicheres Nilbecken vorzufinden.

Contents

Abbreviations and Short Names

ADB	African Development Bank
AHD	The Aswan High Dam
AU	The African Union
B	Beliefs
Br	Britain
C	Capabilities
CFA	Nile Basin Cooperative Framework Agreement
CIDA	Canadian International Development Agency
CNS	Comprehensive Nation of Strategy of Sudan (1992–2002)
CPA	The 2005 Comprehensive Peace Agreement between Northern and Southern Sudan
CWE	China Water and Electric Corp
D3 Project	Component D of the NBRAP; "Regional cooperation" that calls for "the establishment of a basin-wide, multidisciplinary framework for legal and institutional arrangements"
DFID	The UK Department for International Development
DIU	Dams Implementation Unit in Sudan
Do	Downstream riparians
DoP	The Agreement on Declaration of Principles on the GERD
E	Environmental context
EAC	The East African Community
EACC	The East African Nile Waters Coordinating Committee
EAPP	The Eastern Africa Power Pool
EDF	Electricité de France
EEPCo	Ethiopia's Electric and Power Corporation
Eg	Egypt
ENA	The Ethiopian News Agency
ENCOM	Eastern Nile Council of Ministers of Water Affairs
ENSAP	Eastern Nile Subsidiary Action Program
ENTRO	Eastern Nile Technical Regional Office
Eq	Equatorial Lakes riparians

Er	Eritrea
Et	Ethiopia
FAO	The United Nations, Food and Agriculture Organization
FDRE	The Federal Democratic Republic of Ethiopia
Fr	France
GERD	The Grand Ethiopian Renaissance Dam
GT	Game Theory
Hydromet	The Hydrometeorological survey on the Nile Equatorial Lakes (1967–1992)
I	Interests
IBRD	The International Bank for Reconstruction and Development, World Bank
IGAD	Intergovernmental Authority on Development in Eastern Africa that superseded IGADD in 1996
IGADD	Intergovernmental Authority on Drought and Development that was founded in 1986
ILC	International Legal Committee at the UN
IPoE	International Panel of Experts on the GERD (2012–2013)
IWLP	International Water Law Project
IWRM	Integrated Water Resources Management
IWT	International Water Treaty
JMP	Joint Multipurpose Program, ENSAP
MDPIU	The Merowe Dam Project Implementation Unit in Sudan
NBD	Nile Basin Discourse
NBI	Nile Basin Initiative (1999–to date)
NBTF	Nile Basin Trust Fund (2001–2015)
NDFs	Nile Discourse Forums
NELSAP	Nile Equatorial Lakes Subsidiary Action Program
NELSAP-CU	Nile Equatorial Lakes Subsidiary Action Program Coordination Unit
Nile-COM	Nile Council of Ministers of Water Affairs
Nile-SAPs	Nile Subsidiary Action Programs
Nile-SEC	Nile Basin Initiative Secretariat
Nile-SVP	Nile Shared Vision Program
Nile-TAC	Nile Basin Initiative Technical Advisory Committee
NRBAP	Nile River Basin Action Plan
NRBC	Nile River Basin Commission (to be established after the CFA would come into force)
OAU	The Organization of African Unity
Ot	Other Riparians
PD	Prisoners' Dilemma
PoE	Panel of Experts on the CFA (1997–2000)
RBO	River Basin Organization
TECCONILE	Technical Cooperation Committee for Promotion of Development and Environmental Protection of the Nile Basin (1993–1999)

TFDD	Transboundary Freshwater Dispute Database at Oregon State University
TNC	The Technical National Committee on the GERD
SCAF	The Supreme Council of the Armed Forces in Egypt
SCNW	The Supreme Council for the Nile Waters in Egypt
SGB	Sudan's Gezira Board
Sinohydro	Chinese National Water Resources and Hydropower Engineering Corporation
SIWI	Stockholm International Water Institute
SPLA/M	The Southern Peoples' Liberation Army and Liberation Movement
Su	Sudan
SS	South Sudan
SW	Scott Wilson of UK
Ug	Uganda
UNDP	The United Nations Development Programme
Undugu	An initiative on the Nile; the word means "brotherhood" in Swahili (1983–1992)
UNEP	The United Nations Environment Programme
UNEPTA	The United Nations Expanded Programme of Technical Assistance
UNWC	The 1997 UN Convention on the Law of the Non-navigational Uses of International Watercourses
Up	Upstream riparians
WDI	World Development Indicators by the World Bank
WMO	World Meteorological Organization
WSP	Water Sector Policy of Ethiopia

List of Figures

List of Tables

Chapter 1
Introduction: The Nile Development Game

1.1 Background

Although 71% of the earth surface is covered by water, only 0.01% of total water volume on our blue planet is freshwater running in lakes and rivers (Ojakangas 1997: 67). About 60% of this global surface water is shared between two or more countries (UN-Water 2008: 3), creating 276 transboundary lakes and river basins, which host around one-third of the world's population, cover almost one-half of the world's land surface, and cross the borders of 148 countries (Giordano et al. 2014: 245, 246). Because freshwater is the source of life and the essential element that fuels vital sectors in the economy of every nation, these figures can justify the growing concerns associated with various scientific fields studying issues related to transboundary river basins, amongst which are the disciplines of international relations (IR), water governance, and spatial planning. In this respect, it is widely argued that large-scale developmental schemes in transboundary watersheds may provoke conflict or promote cooperation amongst different riparian states sharing international watercourses, often affecting basin communities and ecosystems, especially in those regions of water scarcity including the Middle East and North Africa (MENA) (Fig. 1.1). At the heart of these debates lies the Nile Basin hydropolitics.

The Nile, often considered the longest river in the world,[1] covers 10% of Africa's surface area (Fig. A.1) and crosses the borders of 11 countries (Fig. 1.2): Ethiopia, Eritrea, Sudan and Egypt in the Eastern Nile sub-basin; South Sudan, the Sudd sub-basin; Burundi, Rwanda, the DR Congo, Tanzania, Kenya and Uganda sharing the Equatorial Lakes sub-basin. Despite the fact that more than 2000 BCM of rain falls annually over the Nile Basin on average (Johnston 2012: 71), only 3.9% of this volume runs in streams, which is a considerably low combined runoff coefficient

[1] As reported in the BBC, it has been recently argued that the Amazon River is the longest after an expedition to Peru discovered in 2006 a new source for the river to reach a total length of 6800 km, while the Nile is 6695 km long (Duffy 2007).

M. M. Samaan, *The Nile Development Game*,
https://doi.org/10.1007/978-3-030-02665-3_1

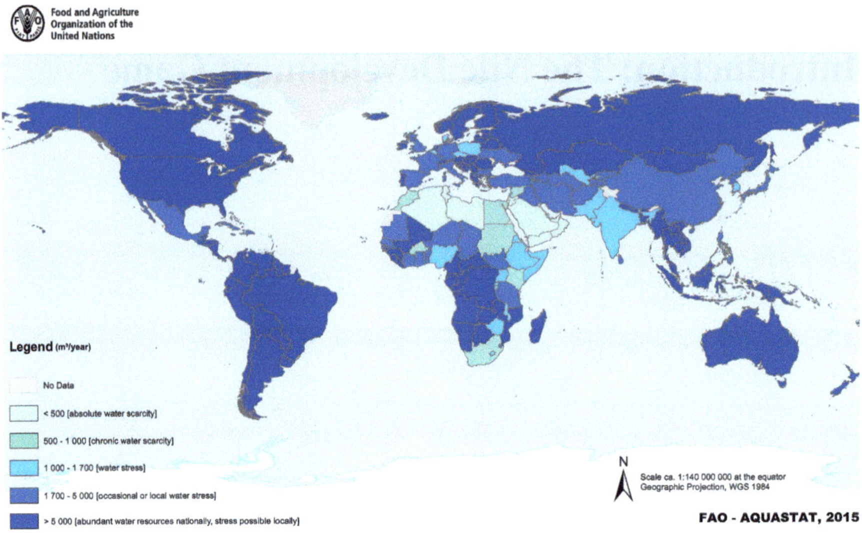

Fig. 1.1 Total renewable freshwater resources per inhabitant. (Source: FAO, Aquastat Database, World thematic maps, available at: http://www.fao.org/nr/water/aquastat/maps/TRWR.Cap_eng.pdf (accessible at July 11, 2018))

compared to other river basins (NBI 2012a: 30). Rainfall over the basin is characterized by highly uneven seasonal and spatial distributions (Figs. A.2 and 3.6), where precipitation reliability and volume decline moving northwards with peak values during the summer months (NBI 2012a: 31). Therefore, the basin includes distinctly different climatic conditions, considering that two-fifths of its total area is hot arid, particularly in Egypt and northern Sudan, contributing almost no runoff. On average, 86% of total runoff reaching downstream originates from the Ethiopian HighlandSwain 1997s: 59% through the Blue Nile (Abay); 14%, Baro-Akobo-Sobat; and 13%, Tekeze-Atbara (: 675). The remaining portion originates mainly in the Equatorial Lakes Plateau (Victoria, Kyoga and Albert Lakes), whereby more than half of the original flow is lost in the Sudd swamps due to high rates of evapotranspiration (Howell et al. 1988: 21). The water flow from the Equatorial Lakes converges with the Baro-Akobo-Sobat River to form the White Nile. This branch of the river meets the Blue Nile in Khartoum, the capital of Sudan, and then combines with the Tekeze-Atbara River, forming what is called the Main River that flows northwards to Egypt and discharges eventually in the Mediterranean Sea through the Nile Delta.

The Nile hydrology has shaped the life of its people throughout history. Since millennia, life in Egypt was established around the Nile Valley, depending on a sole water stream and narrow fertile strip penetrating Egypt's vast areas of deserts. Ancient Egyptians named their country "Kemet," which is thought to mean "black" in reference to the "black land" or the silt deposited by the Nile annual flood (Sambu 2011: 162). They also identified themselves as the "people of the land of the

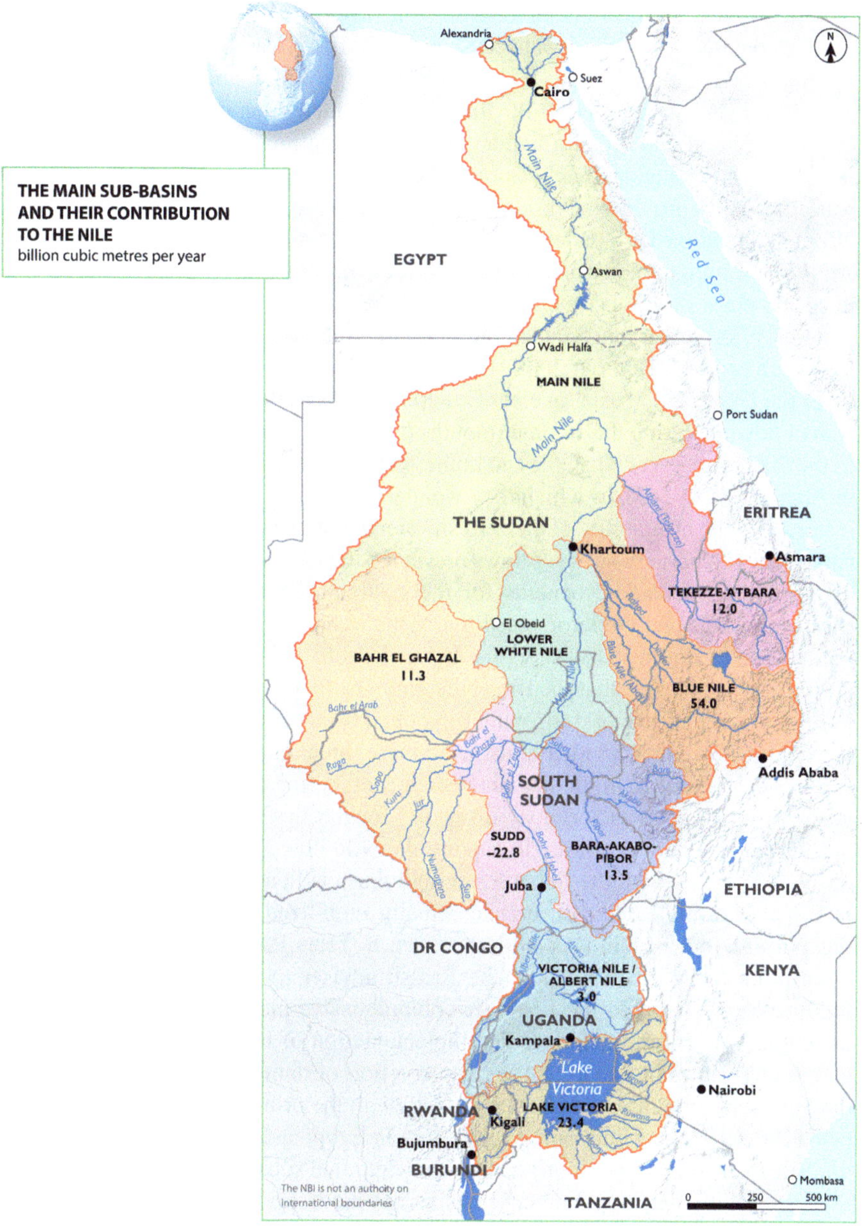

Fig. 1.2 The main sub-basins in the Nile and the contributions of each. (Source: NBI, the State of Basin Report (2012a: 39))

Nile-flood," and they associated the Nile with the God "Hapi," whom they sanctified as the "container of spilling water" (Sharman 2014). It can be said that the ancient Egyptian civilization, which evolved since King Menes unified Egypt into one centralized state around 3,000 BC, had been founded upon the need for managing seasons of flood and drought and administrating the process of irrigation, cultivation, crop harvesting, and yield storage. Moreover, in the age of the Romans, Egypt was the major provider of grain to the Roman Empire and became known as "Rome's bread basket" (Booth 2007: 137). Two centuries ago, when Muhammad Ali Pasha sought to modernize Egypt and transform it into a powerful state, he centered his plans on the prosperity of agriculture and the development of irrigation systems, after which long-staple cotton became the backbone of Egypt's economy for over a century (Dumont 2009: 15; Tignor 2011: 215).

In the Upper Nile basin, ancient Ethiopians viewed the Abay, with its sweeping waters flowing during the summer months downwards through deep steep canyons, as the "father of rivers" that "has no home," according to the popular saying, quoted in Arsano (2010: 163), in which they wondered, "Where does it take all those logs it carries?" The availability of other freshwater resources and the lack of technological capacity to harness the river flows prevented the utilization of the Nile waters for deriving socio-economic benefits for the upstream communities. Yet, successive Ethiopian kings and emperors used the water weapon in dealing with Egypt over political and religious issues. As quoted in Degefu (2003: 150) after Moorehead (1962/1985), "the idea that the Blue Nile might be blocked or poisoned at its source in Ethiopia as a means of destroying Egypt had been canvassed in every age."

Nevertheless, such an idea remained a mere threat beyond the capacity of those who thought about it. In the late nineteenth century, the concept of building a dam upstream of Egypt was seriously discussed for the first time ever in Paris. In the meantime, France was aiming to capture the Nile flow at Fashoda, the confluence of the White Nile with the Sobat River, and thus challenge the presence of Britain in Egypt (Collins 1990: 49). France actually mobilized a military expedition for this purpose, but Britain was able to prevent it. Years later, with a completely different aim, Sir William Garstin, the British adviser to the Egyptian Ministry of Public Works, prepared the first-ever comprehensive plan for the development of the entire Nile Basin, including the implementation of large irrigation scheme for cotton cultivation in Sudan and the construction of dams in the Equatorial Lakes, the Sudd and the Ethiopian Highlands. Although the primary objective of Garstin's plan (1904: 195) was flourishing agriculture in Egypt and Sudan through regulating the Nile flow from where it originates, it inaugurated subsequent plans and negotiations over water development schemes across the Nile. Along the first half of the twentieth century, negotiations had run over the Gezira scheme, the Sennar Dam and the Jabal Auliya Dam in Sudan, the Tana project in Ethiopia, and the Owen Falls Dam in Uganda.

From the 1950s to the 1980s, the basin countries shifted from negotiations over the successive plans prepared by the British experts in Egypt to large national projects. Egypt and Sudan coordinated their projects, namely the Aswan High Dam (AHD), the Roseires Dam, the Jonglei I project and the Peace Canal. Ethiopia invited the U.S. Bureau of Reclamation to conduct an extensive study on the Blue Nile, proposing multiple irrigation and hydropower projects, none of which were implemented at the time, except for the Finchaa Dam. From the 1990s to date, national schemes continued with the Toshka project in Egypt, the Merowe Dam and the Atbara-Setit Dam in Sudan, the Kiira-Nalubale scheme and the Bujagali Dam in Uganda, and the Tekeze Dam, the Beles hydropower plant and the Grand Ethiopian Renaissance Dam (GERD) in Ethiopia. In parallel with these projects, progressive attempts have been made since the 1960s with the support of international donors to develop basin-wide programs, namely the Hydromet survey, the Undugu group, the TECCONILE, the Nile Basin Initiative (NBI), and the Cooperative Framework Agreement (CFA).

Not all of the projects and programs mentioned above were fully carried out, and they have varied considerably in terms of their impacts on inter-state relations in the basin, generating a story of several dramatic scenes of conflict and cooperation over the Nile waters that range from the signature of agreements to the exchange of threats. For instance, the most recent chapter of the story comprises the following main scenes. In May 2010, the riparian states failed to reach a compromise over the CFA, and the upstream riparian countries signed the agreement individually, while Egypt and Sudan suspended their participation in the NBI. In April 2011, Ethiopia's Zenawi inaugurated the construction of the GERD unilaterally without prior consultation with the downstream countries, Egypt and Sudan. In June 2013, Egypt's Morsi threatened Ethiopia that blood would be the alternative to the loss of any drop of water, and Ethiopia's Desalegn responded that no one and nothing would halt the dam construction. In December 2013, Sudan's Bashir announced publicly his country's support for the dam. In March 2015, Egypt's el-Sisi, Ethiopia's Desalegn and Sudan's Bashir celebrated the signature of the Declaration of Principles (DoP) on the GERD in March 2013. In September 2016, the three water ministers from the three countries signed contracts with two French consultancy firms to conduct hydrological, socio-economic and environmental impact studies on the dam. To date, until July 2018, almost 65% of the GERD's construction works have been completed, and still the technical and ministerial committees stemmed from the three countries are still discussing the baseline upon which the consultancy firms can measure the GERD's negative impacts while conducting the required joint studies.

In fact, the essential factor behind the drama of the Nile story is the asymmetric configuration of the basin (Allan 2009), comprising downstream countries that are heavily dependent on water resources that originate outside their borders vis-a-vis upstream riparians that rely on other freshwater resources and do not utilize the Nile waters originating on their territories. The balance among the Nile riparians is not only asymmetric in their dependency on the river, but also in their powers (Zeitoun and Warner 2006): while downstream riparians, particularly Egypt, are more powerful in terms of economic and military capabilities, the upstream riparians possess the power of the geographic position. Given the facts that the Nile represents 97% of Egypt's freshwater resources, that 95% of Egyptians settle in the Nile Valley and Delta (Fig. A.4), and that 95% of Egypt's area is left uninhabited, it can be imagined how Egyptian rulers were extremely cautious about the implementation of dams and irrigation schemes upper the basin. On the other hand, Ethiopia is still suffering from problems of poverty, frequent famines and energy shortages, all of which make hydropower dams and supplementary irrigation schemes are of great importance. What further complicates the story is distrust from the past, generated by the history of using water as a weapon against Egypt vis-a-vis the Egyptian historical attempts to dominate over the upstream riparians. Moreover, the serious challenges of the future overshadow the story, mainly due to the uncertainties of climate change and rapid population growth.

1.2 Aim and Question

Since hydropolitics is "the systematic study of conflict and cooperation between states over water resources that transcend international borders" (Elhance 1999: 3), all three IR branches have found the Nile story a rich and unique case for continuous analysis. The asymmetric powers, the growing needs and the history of mistrust among downstream and upstream riparian countries lead neo-realists to argue that water is a driver for conflict in the Nile Basin (Buzan 1991/2007: 118; Gleick 1993: 111–112; Homer-Dixon 1999: 139; Keith et al. 2013, 2014). Whereas, institutionalists believe that the basin asymmetries reflect the diversity of interests and capabilities that can be a source of cooperation rather than conflict. They analyze the situations where cooperation was established and institutionalized in the basin (Wolf 1999: 23; Metawie 2004; Iyob 2010). They also investigate how benefits can be generated and shared among the riparians to present attractive economic incentives for cooperation (Wichelns et al. 2003; Whittington et al. 2005; Dinar and Nigatu 2013; Tawfik 2015). Constructivists pay special attention to the ideational dimension of power in the formation of hegemonic and counter hegemonic strategies employed by Egypt and Ethiopia respectively (Zeitoun and Warner 2006; Cascao 2008; Ibrahim 2011; Cascao and Nicol 2016). They show how international law and norms can shape cooperation (Brunnee and Toope 2002; Salman 2013) and how conflict and cooperation co-exist in the Nile (Zeitoun and Mirumachi 2008).

In this respect, the three branches of the hydropolitical scholarship have found the Nile an interesting case for analysis. Yet, given the dramatic events and developments witnessed by the basin during recent years, further research is needed to cover the gaps realized in the findings of each branch. In contrast to the neo-realist argument, reality shows that water is not always a driver for conflict and that cooperation based on common interests over shared waters is attainable and feasible. At the same time, reality also shows that the enormous efforts of institutionalizing basin-wide cooperation through the CFA have failed to date and that the riparian states have chosen to implement their large national schemes outside the NBI, which collides with the institutionalist argument. As for the constructivist/critical argument, observations indicate that international norms have not been sufficient to shape cooperation in the basin, since every side sticks to the principles that favor its position and relies on its power to impose its own vision on the other.

Adopting the institutionalist approach, this study believes that the Nile's people and ecosystem deserve better and that the next generations have the full right to find the basin more developed but healthier and more peaceful, which cannot be achieved but through institutionalized cooperation. Yet, a crucial step towards this path is to explain how the basin's historical status quo was changed into the construction of national large-scale water projects in the upstream as well as the downstream riparians, and why they have failed to date to build basin-wide cooperation despite the enormous endeavors invested in the NBI and the CFA. These specific lines of inquiry lead to the more general question of what brought about the change in the historical status quo in the Nile and by what means this was done. The study assumes that the key to answer this question is to further investigate how the events associated with water development schemes in the Nile have been manufactured, contending that all of those events are no more than the tip of the Nile huge iceberg. Thus, behind the research question lies the imperative to dive deep into the Nile story to view the full picture. However, this journey requires a theory, a framework, a hypothesis and a method in order to attain its goal.

1.3 Theoretical Foundations

Because reality comprises complex and intricate details, scholars need a theory to abstract an image focusing on specific outcomes and to analyze the causes that brought it about. This is precisely illustrated in the following quotation. "International relations comprise a plethora of events, issues and relationships which are often enormous in scale and bewildering in their complexity. ... To the scholar of the 'international', theories are unavoidable," as "the events and issues which comprise international relations can be interpreted and understood only by reference to a conceptual framework" (Burchill and Linklater 2005: 16–17). Nevertheless, as described by Holsti (1985), IR is a "dividing discipline" because of the progressive debates and disagreements between different schools of thought studying world politics. Variations do not take place only between different schools, but also among proponents of each. However, such differences have enriched the overall scientific contribution of the field, seeing that "heterogeneity is a strength and an obstacle to ossification" (Burchill and Linklater 2005: 18). Criticisms raised by the exponents of every theory present a powerful motivation for its development. Comparing traditions of international relations, Wight (1992: xviii) remarks that none at all independently presents the complete truth, rather, such continuing deliberative dialogues advance our knowledge about world affairs. Those who extensively reviewed the IR literature often use the term "toolkit" to indicate the diversified roles of various theoretical perspectives in analytic work (Donnelly 2005: 54; Reus-Smit 2005: 208; Ward 2002: 65).

As mentioned in the previous section, this study belongs to the institutionalist branch of hydropolitics, which views states as self-interested rational actors, however it believes that the world is not a pool of "colliding billiard balls" but a "cobweb" of interests related to different stakholders (Reus-Smit 2005: 189; Keohane and Nye 1972). In principle, the Rational Choice Theory (RCT), as summarized by Elster (1989: 22), presumes that "when faced with several courses of action, people usually do what they believe is likely to have the best overall outcome." The action of development in general can be understood through this theory, in that every government contemplates the optimum use of its resources that will bring maximum benefits. There are different theories stemming from the RCT depending on the nature of the situation (Ward 2002): the Decision Theory, when one actor takes one action; the Collection Action Theory, when a group of actors seeks common action; and the Game Theory (GT), when a group of actors takes individual actions and all of them affect one another.

Since water is the fundamental resource that fuels every sector in every country, the process of governing national water resources can be modeled through the Collective Action Theory. In effect, the various stakeholders need to jointly assess the rewards and tradeoffs of the different water uses to reach a unified national vision. At the international level, the absence of an overarching authority having supreme jurisdiction over sovereign states makes the failure of collective action more likely to occur. The contr0ution of the Game Theory, therefore, is substantial in modeling transboundary water governance, taking into account both the sovereignty of each riparian state in utilizing the waters running on its territory and the mutual effects of the decisions taken by the riparians on one another. However, there are different types of games that can be used in modeling transboundary water interactions. Neo-realists adopt zero-sum games in which there is no win-win outcome, as every cubic meter of water used by one riparian is considered a loss for another. The tug-of-war is a clear example of this, as every step gained by one side is lost by the other. On the contrary, institutionalists use positive-sum games, believing that riparians can cooperate to achieve a win-win outcome that benefits all of them, even if in varying degrees. Whereas, all actors in such games are aware that straightforward maximization of their interests is not a rational choice and that the game will come eventually to equilibrium. This is conceptualized by John Nash as the point where every player finds the "best counter-strategy to what one expects the other person(s)' choice will be," as simplified by Chwaszcza (2008: 145).

This study argues that the iterated Prisoners' Dilemma (PD) game,[2] a positive-sum game, perfectly abstracts interactions over transboundary river basins. An analytic framework therefore is constructed for this purpose, elaborating on the models of Axelrod (1984), Axelrod and Keohane (1985), and Keohane and Ostrom (1995), in addition to the general implications made on hydropolitics (Kilgour and Wolinsky-Nahmias 2004; Qaddumi 2008; Anisfeld 2010), and on the particular case of the

[2]As illustrated by Hargreaves Heap et al. (1992) reviewed in Chwaszcza (2008: 145–146), the Prisoner's Dilemma is the story of two members in a criminal gang whom police arrested. The investigator is sure that both are guilty but has no clear evidence. He gives an offer for each to think about individually in their solitary custody. The offer is as follows: if both do not confess, he will give them a minor punishment; if both confess, he will recommend "less than the most severe sentence" for both; and if one confess and the other does not, the confessor will get the "lenient treatment," whereas the other get the "book slapped at him."

Nile Basin (Waterbury and Whittington 1999; Waterbury 2002; Whittington 2004; Madani et al. 2011). Before all, it is important to recall the definition of "cooperation," which is adopted throughout this dissertation. The model of Axelrod and Keohane (1985: 226) represents the situations where there is no "complete identity of interests" among actors, but there exists a "mixture of conflicting and complementary interests." Thus, they define "cooperation" as "when actors adjust their behavior to the actual or anticipated preferences of others" (1985: 226). This definition adequately expresses the meaning of cooperation over shared waters. Water uses of one riparian have a mixture of negative and positive impacts on the other riparian states. For instance, although upstream dams may regulate waters, control sediments, and manage floods downstream, they may alter the quantity, quality and timing of the downstream water flow. Dams also may flood lands and displace indigenous people upstream. Furthermore, large-scale hydraulic projects constructed at any point along the river may have severe adverse impacts on the basin's ecosystems, threatening the livelihoods of local communities (WCD 2000).

Cooperation over transboundary waters then occurs when a riparian seeks to reap benefits from its schemes in such a way as not to adversely affect the other, the core concept upon which international water law is founded. Both the 1966 Helsinki Rules on the Uses of the Waters of International Rivers and the 1997 UN Convention on the Law of the Non-navigational Uses of International Watercourses are based on the principles of equitable and reasonable utilization and causing no significant harm (McCaffrey 1998). The term "equitable" refers to the equal rights of all riparian states to benefit from the shared resources of the basin, while the term "reasonable" indicates that such rights should be exercised with full coordination with others so that none will be significantly harmed. The need for such a balanced perception of water rights is evident in the case of alluvial river basins, like the Nile and the Euphrates. While downstream states had been major water users since millennia, the progressive technology associated with large dams in recent decades has enabled the upstream states to develop their own hydropower and irrigation schemes.

Here lies the Prisoners' Dilemma game: riparian states on international rivers may choose either to cooperate over their water uses or to defect by utilizing the river unilaterally regardless of the harm caused to the other. Consequently, using the language of the two-person PD game, there are four potential payoffs: whether one side utilizes the river solely, tempted to reap maximum payoff (T) leaving the other with the sucker's (S); whether the two sides utilize the river jointly so that both yield rewarding payoffs of cooperation (R); or whether the two sides utilize the river unilaterally and both get punishing payoffs of contention (P). If one side chooses not to act unilaterally, it may be rewarded for cooperation (R-R), but also it may emerge the sucker (S-T). Otherwise, if it chooses to act unilaterally, it may be punished for contention (P-P), but also it may reap the maximum tempting payoff solely (T-S). Ultimately, in case the game is played once, it is more likely that both sides defect and thus be punished.

However, Axelrod (1984) notes that the PD game played in reality among people and states is not a one-move game but an "iterated" game, since they interact for several moves and rounds over their common issues. This is obvious in hydropolitics:

as long as the river flows from its source to its mouth, the riparians will continue to interact over water utilization. Through his extensive research on the iterated PD game, Axelrod (1984) concludes three important remarks: that the longer the game is iterated, the more likely cooperation will emerge and be sustained; that the most fruitful strategy is the one based upon reciprocity and forgiveness; and that actors can form clusters where they cooperate with one another more often than with outsiders. When Axelrod and Keohane applied these findings on the IR arena, they came up with two main points: that the context of the game has a crucial influence on the outcomes; and that institutions are needed to facilitate reciprocal cooperation and to deter defection for the prevention of entering into closed circles of retaliation and counter-retaliation. The context here refers to "issue linkage" and "domestic-international connections" (Axelrod 1984: 239), considering that states are the primary unit of analysis and that they are affected both by the different linked issues and by the interconnected events taking place at the multiple levels.

Nevertheless, this game theoretic model has been criticized for considering players identical in their characteristics and only different in their strategies. For instance, Zagare and Kilgour (1993) and Kilgour and Wolinsky-Nahmias (2004) focus on the case of "asymmetric deterrence," when one player is not satisfied with the "status quo" outcome (challenger) and the other is satisfied (defender). The game equilibrium in this case depends on whether each prefers "conflict" (being "hard") or "capitulation" (being "soft") and on the extent to which each is uncertain about the preference of the other. Moreover, Keohane and Ostrom (1995) introduce the concept of "heterogeneities" among actors sharing a common pool resource (CPR). They show how interests, capabilities and beliefs may be heterogeneous among actors, the third element of which refers to the level of trust/mistrust in entering into binding commitments with others, taking into account the uncertainties about others' intentions and the different perceptions of rights. As defined by Snidal (1995: 62), heterogeneity here describes "variation across actors on some significant attribute." This concept extends the role of institutions to the transformation of heterogeneities among states into a catalyst for cooperation rather than conflict.

Overall, it can be said that what defines the equilibrium point of a game is the level of heterogeneities over interests, capabilities and beliefs among its actors, and that the context and institutions may push the equilibrium either towards or away from cooperation. Such a comprehensive conceptualization, which brings together both rationalist and critical elements of analysis, has been endorsed in further studies by institutionalists (Keohane and Martin 1995; Keohane 1998, 2010; Axelrod 2000). This also led Hasenclever et al. (2000: 10, 11, 25) to highlight the potential of supplementing the rationalist approach with the "minimalist cognitivism," referring to the approach of Goldstein and Keohane (1993), which contends that both beliefs and interests have causal weights in generating human actions and that neither of them shapes individually world politics. In the hydropolitical scholarship, all of Kilgour and Wolinsky-Nahmias (2004), Qaddumi (2008) and Anisfeld (2010) indicate that the GT still has more to contribute in analyzing interactions over shared waters. In specific, the application of the GT by Waterbury on the Nile case emphasizes its usefulness for further investigation over the current changes in the basin.

1.4 Research Design

The analytic framework constructed by this study has benefited from all the elaborations shown above. Interactions over transboundary waters in general and the Nile story in particular are modeled accordingly as an iterated PD game: the riparians are the actors; water utilization is their common issue; the linkage with other political and environmental issues at the multiple levels is the context; the basin asymmetries are the inherited heterogeneities; clusters are formed through agreements; and the changing equilibrium over the game rounds is the generated status quo of every historic period. Accordingly, the research question can be reformulated; from "what brought about the change in the historical status quo in the Nile Basin, and by what means?" to "how the equilibrium in the Nile game is shaped and changed, and what are the process and mechanisms?"

This study points to the game context as the major driver of the equilibrium dynamics. The research hypothesis therefore is that the equilibrium in the Nile game is driven by the changing contexts through their varying influences on the inherited heterogeneities among the riparians, generating different outcomes over rounds and shifting the equilibrium from one point to another.

The research variables are built upon supplementing the PD model of Axelrod and Keohane (1985) and the concept of heterogeneities introduced by Keohane and Ostrom (1995) with the definition of contexts indicated by Cascao and Zeitoun (2010: 40). Thus, the independent variables are the inherited heterogeneities among the Nile riparians over interests, capabilities and beliefs, in that interests refer to the need for utilizing the river, capabilities represent the capacity to utilize the river and justify such utilization, and beliefs describe the perception of water rights and the level of trust based on the history of interactions and the information about the others' future plans. The dependent variable is the game equilibrium generated through the accumulation of outcomes. As mentioned earlier, according to the PD framework adopted by the study, there are four possible outcomes: sole utilization, when either is tempted to reap maximum benefits by utilizing the river solely and leaving the other with the sucker's payoff (T-S & S-T); cooperation, when both utilize the river jointly assuring rewarding payoffs (R-R); and contention, when both are punished due to consequent negative impacts of their unilateral projects (P-P). In addition, the intervening variable is the changing contexts encompassing the concurrent political and environmental events at the domestic, regional and global levels. The significance of validating this hypothesis with respect to institutions is to revisit the philosophy of international development partners in designing the institutional settings in the basin for decades. In other words, if the political and environmental context is truly the major driver behind every change in the game equilibrium, it is essential to investigate whether the institutions have considered such contextual influences on the game or not.

To make it crystal clear, the role of theory in this research is mainly to explain the complex interactions among the riparian states in transboundary river basins. This is done through modeling the riparian countries as rational state actors playing an iterated PD game over the development of the river basin. The decisions of each are based on interests, capabilities and beliefs. The variations or heterogeneities among these three attributes influence the game outcomes. The accumulation of outcomes in every historic period defines the round's point of equilibrium. According to this model, it is argued that the changing contexts of the game drive the influences of heterogeneities to generate different outcomes and thus induce the equilibrium in the game over sequential rounds. It is crucial here to identify what is context and what is not. The context plainly is the group of events occurring at global, regional and domestic levels concurrently with the game interactions. The contextual event becomes relevant if the following condition applies: when the progression path of water-related interactions among the riparian states is altered under the influence of different linked issues or intervening actors (Table 1.1). From this perspective, the study looks at the vector and magnitude of the contextual effects and consequences on the inherited heterogeneities among different riparian states over interests, capabilities and beliefs, driving, as a result, the equilibrium. Also, of great importance is to determine the domain of linked issues and intervening actors in the context. Since the framework regards water as the central issue of the game and the riparian states as the primary actors, the context involves distinguished linked issues and intervening actors of political or environmental nature at the global, regional or domestic level. Political issues may include foreign occupation, change in the world order, border dispute, regional conflicts, civil wars, and regime transformation, while environmental issues may include floods, droughts, famines, and climate change. Similarly, political actors involve global superpowers, international institutions (not only organizations but also norms and principles), neighboring countries, and internal insurgent groups, while the climate, the river and its ecosystems (Fig. A.6) are considered the environmental actors.

1.5 Methodology

The method of process tracing is used to validate the research hypothesis, that every change in the contexts affects the game outcomes, and more important to explore the mechanisms of such effects. Process tracing is a significant qualitative method for generating in-depth knowledge and validating a specific theoretical hypothesis through the analysis of a single case study (Bennett and Elman 2006; Collier 2011; Mahoney 2012; George and Bennett 2005). As stated in Mahoney (2012: 1), process tracing is "a methodology for testing hypotheses in social sciences" by which "the analyst combines preexisting generalizations with specific observations from within a single case to make causal inferences about that case." Bennett and George (1997: 5) metaphorically demonstrate the exploration of "causal mechanisms" that link an assumed condition with a given outcome as a line of dominoes placed on a table with a blind that permits to see only the first and the last pieces, contemplating the path and procedures through which the fall of the first piece has led to the fall of the last one. In other words, the use of process tracing is not confined to proving that

Table 1.1 The contextual events and intervening actors along the four rounds of the Nile game

Round	Global level	Regional level	Domestic level
Colonial round (1860s–1940s)	American Civil War and rise of international cotton prices in 1860s; opening the Suez Canal in 1869 (Britain, France and Ottoman Empire); European rivalry and the first and second World Wars (Britain, France, Italy, Germany and Belgium); world economic depression in late 1920s (Britain and Ethiopia)	Egyptian military expeditions in Africa in the nineteenth century (Mohammed Ali Dynasty); low-flood years of 1910s (British experts in Egypt and Sudan); assassination of the British general of Sudan in Cairo (Britain, Sudan and Egypt); claims of Egypt over Sudan and of Ethiopia over Eritrea	Financial/political crises leading to the British occupation of Egypt in 1882 (Ottoman governor, Egyptian nationalists, and Britain); nominal independence of Egypt in 1922; internal feudal system in Ethiopia in 1920s (central government and local governors); Italian invasion of Ethiopia in 1935–1936
Cold War round (1950s–1980s)	The Cold War (the USA and the Soviet Union); the role of the International Bank for Reconstruction and Development (IBRD); the Suez Crisis in 1956 (Egypt, Britain, France, Israel the USA, the Soviet Union, and the UN); the collapse of the Soviet Union in 1991	Hala'ib Triangle dispute (Egypt and Sudan); Arab-Israeli Conflict; Ethiopian-Eritrean War; Ethiopian-Somali War; the leadership of Egypt's Nasser in Africa and the Arab World; Great rains of 1961–1964; October 1973 war; severe drought of 1983–1985	Independence of states in African in 1950s and 1960s; the 1952 revolution in Egypt; the spark of the first and second civil wars in Southern Sudan in 1955 and 1983 respectively; military control over Sudan in 1958; overthrown of Haile Sellasie and Mengistu in 1974 and 1990 respectively
Post-Cold War round (1990s–2000s)	Major role of international donors in funding development project in 1990s; emergence of China as an alternative source of financial and technical assistance in 2000s	The Gulf War in 1990–1991 (Kuwait, Iraq, Egypt and Sudan); assassination attempt against Egypt's Mubarak in Addis Ababa in 1995 (Egypt, Ethiopia and Sudan); Ethiopian-Eritrean War in 1999–2001	New regimes in Sudan and Ethiopia in 1989 and 1990 respectively; dropping external debts from Egypt in recognition of its role in the Gulf War; imposing sanctions on Sudan for its support to terrorists; Ethiopia's relative economic growth and political stability
Post-2011 round	Investment opportunities in the agricultural sector after the international food crisis in late 2000s; uncertainties of climate change	The so-called Arab Spring uprisings in 2011; Hala'ib Triangle dispute (Egypt and Sudan); tensions between Egypt and Saudi Arabia over the Syrian problem and the Tiran and Sanafir islands in the Red Sea	The Egyptian January 25 revolution in 2011; the independence of South Sudan in 2011; ethiopian strategy to implement large national projects with national funds; the Egyptian June 30 revolution in 2013; crash of the Russian plane in 2015; mysterious killing of the Italian researcher in 2016

condition X leads to the occurrence of outcome Y, but rather it aims at exploring the process and mechanisms linking both of them. This method perfectly fits the scope of the research. In the institutionalist hydropolitical scholarship, Mostert (2003), Yoffe et al. (2003), Delli Priscoli and Wolf (2009), Brochmann and Hensel (2011), Schmeier (2013), and Giordano et al. (2014) conducted multi-case studies to explore correlations between characteristics of river basins and riparian states, on the one hand, and the signature of international water treaties and establishment of river basin organizations, on the other. These studies illustrate that the influences of asymmetries between riparians and the effectiveness of various structural and procedural design of river basin institutions differ from one case to another, opening the way for closer investigations in single-case studies. In addition, the study aims not merely to prove that the context affects the outcomes in the Nile game, but to explore how the changing political and environmental contexts at the multiple levels have influenced the heterogeneities among the riparian states and shaped the equilibrium in the game.

Moreover, process tracing, or "colligation" as Roberts (1996) chose to call it, works as historical explanation in tracing and discussing a complex consequence of events in a "narrative" form. Yet, the use of process tracing is restricted by "analytic" and "theoretical" constraints (Bennett and George 1997: 7). Such constraints cover two important aspects of analysis with the purpose of either testing or generating a theory. First, process tracing analyzes predefined sets of conditions and outcomes across the whole consequence of events, the restriction to which historians do not always oblige themselves. Second, process tracing seeks to explore the causal mechanisms linking the conditions with the outcomes, the objective of which historians may target more implicitly rather than in a systematic and explicit way. In the literature on the Nile, many scholars have adopted the historic approach (Collins 1990; Tvedt 2004; Carlson 2013; Salman 2016), but this study follows the path of El-Atawy (1996) and Waterbury (2002) who traced the Nile story within a constrained framework. Furthermore, Bennett and Elman (2006: 459), and Mahoney (2012) define the conditions that should be fulfilled while using the method of process tracing, which can be summarized in three points as follows. The first is to select a case of rich details about both the conditions and outcomes and at the same time, with an adequate degree of causal complexity. The second is to cover adequately the historic path of the process, specifying the appropriate starting point of analysis on the case timeline and securing the continuity of the path with no temporal breaks. Finally, the third is to diversify the pieces of evidence and observations upon which the hypothesized explanation is validated and the alternative ones are eliminated.

These three conditions have been fully respected while selecting and analyzing the Nile case, as follows. First, the Nile is obviously a rich case with a long history of hydropolitical events and interactions among different actors from inside and outside the basin. The case can be also considered a deviant case for the institutionalist theory to which the study belongs, since institutional basin-wide cooperation has been hindered for long times, in contrast to other cases where river basin institutions have been developed. Second, the study covers the historic progression of the

Nile Basin along the twentieth century and up to date, taking into account the contexts in which the water-related events have occurred and progressed on the case timeline with no temporal breaks. This time span is divided into four sequential phases (rounds of the iterated game), each of which has different external actors and comprises sets of various events at the domestic, regional and global levels. Briefly, these phases are the colonial round until the mid-twentieth century, the Cold War round until 1980s, the post-Cold War round until 2000s, and the post-2011 round that covers the current events and the shadow of the future. Third, the study collects evidence and observations that test the research hypothesis, showing that the game outcomes have varied between sole utilization, cooperation and contention based on the changing influences of the contextual events on the heterogeneities among the riparians over interests, capabilities and beliefs. Doing so not only validates the research hypothesis, but also eliminates the neo-realist argument contending that conflict has prevailed and will continue to prevail because of the scarcity of water resources in the Nile, in which all the riparians are placed under the huge pressure of increasing water, energy and food needs of the rapidly growing populations across the basin. The study also counters the other critical argument suggesting that general norms and principles of international water law are sufficient to generate sustainable cooperation between the river basin's riparian states.

However, there are different techniques for employing process tracing. The study uses implicitly the technique of combining the hoop and smoking gun tests that is developed by Mahoney (2015) to be applied in case of adopting a deductive methodology, in which the research hypothesis is constructed after a general theory. The concept behind these tests can be simplified using the illustration of Van Evera (1997: 31–32), which describes the researcher as an investigator collecting evidence to prove that the accused is guilty. In the hoop test, the investigator needs to show that the accused was in the state not abroad at the time of the murder. Passing the test weakly supports the hypothesis but does not validate it, while failing to pass the test proves it invalid. In the smoking gun test, the investigator needs to show that the accused was seen with a smoking gun on her/his hand after committing the murder. Passing the test strongly validates the hypothesis, while failing to pass the test does not prove it wrong. In this picture, the accused refers to the independent variable or the condition, and the murder to the dependent variable or the outcome, while the smoking gun to the intervening variable or the causal mechanism.

Both the hoop and smoking gun tests are then used respectively as negative and positive tests while tracing the Nile game. This is done throughout the narrative analysis in two subsequent steps: to show that every change in the game outcomes and thus the overall equilibrium is associated with a change in the context (passing the hoop test); and to explain the complete steps through which the context drives the heterogeneities among actors to change the game equilibrium (passing the smoking gun test). The application of this technique can be demonstrated in the following example. The cooperation outcome of the 1929 Agreement, through which Egypt and Sudan reached a compromise over their water allocations, took place in light of a profound change in the game context, as the entire Nile Basin had been under the British control, either by direct colonization or by signing treaties with other European powers. The hoop test is therefore passed by showing that the outcome is associated with a change in the game context. Yet, it is necessary to trace the complete steps linking both of them in order to pass the smoking gun test, as follows. Opening the Suez Canal in 1869 and the subsequent financial and political disturbances in Egypt led Britain to occupy the latter in 1882 to secure the new strategic sea-route to its Indian colonies. Britain realized that the security of the Suez Canal required the stability of Egypt, which fundamentally relied on maintaining the Nile flow downstream. The French attempt to invade Fashoda in 1898 led Britain to occupy Sudan with an Anglo-Egyptian expedition. After the signature of several treaties with Germany, Belgium, France and Ethiopia to secure the Nile sources, the implementation of large-scale irrigation schemes in Sudan soon became a priority for Britain as part of its plan to expand cotton cultivation across its colonies. Britain then sought to convince and even pressure upon Egypt to accept, for the first time ever, the utilization of the Nile waters and construction of large dams upstream of Aswan. Britain's desire to maintain its interests in both of Egypt and Sudan led it to moderate between the two countries until the 1929 Agreement was successfully signed.

1.6 Resources and Limitations

Employing the method of process tracing and passing the combined hoop and smoking gun tests for all the moves along the Nile game require gathering available data about the water agreements, developmental schemes, and the contextual events associated with every round. Hence, the study relied on different resources in doing so. Initially, original technical reports were analyzed to identify the benefits and costs of every scheme and thus to define the payoff gained by relevant riparians. Examples of these are Garstin (1904), Peel (1904), USBR (1964a, b), Howell et al. (1988), EDF and SW (2007), Blackmore and Whittington (2008), EAPP and EAC (2011a, b, c), IPoE (2013), and MIT (2014). In addition, data sources included official statements by water and foreign ministries from different riparian states, non-governmental reports from international donors (e.g. the World Bank), institutions in the Nile Basin (e.g. the NBI), and environmental organizations (e.g. International Rivers). Another main resource, this research relied on, is the collection of the historic statements quoted after the public figures who took part in the studied events, mainly heads of state/government and ministers. Moreover, water and contextual events were extracted from different sources of historical literature, the Transboundary Freshwater Dispute Database (TFDD), media reports, and biographical notes. Statistical data was also used in reference to the World Development Indicators (WDI), the Aquastat Database of the Food and Agricultural Organization (FAO), and the UN World Population Prospects Datasets. Furthermore, during the 2016 Stockholm World Water Week, where the author delivered a presentation at the water security session, several face-to-face unstructured interviews were conducted with the top officials managing the water sector in Egypt, Ethiopia and

Sudan, with the then executive director of the NBI, and with the former president of the Nile Basin Discourse (NBD). Additional interviews were carried out with international experts on water, legal and engineering issues, namely Professor Owen McIntyre, Professor Ashok Swain, Dr. Anders Jaegerskog, and Dr. Ana Elisa Cascao. The purpose of these interviews was to investigate the missing details behind the current events and to provide the author with deeper insights concerning the ongoing interactions in the Nile. All the interviews are recorded in audio files following the permission of the interviewees in line with research ethical guidelines. Although those top officials gave approval to use their comments where appropriate in the text, they are referred to by initials only, not their full names, so as to keep the focus on the value of their statements and not to affect, in any degree, the ongoing talks on the GERD.

It is worth emphasizing that in spite of the interdisciplinary scope of analysis, this study is specialized in the discipline of International Relations. Thus, the study should not be expected to make decisions about sophisticated technical issues related to the Nile hydrological projects. At the same time, the analysis depends on the data provided by the reliable international studies that evaluate such projects. In addition, given that the primary unit of analysis in this study is the riparian states, the aim is not to compile an encyclopedia of the Nile modern history, but to analyze how the contextual events have affected the decision-making process during every round, leading to different outcomes over the studied schemes. It is also important to stress that this study does not aim to judge the actions taken by each of the riparian states. Rather, it looks into the justifications made by the riparian states along the game, since the model deployed by the study requires deep understanding of the interests, capabilities and beliefs of every actor as seen by the actors themselves. In terms of reflexivity, the author of this dissertation is from Egypt. This benefited the research due to his awareness of the complex backgrounds and updates of the Nile story, besides his ability to use Arabic sources in the study. Against the critical notion of Cox (1981: 128) that "knowledge is always for someone and for some purpose," this study, following the rationalist approach, believes that science should be blind and unbiased enough to come up with reliable findings that may be useful for the targeted beneficiaries of the research. These are, in the first place, the next generations in the entire Nile Basin and certainly those in Egypt.

1.7 Book Structure

After introducing the research design and the theoretical and methodological foundations of the study in this chapter, the second chapter builds a generalized conceptualization for the puzzle of transboundary river basins and constructs the analytic framework that models the dilemma of transboundary waters as an iterated PD game. To show the validity of the framework, different examples are given from the Aral Sea, the Jordan, the Indus, the Mekong, the Orange, the Rhine, and the Tigris-Euphrates, in addition to the Nile.

The empirical analysis of the Nile game begins in the third chapter with the colonial round. It models the basin before and after the British occupation of Egypt in 1882, investigating how the presence of one superpower mastering the Nile affected the development of the basin and why some projects were successfully implemented whereas others were not. The Cold War round is covered in the fourth chapter, focusing on the global rivalry between the U.S. and the Soviet Union that developed after the Second World War and persisted until 1991, besides the regional conflicts between the Arab and Israel and between Ethiopia and Somalia, and the civil wars in Ethiopia and Sudan. The role of institutions in the Nile game begins in this round. The chapter therefore tracks the influences of the concurrent political and environmental circumstances on their involvement and effectiveness. The fifth chapter analyzes the post-Cold War round, in which the game context shifts from the global bipolar rivalry to the growing role of international development partners that sponsored the establishment of an inclusive basin-wide platform. The chapter addresses the emergence of new alternatives of technical and financial assistance for the Nile riparians to implement their national developmental plans outside the basin institutional umbrella, such as China and the Gulf States. The analysis then tracks the multilateral and unilateral paths that emerged in the mid-1990s and converged in 2010 with the complete deadlock on the CFA negotiations and with the commencement of large-scale national water projects, particularly in Ethiopia and Sudan. In the sixth chapter, the post-2011 round starts with the independence of South Sudan, the Egyptian Revolution, and Ethiopia's unilateral decision to construct the GERD. It shows how regime transformation in Egypt and the new conditions in Sudan affected their longstanding positions on the Nile. The chapter covers the protracted tripartite negotiations over the GERD, running since May 2011 until the finalization of this book in mid-2018, and it ends with rendering the shadow of the future in light of the rapid population growth, climate change risks, and the ambitious national plans of the riparian states. These four empirical Chaps. (3, 4, 5, and 6) are structured to define the contexts of the round at the beginning (Table 1.1), then trace the events of the round in a narrative form, and eventually analyze how sole utilization, cooperation, and contention outcomes are driven by the multilevel changing contexts through their influences on the inherited heterogeneities amongst different riparian states. Finally, the seventh chapter concludes by presenting the study's main findings and key messages, as well as the policy, theoretical and methodological implications, paving the way for future research.

Chapter 2
The Analytic Framework: Modeling the Dilemma of Transboundary River Basins as an Iterated PD Game

The framework is broad enough to encompass not only people but also nations and bacteria. Nations certainly take actions which can be interpreted as choices in a Prisoner's Dilemma ...

Robert Axelrod (1984: 18)

2.1 Conceptualizing the Puzzle of Transboundary Water Management

2.1.1 Governing the Freshwater Commons Under International Anarchy

One way of analyzing different forms of managing freshwater is to consider it an economic good (Menzel 2014). This is because it passes along its lifecycle through a group of economic processes: collection, production, drainage, sanitation, treatment and reuse. These processes require a set of policies that define priorities of supplies, regulate pricing, and control expected impacts on society and environment. Cornes and Sandler (1986) classified economic goods into four groups, based on the factors of "exclusion" and "rivalry": private, public, club and common goods (Table 2.1). The factor of "exclusion" refers to who can get the service, while "rivalry" indicates how the usage of someone affects the ability of others to use the service. According to this classification, freshwater can take one of those four forms: a bottle of water in supermarket (private); water pipelines constructed by public authorities to supply houses (public); a well in the desert drilled by a tribe (club); or rivers and lakes (common). The most complicated problems are related to the fourth type, since leaving common water resources unregulated and unmanaged, regarded as the "gift of nature," entails severe threats to both the resources and communities dependent on them. In contrast, each of the other types usually has solid mechanisms of regulation and control, even if with different levels of effectiveness and efficiency.

M. M. Samaan, *The Nile Development Game*,
https://doi.org/10.1007/978-3-030-02665-3_2

Table 2.1 The classical classification of economic goods

Factors		Rivalry: Yes	Rivalry: No
Exclusion	Yes	Private goods	Club goods
		Private company (regulated by market: who cannot pay is excluded and the good consumed by one person can no more be consumed by another one)	Association (regulated by its statute: non-members are excluded and using the facilities of club by one member does not prevent others to use them)
	No	Common goods	Public goods
		A gift of nature (regulated by traditions: no one is excluded and the utilization of one person affects others)	Public authority (regulated by government: no one is excluded and there is no rivalry between consumers)

Source: Created by the Author after Menzel (2014: 55–64) citing Cornes and Sandler (1986)

In general, every common-pool resource may witness a "tragedy" when those, who are sharing it, do not assure that rights are distributed in a fair and equitable manner (Hardin 1968). No one will wait to see others benefiting from this shared resource without pursuing self-interests. Everyone, therefore, will "free-ride" in securing her/his maximum benefits (Albanese and Van Fleet 1985), which leads at the end to a situation where the resource is overexploited or depleted and the whole group is worse off (Hardin 1968). The conventional alternative to the "tragedy of the commons" is the transformation either into centralization or into privatization (Hardin 1978: 314). However, both are substantially challenged, due to probable errors in monitoring information, enforcing rules, and sanctioning defectors in the former case, and because of the difficulty to divide the common property into private rights in the latter case (Ostrom 1990: 10, 13). Another alternative is represented in self-governance, in which those sharing the resource negotiate and agree upon the way they will manage it and define the appropriate system of enforcing rules and imposing penalties, either by assigning these administrative tasks to an external body or by creating an internal one (Ostrom 1990: 16). To reframe this debate in the terms of Cornes and Sandler (1986), it can be said that Hardin (1978) discusses the transformation of the "commons" into "public" or "private" goods, while Ostrom (1990) shows how "club" goods of self-organization and self-governance represent the optimal form of institutionalizing the use of "commons."

There are many examples of transforming common freshwater resources into other types of goods. Millennia ago, centralized states were established around alluvial river basins, such as the Nile, Mekong, Euphrates and Indus basins, where solid mechanisms have been developed in order to regulate irrigation, navigation, and risks of floods and droughts (Sadoff and Grey 2002: 391–392). In such nations, governing common waters as a public good by a powerful centralized state gave profound significance to hydraulic control over waterways, the condition of which Wittfogel (1957) calls "hydraulic monopoly." Besides, a traditional example of managing freshwater as a club good is found in pastoral communities relying on rainfall or groundwater. In addition, handling common waters as a private good is often manifested in companies obtaining concessions to sell freshwater, regulated and priced through market mechanisms. Whatever the form into which the common water resources are transformed, the process necessarily requires a supreme jurisdiction, represented in a public authority, a private arbitrator or a users' association, which can be attainable at the local level, not at the global one.

The common feature of world politics upon which different IR schools of thought have agreed is that sovereign states interact together under "anarchy" (Reus-Smit 2005: 190). This indicates specifically the absence of an overarching authority and the lack of solid hierarchical structures similar to those developed at the national and sub-national levels. Since transboundary water resources are shared between groups of sovereign states, the absence of such an ultimate body imposes an additional dimension to the problematic of governing the commons in international river basins. This is evident in the fact that there is no international water court or international water institution exercising supreme jurisdiction over transboundary water conflicts (Wolf 2007: 3.18). Even international water law presents general principles and norms rather than concrete conflict-resolution procedures and mechanisms (Biswas 1999: 429), putting transboundary water interactions under the pressure of international anarchy. Accordingly, the three main schools of thought in IR have had different positions on the consequences of such a configuration (Fig. 2.1). From the neo-realist perspective, the feature of international anarchy leads states to prioritize the use of power in securing relative gains, provoking conflicts in a world of scarce resources (Waltz 1979/2010: 114, 146). In hydropolitics, neo-realists have adopted this general concept, introducing the thesis of "water wars," especially where freshwater is particularly scarce and greatly needed (Starr 1991; Gleick 1993).

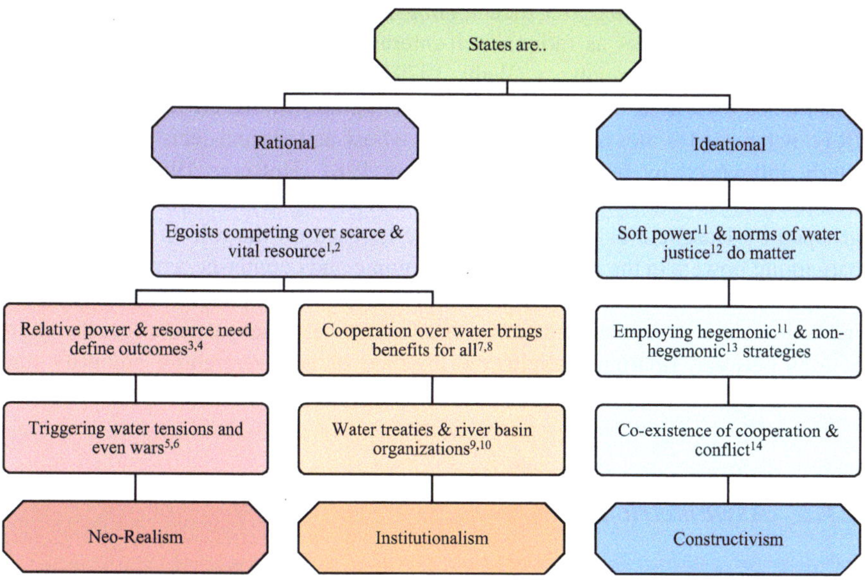

Fig. 2.1 The hydropolitical branches of the major IR schools of thought. (Created by the author after: (1) Elhance (1999: ix, 3); (2) Wolf (2007: 3.5); (3) Lowi (1993: 11); (4) Homer-Dixon (1999: 139); (5) Westing (1986: 9); (6) Starr (1991: 19); (7) Sadoff and Grey (2002); (8) Phillips et al. (2008); (9) De Stefano et al. (2010); (10) Giordano et al. (2014); (11) Zeitoun and Warner (2006); (12) Zeitoun et al. (2014); (13) Cascao (2009); and (14) Zeitoun and Mirumachi (2008))

On the other hand, institutionalists draw close attention to "interdependence" among states and seek to present it as a catalyst for cooperation based upon common interests (Keohane and Nye 1972). Moreover, they argue that international cooperation is achievable under anarchy through reciprocal rather than hierarchical relations, highlighting the role of international institutions in facilitating cooperation among states (Krasner 1983; Keohane 1984/2005; Axelrod and Keohane 1985). As stated by Keohane and Martin (1995: 42), governments "attempt" to construct international institutions when they "can jointly benefit from cooperation," because institutions will then "provide information, reduce transaction costs, make commitments more credible, establish focal points for coordination, and in general facilitate the operation of reciprocity." The institutionalist approach is also applied to governing common natural resources in general (Haas et al. 1993; Keohane and Ostrom 1995) and freshwater bodies in particular (Mostert 2003; Delli Priscoli and Wolf 2009; Schmeier 2013). Scholars that belong to the institutionalist branch of hydropolitics show through empirical analysis how international cooperation over waters exists and takes various forms, such as signing water treaties, establishing river basin organizations, and implementing collaborative hydrologic projects (Delli Priscoli and Wolf 2009; Scheumann et al. 2014; Schmeier 2010). Several case studies have also shown how water institutions can work on sharing information among riparian states, implementing monitory schemes, proposing cooperative projects that benefit all parties, and assuring that good faith of one riparian regarding a certain issue will be then reciprocated in the future by others (Mostert 2003; Schmeier 2013; Hooper and McWilliams 2010).

Hence, despite the fact that the neo-realist and the institutionalist schools of thought differ profoundly over their findings, both build their arguments upon the representation of states as rational self-interested actors, adopting the microeconomic theory in their analyses (Waltz 1979/2010: 89; Keohane 1984/2005: 29). Established on a radically different ontological platform, the critical/constructivist IR branch regards states as ideational actors whose actions and decisions are significantly influenced by beliefs and ideas (Cox 1981; Hoffman 1989; George and Campbell 1990). The hydropolitical studies that belong to this cognitivist family, as grouped by Hasenclever et al. (1997), have focused on the role of ideational and bargaining powers in the formation of hegemonic and counter-hegemonic strategies by riparian states (Zeitoun and Warner 2006; Cascao 2008, 2009). Thus, they examine how various principles and norms of international water justice affect the interactions between riparians despite the absence of an overarching authority that enforces rules among them (Mirumachi et al. 2013; Zeitoun et al. 2014).

2.1.2 *Asymmetric Interdependence*

The problematic issues of transboundary river basins are not limited to governing the commons under international anarchy, but also involve the so-called "asymmetric interdependence" in upstream–downstream constellation (Dinar 2006;

Scheumann et al. 2014; Ovodenko 2014). Shared water resources in transboundary river basins create a sort of interdependence among riparian states but in asymmetric form. This is due to the hydrologic nature of international rivers flowing across political borders, whereby the stream defines which states are located upper or lower the river basin. Such a configuration is characterized by asymmetric consequences of each riparian's actions on the others, which is not necessarily the case in other shared water bodies, such as international lakes, seas, and border rivers (Ovodenko 2014). This asymmetric interdependence, shaped by nature, provides upstream states with a significant sort of power gained by its geographic position (Zeitoun and Warner 2006). Conflicts, therefore, often arise between upstream and downstream states over projects constructed upper the river, especially large-scale dams and irrigation schemes, either if the upstream state implements such projects without taking into account the negative impacts created lower the river or if the downstream state attempts to halt them in every way to prevent any potential risks (Scheumann et al. 2014).

This upstream–downstream configuration is further complicated by the controversial conceptions of extreme water rights adopted by each side. In most cases, the upstream riparian asserts the absolute right of "national sovereignty" to utilize the waters flowing across its territory, while the downstream riparian advocates the absolute right of "riverain integrity" to receive its historic and acquired share of waters without any reduction or alteration (Wolf 1999: 5, 8). These controversies in the upstream–downstream extreme water rights, also referred to as the debate over "hydrography versus chronology of use" (Delli Priscoli and Wolf 2009: 70), have shaped the foundational principles of "reasonable and equitable use," "prior notification," and "causing no appreciable harm" in international water law. These principles recognize the rights of both sides and at the same time define the limits for those absolute conceptions by putting obligations on each side to respect the others' needs and rights (Wolf 1999: 5; Anisfeld 2010: 265). Based on these, the downstream state should share the watercourse in an equitable way with its co-riparians, on the one hand, and the upstream state should cause no harm to others and notifies them in advance about its potential projects, on the other. However, such principles are considered "broad" and "vague," due to the open and varying interpretations about what defines "equitable" sharing and "appreciable" harm (Biswas 1999: 429).

2.1.3 Vitality, Variability and Vulnerability of Freshwater Resources

While freshwater is regarded as the source of life, it is also a source of risk and vulnerability (Watkins 2006). Freshwater is vitally needed for fueling various sectors of development in every nation. Besides its basic use for drinking and sanitation, it is a fundamental element in the fields of food production, power generation, industry, navigation, and tourism (Phillips et al. 2008). Such multiple uses define water management as a multi-objective and multi-stakeholder process (Wolf 2007: 3.5), creating a "nexus" of assorted tradeoffs and synergies with other sectors, notably food, energy, land, and climate (WEF 2011; Bazilian et al. 2011; Lawford et al. 2013). In addition to vitality, freshwater is a non-stationary and variable resource in both space and time, and it is vulnerable to any sort of fragmentation in both quantity and quality (Anisfeld 2010: 12–16). Currently, the variability of freshwater resources is subject to uncertainty under climate change risks, in addition to the increased vulnerability associated with land fragmentation, which has been further exacerbated due to the growing rates of urbanization, industrialization, and construction of mega hydraulic projects. The characteristics of vitality, variability and vulnerability place heavy burdens, not only on the resource itself, but also on the ecosystems and communities dependent on it. States, therefore, seek to develop their national water policies in order to prioritize and balance the wide range of benefits and costs associated with water resources development projects under various challenges and changes, and to coordinate different stakeholders' needs in an integrative perspective (GWP 2010). Moreover, international institutions have developed a group of norms in order to define "good governance" in general, including accountability, transparency, public participation, and rule of law (World Bank 2013a). Furthermore, particular procedures have been recommended for considering environmental and social sustainability when constructing large-scale dams.

The most significant framework in this regard is the one provided by the World Commission on Dams, conducted in the late 1990s, which presented an extensive study to evaluate the processes upon which dams were constructed throughout the second half of the twentieth century. Its published report (WCD 2000) listed several social and ecological expensive costs that are often generated because of prioritizing the dams' economic benefits of hydropower, year-round agriculture, flood control, and water storage. Examples of those expensive costs are altering water flows in quantity, quality, and timing, devastating ecosystems and biodiversity, emitting greenhouse gases due to the submergence of vegetation cover under dams' reservoirs, eroding riverbeds, banks and deltas, displacing local people, and threatening the livelihoods dependent on fisheries and traditional floodplain agriculture. In order to balance these very complicated and interconnected benefits and costs, the WCD developed a framework for decision-making founded on a participatory approach. The framework emphasized the comprehensive involvement of various stakeholders in projects' design and planning, evaluation of alternatives, and setting mechanisms of compliance and accountability. All these considerations aimed at securing the interests of vulnerable groups and assuring the fair distribution of benefits and costs among stakeholders.

Nevertheless, when those water-related projects are constructed on internationally shared waterways, their benefits and costs are distributed among more than one sovereign state, triggering conflict or cooperation among them, based on the net gains brought to each (Sadoff and Grey 2002). Thus, transboundary water management becomes a more complicated process involving numerous users and stakeholders from different sectors and at multiple levels (Mostert 2003). In other words, governing transboundary river basins is an intricate process in which the interests of relevant state and non-state actors are obviously varying, from harmonization and integration to tensions and disputes. The geopolitical dimension is then added to the biophysical and socio-economic dimensions (Brown et al. 2009), challenging the effectiveness of norms defining good governance and sustainability in transboundary basin management. As for international water law, specifically the UN 1997 International Watercourses Convention (UNWC), a set of principles are devoted to urge riparian states to give "special regards" to "the requirements of vital human needs," and to consider the "balance of uses," taking into account environmental, social and economic impacts of current and potential plans. Yet, these principles are again general and subject to opened and varying interpretations (Anisfeld 2010: 265).

2.2 Setting the Game of Development on Transboundary River Basins

2.2.1 Mixed Motives over Transboundary Waters

In light of the conceptualization discussed above addressing transboundary water management, this study identified Game Theory (GT) as a holistic approach for modeling hydropolitical interactions in internationally-shared basins. As discussed briefly in the introductory chapter, riparian states, in the GT perspective, are considered a group of players, each of which believes that straight-forward maximization of utility is not a rational choice, as the consequences of one's course of action partly depend on the decisions of others (Ward 2002: 66). However, Kilgour and Wolinsky-Nahmias (2004) distinguish cooperative from non-cooperative games. In the former, riparian states are able to communicate and work together seeking fair distribution of benefits generated from their co-management of the river (Whittington et al. 2005; Wu and Whittington 2006), while in the latter, they compete with one another aiming to achieve maximum possible benefits (Waterbury and Whittington 1999; Madani et al. 2011). In addition, there are two different types of non-cooperative games: zero-sum and non-zero-sum games. The first type models those situations where everything gained by one player is certainly lost by the other (Chwaszcza 2008: 144). Thus, outcomes of such games are either win-lose or lose-lose scenarios. In contrast, the second type addresses those situations where the interests of players are neither completely opposing nor harmonized with one another, but a "mixture of both conflictive and complementary interests" exists (Axelrod and Keohane 1985: 226). Therefore, this type is also called "mixed-motives" game, in which there is a chance for both players to achieve a win-win outcome through cooperation (Chwaszcza 2008: 145). Limiting transboundary water management to quantitative water allocation leads to zero-sum outcomes, since every cubic meter of water allocated to one state is subtracted from others.

Table 2.2 Possible scenarios and payoffs generated according to the decision of each players

Scenarios		Player Y Cooperative	Defect
Player X	Cooperate	Cooperation (R-R) X and Y cooperate together Both get rewarding payoffs	One-sided exploitation (T-S) X is exploited by Y X gets the sucker's payoff Y gets tempting payoff
	Defect	One-sided exploitation (S-T) X exploits Y X gets tempting payoff Y gets the sucker's payoff	Contention (P-P) X and Y defect together Both get punishment payoff

Source: Created by the author after Axelrod (1981: 306)

Taking into account the upstream--downstream debate on extreme water rights, such a zero-sum game turns ultimately into an exhausting tug-of-war. Contrarily, choosing to share benefits, rather than provoking conflicts over water allocations, creates a range of common interests, transforming the model's interactions into a mixed-motive game (Phillips 2005).

2.2.2 *Choices and Preferences in the Hydropolitical Realm*

As illustrated by Anisfeld (2010: 271–272), a perfect abstraction of the transboundary water rights puzzle is the two-person Prisoners' Dilemma (PD) game (see fn. 2). Each riparian in this game has the choice whether to pursue its interests only in coordination with the other (to "cooperate") or to implement its plans unilaterally paying little attention to the interests of the other (to "defect"). The combinations of the choices made by each side, to cooperate (C) or to defect (D), generate a matrix of four scenarios, as plotted in Table 2.2. These four scenarios are sole utilization by either (DC or CD), cooperation (CC) and contention (DD). Each riparian will sort its preferences differently under these four scenarios as follows. The best scenario at all for each is to utilize the basin resources alone, yielding maximum benefits from the river without being harmed by the uses of the other (DC). Conversely, the worst at all for each is to be unable to pursue its water rights while the other utilizes the basin resources solely (CD). In addition, the scenario of cooperative management of the basin (CC) remains better than the simultaneous implementation of unilateral projects by both (DD). Accordingly, these scenarios are arranged from the best to the worst as DC, CC, DD and CD (Table 2.3), which is the typical order of preferences in the PD game (Axelrod and Keohane 1985: 229).

There are other examples of mixed-motives games with a different order of preferences. However, this study argues that the abstraction of the PD game represents the typical dilemma of transboundary water interactions, better than the Stag

Table 2.3 The order of preferences in the PD game and other mixed-motives games

Preferences	Best		Better		Worse		Worst
Payoffs	Tempting (T)		Rewarding (R)		Punishment (P)		The Sucker's (S)
Order	DC	>	CC	>	DD	>	CD
Other games	CC> DC **Stag Hunt** Cooperation is the best		DD > CC **Deadlock** Contention is better than cooperation		CD > DD **Chicken** Contention is the worst		

Source: Created by the author after Axelrod and Keohane (1985)

Hunt,[1] the Chicken,[2] and the Deadlock games. In the Stag Hunt game, each side believes that cooperation is the best scenario at all, even better than sole defection (CC > DC > DD > CD). This is a symmetric game, where choosing to cooperate brings either the best or the worst scenario at all (CC or CD). Yet, in the realm of hydropolitics, where there are always transboundary tradeoffs, each riparian often associates cooperation over shared resources with the necessity to make concessions and to respect constraints. Thus, typically, it is not a Stag Hunt game, since the scenario of absolute unilateral utilization is still more preferred, from the national perspective as opposite to making concessions to reach cooperation. In the Chicken game, each side believes that the worst scenario at all is contention, even worse than being exploited by the other (DC > CC > CD > DD). This is also a symmetric game, where choosing to defect brings either the best or the worst scenario at all (DC or DD). Looking back to the hydropolitical realm, one can deduct that nothing is worst for a riparian than being not able to obtain any of its water rights while the other yields all possible benefits of the river unilaterally. Therefore, it is also not a typical Chicken game. In the Deadlock game, each side believes that contention is better than cooperation (DC > DD > CC > CD). This game drives nowhere but to a deadlock since choosing to defect brings the two most preferred scenarios for each (DC & DD). Thus, both will definitely choose to defect. Nevertheless, cooperation over transboundary waters remains more preferred than contention, as the simultaneous implementation of unilateral projects by each riparian will lead to considerable costs for both, those of which could have been mitigated in case of cooperation. Overall, it can be argued that none of these games can model the hydropolitical interactions more accurate than the PD game.

[1] The Stag Hunt game: "Two hunters can either jointly hunt a stag (an adult deer and a rather large meal) or individually hunt a rabbit (tasty, but substantially less feeling). Hunting stags in quite challenging and requires cooperation. If either hunts a stag alone, the chance of success is minimal. Hunting stags is collectively most beneficial but requires a lot of trust among the hunters." The game is named after a famous passage in Rousseau, as reviewed in Chwaszcza (2008: 154–155).

[2] The Chicken game: "Two adolescents decide to resolve a dispute by riding towards each other down the middle of a road. The first to turn away loses. If both continue straight ahead, they will crash and risk serious injury," as illustrated by Hargreaves Heap et al. (1992) reviewed in Chwaszcza (2008: 160).

2.2.3 Scenarios of Utilizing the River and the Generated Payoffs

Each of the four scenarios in the PD game brings different payoffs to the involved riparian states. As stated by Axelrod (1981: 306), the highest payoff is the "temptation" to defect unilaterally (T), leaving the other with the "sucker's" payoff (S), which is the lowest. In this case, it is said that the cooperator is "exploited" by the defector. Whereas, contention brings a "punishment" payoff (P) for both, which is lower than the "reward" for cooperation (R). One additional statement about the payoffs in the PD game is that R remains higher than the average of both T and S, which makes cooperation still better than alteration between exploiting and being exploited by the other.

The application of this payoffs' structure on transboundary river basins can be simply exemplified through the concept of compensation (i.e. why a riparian would compensate another). Cases of dams constructed on the Columbia River and the Lesotho Highlands Water Project (LHWP) are good examples in this regard (Scheumann et al. 2014: 112; Wolf and Newton 2009: 214–216). In the former case, USA compensated Canada, based on the 1961 Columbia Treaty, by paying half the cost of the flood protection and electricity generation. In the latter case, South Africa buys water from Lesotho and pays costs of water transfer based on the 1986 LHWP Treaty, while Lesotho, supported technically and financially by the World Bank among other donors, gets a share from the generated hydropower. Hypothetically, Canada and Lesotho could have carried out unilateral projects without minimizing potential risks and sharing benefits with their downstream co-riparians. Therefore, all sides avoided getting the sucker's payoff (S) generated in case if one side exploited the other (CD). This is done through cooperative implementation (CC) of the project alternative that harms neither side (P), but rewards benefits for both (R). At the same time, the upper riparians are compensated for limiting their potential benefits, comparing the tempting payoff (T) of utilizing the river solely (DC) against the rewarding payoff (R) of the cooperative scenario (CC) that makes both sides better off.

Furthermore, Axelrod (1984: 17) introduced a broad perception of this structure, arguing that those payoffs need not be "comparable" nor "equivalent.[3]" Obviously, such characteristics are deeply significant when modeling hydropolitical games. Notably, the payoffs generated from utilizing the basin resources vary in their values and forms. This is referred to, in the hydropolitical scholarship, as the "basket of benefits" (Sadoff and Grey 2002; Wichelns et al. 2003; Phillips et al. 2006, 2008; NBI 2009). For instance, Sadoff and Grey (2002) put forward four categories of benefits to, from, because of and beyond the river, indicating respectively the environmental aspect (e.g. biodiversity and water quality improvement), the direct socio-economic aspect (e.g. agriculture, flood control, and hydropower), the

[3] This means that if players "X" and "Y" are playing together, the payoffs T_X, R_X, P_X and T_X need not be equal to T_Y, R_Y, P_Y and S_Y.

hydropolitical aspect (conflict resolution), and the indirect economic and geopolitical aspect (e.g. regional economic integration). Besides, the distribution of payoffs among riparians need not be symmetric in magnitude, depending on the geographic position and power of each, as will be discussed later in this chapter. In particular, the punishment and the sucker's payoffs (S) differ substantially between upstream and downstream riparians. On the one hand, when an upstream riparian defects by implementing unilateral projects, the downstream riparian suffers tangible risks, such as the alteration of flows and the devastation of ecosystems. On the other hand, unilateral downstream projects may foreclose future development upper the basin, because they create new facts on the ground that redefine potential significant harm to be caused by prospective upstream projects, which is referred to as the "opportunity cost" paid by the upstream riparian (Sadoff and Grey 2005: 424). It is worth noting, however, that Articles 11–19 in Part III of the 1997 UNWC and the Operational Directive 7.50 of the World Bank for Projects on International Waterways do not distinguish upstream and downstream riparian states in giving both the right of being consulted prior to construction of any water scheme by the other (Salman 2009: 115–117, 246–252).

Hence, it is important to stress here that payoffs need not be uniform, considering that hydrologic projects generate different environmental and socio-economic externalities. These are not always distributed fairly among various stakeholders. In many cases, riparians focus on the economic benefits and overlook social and ecological aspects. As a result, vulnerable groups may suffer from intolerable costs, even if the project is an outcome of cooperation. An important example is the Senegal River Basin (Niasse 2010; Wolf and Newton 2009: 239–243). Although hydraulic infrastructure, namely the Diama and Manantali Dams, were constructed through collaborative mechanisms, their plans did not consider those potential adverse impacts on local communities and ecosystems. Degradation of ecosystems, erosion of banks, and spread of water-related diseases were some of the negative impacts caused by these dams. Serious efforts were taken by the end of the 1990s to mitigate such costs and promote the involvement of various stakeholders in the decision-making process. Recently, there have been growing concerns for stakeholders' involvement, as this has become an international norm (Sect. 2.1.3). Examples of this are the cases of the International Commission for the Protection of the Danube River (ICPDR), the US-Canadian boundary waters' International Joint Commission (IJC), the La Plata Intergovernmental Coordinating Committee (CIC), and the Orange-Senqu River Commission (ORASECOM) (Hooper and McWilliams 2010).

2.2.4 Not a One-Move Game But an Iterated Game

If the PD game is played once (one-move game), the dominant choice for both players will be "to defect" (Axelrod 1981: 306–307). As shown in Fig. 2.2, in case the competitor chooses to cooperate, deciding to defect will be the best choice, since unilateral defection (DC) is more yielding than cooperation (CC). Whereas, in case

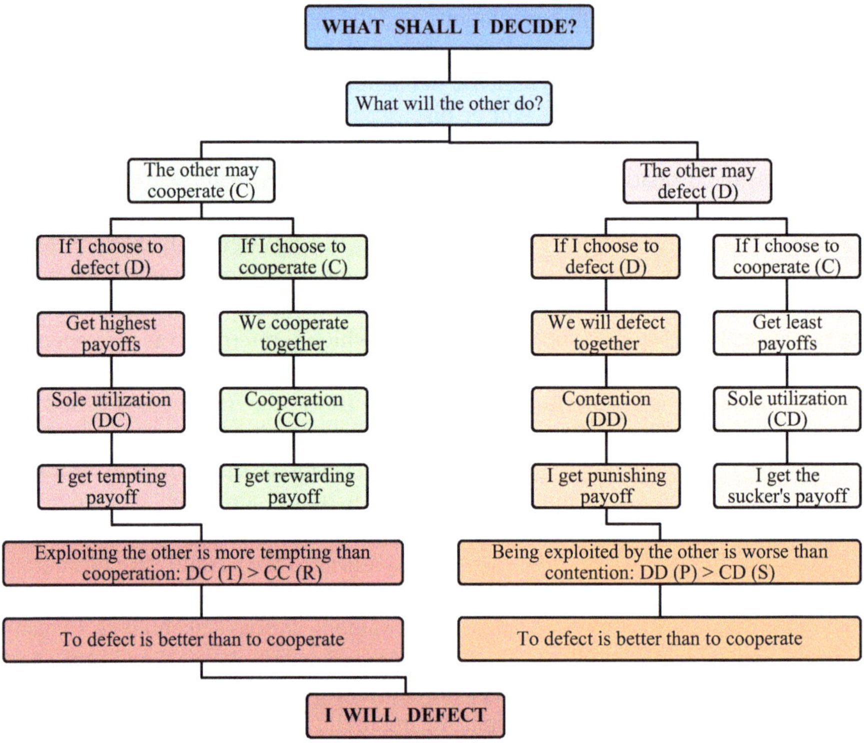

Fig. 2.2 The logic each player will follow while taking its decision in the one-shot PD game

she/he chooses to defect, responding to defection by defection will bring the higher payoff, as contention (DD) is better than being exploited by the other (CD). This means that whatever the other decides, it is better to defect. Because both build their decisions in the same logic, contention (DD) will be the game outcome, by which both will be worse off than if they had mutually cooperated. Moreover, a finitely repeated PD game will not lead to a different outcome, thinking about the last move and coming the whole way round (Luce and Raiffa 1957: 94–102). Still, if the game is played with a high probability of iteration, there will be a chance for different strategies competing to obtain the greatest accumulated payoffs. Thus, in every move, players will not only think about the payoff to be obtained in the present, but also about the cumulative sum of payoffs in the future, which is described by Axelrod (1984: 12) as the future casting a "shadow back upon the present.". However, as he argued, it is a fact that "the future is less important than the present" because "players tend to value payoffs less as the time of their obtainment recedes into the future" and because "there is always some chance that the players will not meet again" (1984: 12). He therefore introduced the W discount parameter that represents the weight or importance of the next move relative to the current move (between 0 and 1) and thereby determines the cumulative sum of payoffs for a whole sequence of moves (1984: 13). A general finding of his work is that the larger this W parameter, the higher the probability of cooperation to emerge and be sustained (1984: 59, 207). It is noteworthy here that the iteration of moves does not necessarily mean that these are played one after another. Axelrod (2000) asserts that the moves in an iterated game can take place either consequently or simultaneously.

Apparently, hydropolitics on international rivers can be modeled as an iterated open game with a high value of the W parameter because of its natural hydrologic cycle, in that as long as the water stream flows, the hydropolitical game will keep iterating. Such a long "shadow of the future" provides the game with the conditions of "long time horizon" and "regularity of stacks," which are the most significant pre-requirements for establishing cooperation (Axelrod and Keohane 1985: 232). Reality, as surveyed through the TFDD,[4] proves that the majority of water events in transboundary river basins tend to be more cooperative than conflictive. The overall theme of the project, as stated by Hamner and Wolf (1998), points to the fact that "the weight of historic evidence tends to favor water as a catalyst for cooperation," and that "water, by its very nature, tends to induce even hostile co-riparians to cooperate." In addition, the TFDD massive collection hints that every hydropolitical game comprises a series of subsequent and simultaneous moves (i.e. water agreements and developmental schemes), each of which differs in its circumstances and outcomes.

An explicit example of the influence of the iteration feature that characterizes hydropolitical games is the case of the Rhine River Basin. As reviewed in Sadoff and Grey (2002: 394), the growing populations and the over-contaminating industries in the Rhine Basin led to the complete extinction of salmon in the river by the 1920s. By the mid-twentieth century, the river was known as the "sewer of Europe." In 1987, the ministers of the Rhine countries agreed on a plan to reduce the chemical contaminants of the river to the level that bring salmon back to life by 2000. The plan, which was named "Salmon 2000," achieved its goals successfully mainly due to the intensive international cooperation, private investments and public support. More schemes of cooperation have been undertaken afterwards in the Rhine for further coordination about flood control and environmental issues. In this example, riparians decided to shift their strategic choice from defection into cooperation as an attempt to rescue the river's ecosystems, when everyone realized that iterated contention (DD) is bringing punishment payoffs (P) for the whole group. Eventually, successful and fruitful cooperation (CC) is being iterated instead.

The Aral Sea Basin is another example, as reviewed in Glantz (2007). Since the 1950s, during the Soviet era, the lake has witnessed severe environmental problems because of the extensive cultivation of cotton. Within 50 years, about three quarters of the lake's total volume diminished. Water quality and quantity have been dramatically deteriorated and the lake has been broken into several

---

[4]The Transboundary Freshwater Dispute Database (TFDD), provided by the Oregon State University, is one of the most important projects addressing empirical data about cooperation and conflict in international basins. The TFDD comprises, among others, massive collections of international freshwater-related events and treaties, in addition to studies analyzing both datasets. All related datasets published by the TFDD are available at http://www.transboundarywaters.orst.edu/

shrunken parts, given the unrestricted water withdrawal from both the Amudarya and the Syrdarya rivers, the main tributaries feeding the Aral Sea, in addition to the excessive use of pesticides and fertilizers. Environmental degradation of the Aral Sea resulted in wide adverse impacts on biodiversity and human health and livelihoods in the disaster zones. Recently, serious efforts have been made by riparian states for the partial restoration of the lake. However, this is significantly challenged due to the expensive political and socio-economic sacrifices that should be made on local and regional levels. The Aral Sea's case embodies, in brief, "what can happen if one does not respect the limits to the exploitation of nature" (Glantz 2007: 325). Moreover, this example illustrates how iterated contention can even shorten the shadow of the future in such hydropolitical games as a result of threatening the physical presence of the shared water resource through riparians' unsustainable uses.

A third example of iterated contention is the series of unilateral actions carried out by water-scarce riparian states sharing the Jordan River Basin[5] in the 1960s, including the Israeli National Water Carrier (NWC), the Syrian dams and diversion plans on the tributaries of the Yarmouk, and the Jordanian King Abdullah Canal. Anisfeld (2010: 276) specifies three main negative consequences of such "unilateralism": threats of violence and military actions between Israel and Syria; severe economic and hydrologic costs due to the lack of water resources co-management; and the degradation of the Lower Jordan River ecosystem. These adverse outcomes represent punishment payoffs (P) for all; though distributed in an asymmetric manner among riparians and in different forms. In spite of extended different international efforts, the complexity of political tensions between Arab riparians and Israel formed a massive obstacle toward the achievement of multilateral cooperation in the basin.

The Senegal River Basin is a fourth example that shows how the vulnerability to climatic variability may catalyze the establishment of multilateral plans for the welfare of the whole group (Niasse 2010; Wolf and Newton 2009: 239–243). During the 1960s and the 1970s, the dramatic drop of rainfalls led the newly independent riparians, Guinea, Mali, Mauritania and Senegal, to launch an integrative basin-wide management. In 1972, the Senegal River Basin Development Organization (OMVS) was established without the full membership of Guinea due to regional political tensions, particularly with Senegal. However, the OMVS has been able to develop various legal and institutional mechanisms to govern the basin shared resources. Receiving international funds and loans, the riparian states have constructed a series of water developmental projects through the conceptualization of multilateral ownership and partnership. Those comprise the Diama and Manantali Dams, constructed in the 1980s for irrigation and hydropower purposes. In general, this example reflects how the collective awareness about the importance of the future stimulated the choice of the riparians to cooperate.

---

[5] Refer to Khaled Abusamhadaneh, (2012). Transboundary water management in the Western part of Jordan: Challenges and solutions – M.A. thesis, Institute of Social Sciences, TU-Braunschweig, which was funded by the EXCEED project that is also sponsoring this doctoral dissertation.

2.2.5 Bilateral Relations Between Riparians and the Formation of Clusters

The PD model, used in Axelrod's paradigm, is a two-person game, in which several players interact together in pairs. However, the concept of "clusters" was introduced to consider the possibility that sub-groups can be formed gathering individuals that meet with each other more often than with those outside the cluster (Axelrod 1981: 315). These two features, pairwise interaction and cluster formation, allow for the consideration of bilateral relations between riparian states in addition to the creation of partial coalitions. Such clusters are formed in transboundary river basins due to either hydrogeographic or political reasons. For instance, in the Mekong case (Sokhem 2010; Wolf and Newton 2009: 216), the lower riparians established a cluster for promoting cooperation among one another. The structure and membership of the cluster have changed since its formation in 1957. Currently, the Mekong River Commission (MRC), launched in 1995, involves Cambodia, Laos, Thailand and Vietnam, while China and Myanmar (the upper riparians) participate as dialogue partners. Nevertheless, the basin-wide developmental schemes of MRC are substantially challenged by both the restrictions of the members' national rules and the unilateral plans adopted by China. These problems show how national agendas of riparians inside and outside a cluster can limit its efficacy.

In the Nile case, two forms of clusters are apparent. The hydrologic characteristics of the basin create two sub-basin regions, the Eastern Nile Basin and the Equatorial Lakes (NBI 2012a), which shapes the administrative structure of the NBI (Sect. 5.2.2). Besides, the legal regime governing the Nile waters, which has been developed in the basin for over a century, resulted in another two clusters. On the one hand, the downstream countries, Egypt and Sudan, advocate their historical and natural water rights based on the 1929 and the 1959 treaties (Sects. 3.3.2 and 4.2.2). On the other hand, those agreements are viewed by the upstream countries as unfair (Degefu 2003) and overlooking the principle of equitable use (Ibrahim 2011). This explains the deadlock at which the two clusters arrived in 2010 over the CFA after more than a decade of protracted multilateral negotiations with the support of several international developmental partners (Sect. 5.4.3). Most of the upstream riparians eventually signed the disputed draft separately despite the rejection by Egypt and Sudan because their current water rights and uses are not recognized, in addition to the absence of the prior consultation principle as stated in the 1997 UNWC (EMFA 2010). Nevertheless, at the bilateral level, different agreements were signed in the early 1990s (Sect. 5.2.1) between Egypt and Uganda, Egypt and Ethiopia, and Ethiopia and Sudan. Accordingly, the framework is able to recognize the bilateral relations between every pair of riparians and the interactions between clusters of states formed due to natural or political reasons.

2.2.6 The Hydropolitical Game Running Under Anarchy

A fundamental feature of the game is that the interactions are running under "anarchy," the concept on which Axelrod and Keohane (1985: 226) emphasized to indicate the absence of central authority forcing rules or norms among players. This is evident in the case of international freshwater resources, in that there is no international water court or any other international water institution of supreme jurisdiction over sovereign states (Sect. 2.1.1). Thus, the 1997 UNWC general principles of equitable and reasonable utilization, causing no significant harm, and prior consultation are left to the consideration of every riparian.

An informative example is the dispute between Hungary and Czechoslovakia (now Slovakia) over the Gabcikovo-Nagymaros Project on the Danube River, which is the first test case for international water law over an absolute water dispute. As illustrated by McIntyre (2010), both countries signed in 1977 a treaty on the construction and operation of the project, which includes two dams and systems of locks in both territories. Environmental lobbies in Hungary objected the project for its serious potential risks on water quality, aquifers' recharge, and ecological wetland areas. Years after, the Hungarian government took its decision to halt the construction work until a full environmental impact assessment is processed. In response, the Slovakian government decided to carry out unilaterally an alternative solution called "Variant C," diverting the Danube into its territory, the scheme of which significantly reduced the water flows running to Hungary. The latter, as a result, terminated the 1977 Treaty. Failing to resolve the dispute by negotiations, both agreed on submitting the case to the International Court of Justice (ICJ). In 1997, the Court's judgment came to vindicate neither side, indicating that Hungary violated the 1977 Treaty and that the Slovakian unilateral implementation of "Variant C" was illegal. The ICJ eventually recommended that both should re-establish cooperative management of their shared resources. A final solution is not yet achieved. However, attempts to resolve this conflict cooperatively have proceeded under the umbrella of the EU's Water Framework Directive. In short, "during the course of the Gabcikovo-Nagymaros case, it became apparent that the court was not a particularly effective tool for dispute resolution" (McIntyre 2010: 229).

Nevertheless, there are other cases showing that arbitration or involvement of a third party may play a positive role in settling water conflicts in a world governed by anarchy. Three examples are illustrated in this regard: the role of arbitration between France and Spain in signing the 1958 Lac Lanoux Treaty, the role of World Bank's mediation between India and Pakistan in signing the 1960 Indus Water Treaty, and the facilitation of Saudi Arabia to settle down the dispute between Syria and Iraq over the Euphrates in 1975.

As reviewed in Wolf (1999: 8), the dispute between France and Spain over the Carol river arose when the former decided to divert the waterway for hydropower generation and compensate the latter monetarily. France, on the one hand, asserted its right of absolute sovereignty as an upstream state, while Spain, on the other, claimed as a downstream country the principle of absolute riverain integrity and rejected the French decision. The arbitration tribunal negated both absolute rights, paving the way to the signature of 1958 Lac Lanoux Treaty, which allowed France to divert a portion of the river's water for hydropower and to return the same quantity back to the stream before reaching the Spanish territory. The significant role of the tribunal decision is not limited to promoting cooperation between both sides, as it contributed to establishing the principle of equitable and reasonable use, allowing upstream water development without causing appreciable harm to downstream communities.

In the second example of the Indus Basin, the World Bank played a crucial role as a mediator between two countries that experienced a long history of political crises. As a neutral body, the World Bank succeeded to bridge the gap between India and Pakistan. It proposed splitting the basin's Eastern and Western tributaries between India and Pakistan respectively (Wolf and Newton 2009: 190–195), which was considered the best possible solution in such complicated circumstances. The Bank assisted both in raising funds from the international community for the implementation of new water projects and in setting adequate mechanisms for conflict resolution. Consequently, India and Pakistan were able to sign the 1960 Indus Water Treaty.

Saudi Arabia, in the third example, facilitated the avoidance of war between Iraq and Syria in 1975, when the water of the Euphrates flowing from the latter into the former was reduced dramatically due to the reservoirs' filling of the Turkish Keban Dam and the Syrian Tabqa Dam during dry years (Leb 2010). Supported by the Saudi Arabian mediation, Syria, the midstream riparian between Turkey and Iraq, agreed to keep 42% of the flow and release the remaining amounts to run downstream towards the Iraqi territory (Wolf and Newton 2009: 243–246).

Yet, all the examples shown above emphasize an important fact that arbitration or third-party involvement requires the acceptance of all concerned riparians and that the effectiveness of the step depends on their willingness to resolve the dispute in good faith.

2.3 The Game Equilibrium

2.3.1 Strategy of Reciprocity

Following the discussion presented in the previous section concerning the situational settings of the hydropolitical PD game, this section addresses the way in which every riparian builds its strategy. Based on Axelrod's game theoretic perspective (Axelrod 1984: 14), a "strategy" is defined as a "decision rule" to specify "what to do in any situation that might arise," analyzing the "history of the game so far." The adopted strategy by a player defines her/his choice whether to cooperate or to defect in every move along the game. Players develop their strategies with the purpose of reaping the maximum possible payoffs in accordance with the performance of the others, since all of them are aware that the outcome depends partly on one's own decision and partly on the competitor's decision. Thus, in every move, each player expects what will be the other's choice and what is the other's expectation about her/his own choice, resulting in the so-called "mutual dependency of choices" or "circular expectations about expectations" (Chwaszcza 2008: 144; Ward 2002: 66–69). Development of strategies continues until the game arrives at an equilibrium point. The most widely accepted understanding of equilibrium in mixed-motives games is the one developed by John Nash, which states that equilibrium takes place when every player finds "the best counter-strategy to what one expects the other person(s)' choice will be," as simplified by Chwaszcza (2008: 145). This means that at equilibrium, no one can be better off by only changing her/his own strategy while the others' strategies are remained unchanged.

Axelrod (1980a, b) and Axelrod and Hamilton (1981) ran a series of real and hypothetical computer tournaments for an iterated two-person PD game, in addition to processing ecological simulation, in order to compare the fruition of all possible strategies. It was found that Tit for Tat (TFT), a "strategy based upon reciprocity," is the one that did the best among others. As defined by Axelrod and Dion (1988: 1385), Tit for Tat, simply, is a strategy that never defects first ("nice"), responds with defection when the other defects ("provokable"), punishes once and then cooperates again if the other returns to cooperation ("forgiving"), and adheres to these rules and is easily understandable for others ("clear"). The TFT obtained the maximum accumulated payoffs along the sequential moves of the computer tournament, and it proved the fittest to survive among the population of competing strategies in the ecological simulation. Moreover, Axelrod (1981) showed via theoretical reasoning that when a cluster of TFT players is created, regardless of their proportion among the total population and whatever defective strategies are employed by the others, cooperation can be collectively stabilized as the Nash equilibrium point of the game.[6] Despite its very simplicity, beginning with cooperation and doing whatever the other did in the preceding move, empirical and theoretical analysis confirmed that "if everyone else is using Tit for Tat, you can do no better than to do the same, and cooperate with them," provided that the shadow of the future is sufficiently long (Axelrod 1984: 59). This means that a group of players employing TFT can maintain the game equilibrium at cooperation, and no outsider can invade them employing another strategy.

When applying these findings to different games of world politics, Axelrod and Keohane (1985) indicated that the emergence and sustainability of cooperation based upon reciprocity is greatly challenged by the problems of identifying, deterring and sanctioning defection. This is evident in hydropolitical games. When a

[6]A strategy is "collectively stable," if "no strategy can invade it," where a new strategy "A"—employed by an incoming player—is said to "invade" a native strategy "B," if "the newcomer gets a higher score with a native than a native gets with another native" (Axelrod 1984: 56). This condition of "invasion" can be written in the following inequality: $V(A/B) > V(B/B)$, whereas $V(A/B)$ is the accumulated payoffs gained by player using strategy "A" when playing with another using strategy "B," and the same for $V(B/B)$ for two players using strategy "B." In addition, a collectively-stable strategy is said to be in Nash equilibrium with itself (Axelrod and Dion 1988: 1388).

riparian implements a unilateral project that may harm others, it may consider this merely an exercise of its water rights, not a "defection." Because rights are viewed from different points of view by upstream and downstream riparians (Sect. 2.1.2), defection is controversially identified. Moreover, when defection is not deterred, the game is driven either to contention, in which both sides utilize the river unilaterally and thus may harm one another, or to sole utilization, in which the sole user is said to be exploiting the one that is not benefiting from the river. In both outcomes, those ones harmed or exploited by the other will seek retaliation, particularly if defectors are not sanctioned. Such pitfalls of identification, deterrence, sanctioning and retaliation are apparent in hydropolitics, since international water law presents general principles and norms and there is no international court having supreme jurisdiction over sovereign states, in that arbitration and mediation require the consent of all concerned sides. Nevertheless, the only sanction imposed against unilateral projects on international waterways is losing the financial support of international development partners (Sect. 2.2.6).

2.3.2 *Heterogeneities Over Interests, Capabilities and Beliefs*

The basic criticism raised against such a game theoretic model is the assumption that states are identical actors and only differ in their adopted strategies. For instance, Zagare and Kilgour (1993) and Kilgour and Wolinsky-Nahmias (2004), analyzed respectively IR and hydropolitical affairs by illustrating the case whereby states may be "hard" (prefer "conflict") or "soft" (prefer "capitulation") either in challenging or defending the "status quo" outcome. Introducing a broader and more comprehensive framework, Keohane and Ostrom (1995: 6–10) argued that the characteristics and conditions of actors in common-pool resources (CPR) games are not always homogeneous, but may be heterogeneous. They categorize such heterogeneities into three aspects: capabilities owned and utilized by each actor in order to achieve its goals; interests and preferences for desirable outcomes; and beliefs shaped by level of mutual trust/mistrust that reflects the history of interactions and uncertainties over the future. From this perspective, employing the strategy of reciprocity is challenged by heterogeneities among states over their interests, capabilities and beliefs. Two specific questions can be stated here, as follows. Do all the states in the game possess the capability to defect in order to reciprocate defection? Do all of them trust each other in order to reciprocate cooperation?

Accordingly, in hydropolitical games, not all riparian states are capable of implementing their developmental projects on the river unilaterally. Thus, the decision to defect may be an unavailable option. Internal and regional political instabilities in addition to social, economic and environmental pressures can disturb such a capability. External actors can strengthen a riparian in a basin and weaken another. Alternatively, the basin's strong riparian can use its power to prevent unilateral actions of others in both direct and indirect ways. Furthermore, the uncertainty about others' decisions and the history of mutual mistrust can hinder cooperation.

In short, the capability of a riparian to act unilaterally and its trust in others' choices constrain the maximization of overall interests in a transboundary river basin. These elements (interests, capabilities and beliefs) are three overlapping factors that complicate hydropolitical games and define equilibrium. First, each riparian evaluates the potential benefits associated with various options of utilizing the basin resources in correspondence to its national needs. Second, each one chooses the best possible option depending on its level of capability, which may oppose the interests of the others. Third, long history of conflictive interactions creates mutual mistrust, leading the riparians to be suspicious about cooperation with one another, even if they will be rewarded by adequate absolute gains. The riparians' decisions are then dependent on these overlapping factors, defining the outcomes of consequent moves. The generated payoffs in every move return to affect the interests, capabilities and beliefs of each riparian. The moves go on and all riparians develop their own strategies until the game reaches an equilibrium point, at which everyone realizes that there is no better strategy to be adopted. Nevertheless, such an equilibrium point is not static, but dynamic, since the interests, capabilities and beliefs of the riparians are subject to changes due to internal or external influences. The riparians that are not satisfied with the equilibrium will take any chance to challenge the status quo, while in the case of those where the equilibrium has been in their favor, they will seek to maintain it. Accordingly, the equilibrium is developed in a cyclic manner as shown in Fig. 2.3. Yet, it is argued that the balance of power among riparians decisively controls the flow of such process. Two cases can be distinguished: symmetric versus asymmetric balance of power, as grouped in Fig. 2.4. This is discussed in detail in the following two subsections.

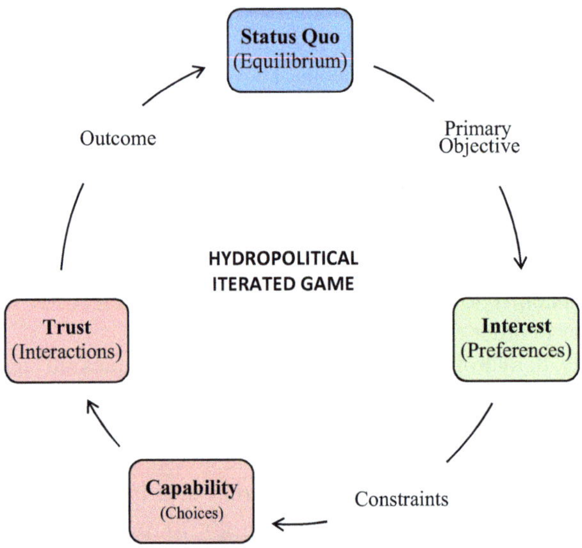

Fig. 2.3 The cyclic progression of interest, capability, and trust in hydropolitical games

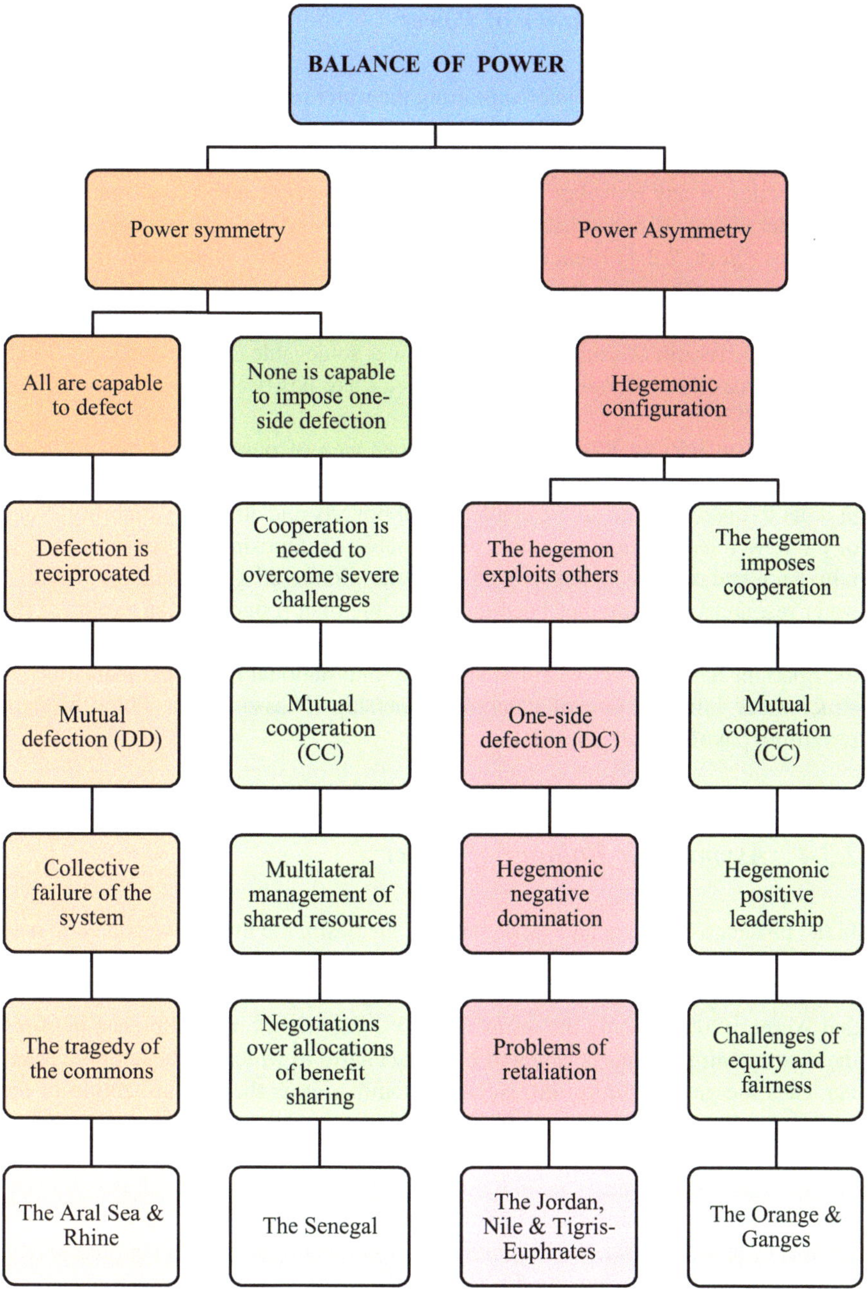

Fig. 2.4 Generated transboundary water stalemates according to the balance of power

2.3.3 Symmetric Balance of Power

If all the riparians are capable of exploiting the water resources flowing on their territories and choose to act unilaterally, ignoring the sustainability of the basin, a collective failure of the system will eventually take place (Sect. 2.1.1).In this case, the defection of any provokes the counter defection of the others, and then contention is iterated. As discussed earlier, the Rhine River and the Aral Sea Lake are clear examples (Sect. 2.2.4). Successful shifting from iterated contention to iterated cooperation relies on the level of environmental degradation, the willingness of riparians to cooperate, and the contextual influences from outside the basin. For instance, the rebirth of salmon in the Rhine was achievable (Sadoff and Grey 2002: 394), whereas the restoration of the Aral Sea Lake has been seriously challenged (Glantz 2007).

The Senegal River Basin provides another face of power symmetry in transboundary river basins, in which all riparians are aware that they have to cooperate in order to overcome the severe challenges of poverty and climatic variability. None of the newly independent riparians was capable of imposing the scenario of sole utilization and reaping highest payoff (T) alone. Cooperative exploitation of shared water resources therefore has been more feasible than if one riparian exploited the others by benefiting from the river solely. The multilateral approach employed by the riparians has been successful in attracting international funds and loans to construct water infrastructure of common ownership (Niasse 2010), which brought rewarding payoffs to all sides (R).

2.3.4 Asymmetric Balance of Power

In the presence of strong and weak states in the basin, the most powerful state may play different roles. It is useful here to borrow the distinction between positive leadership and negative domination from the theory of "hydro-hegemony" by Zeitoun and Warner (2006). Using the terms of the PD framework, the difference between these two configurations is centered on the scenario to which the hegemonic riparian leads the game equilibrium. Negative domination is the sole utilization of the basin resources by the hegemon (DC), while positive leadership takes place when the hegemon imposes cooperation on the others (CC). In both configurations, the hegemon builds its strategy upon the fact that those scenarios of high payoffs (DC and CC) are only achievable when the other does not choose to defect or is not capable of defecting. Yet, the hegemon, in the first configuration, is tempted to reap the ultimate payoff (T), leaving those exploited riparians with the lowest payoff (S) of the worst scenario (CD). In the second, the hegemon shares benefits of cooperation with the others, and thus all reaps rewarding payoffs (R). As mentioned earlier in Sect. 2.2.3, such payoffs need be neither comparable nor equivalent, in that they reflect the positions of the hegemonic and counter-hegemonic riparians in the basin. Thus, the substantial difference between positive leadership and negative domination is the reaction of the counter-hegemonic riparians on the game equilibrium. In the latter, they call for more fair distribution of cooperation benefits, while in the former, they seek to challenge the status quo and wait for the chance to retaliate, turning the equilibrium point from sole utilization of the hegemon to contention driven by unilateral projects.

Zeitoun and Warner (2006: 446) argued that the Jordan, the Nile and the Tigris-Euphrates are examples of negative domination, while the Orange-Senqu River is an example of positive leadership. In addition, Mirumachi et al. (2013: 9–10) mentioned the Ganges-Brahmaputra-Meghna River as an example of the latter configuration. Power asymmetries and water dependency have shaped the hegemonic position of Israel, Egypt, Turkey, South Africa and India in those five basins respectively.

In the examples of the Jordan, Nile and Tigris-Euphrates, the hegemonic riparians employed the strategy of "resource capture" through implementing unilateral water projects: the National Water Carrier (NWC) by Israel, the Aswan High Dam (AHD) by Egypt, and the South-Eastern Anatolia (GAP) multi-dam scheme by Turkey (Zeitoun and Warner 2006: 444). On the one hand, those hegemonic riparians regard these projects as a vital necessity to secure their water rights and meet their urgent needs. On the other hand, the other riparians in each basin consider that the generated outcome is the sole utilization of the river by the hegemonic state (DC) through getting maximum benefits (T) and leaving them with the lowest possible payoffs (S).

The mechanisms through which the hegemon maintains the status quo in its favor differ based on its geographic position. When the downstream riparian is the hegemon, its domination depends on the inability of the others to implement their unilateral projects, either because they lack the financial and technological capacity or they are prevented by force from acting unilaterally. The occupation of the Syrian Golan and the West Bank by Israel in 1967 took place under the pretext of the Arab attempts to divert waters reaching Israel (Wolf 2000). The regional and civil wars in upstream riparians was the main reason behind hindering the construction of large dams upper the Nile Basin until the early 1990s (Waterbury 2002). On the other hand, the ability of the upstream hegemon to undertake its unilateral projects on its territory despite the objection of the downstream riparians is enough to maintain its hegemonic position in the basin. In the Tigris-Euphrates example, Turkey has been able to proceed with its GAP project, which is a "gigantic" integrated water development scheme of 22 multi-purpose dams and 1.7 million hectares of irrigable lands (Leb 2010: 3). Filling the reservoir behind the Ataturk Dam, as part of the GAP scheme, Turkey was able to stop the water flowing to Syria and Iraq for around 1 month in 1990, leading to significant harm in hydropower and agricultural sectors downstream (Zawahri 2006).

As noted above, the counter-hegemonic riparians in the configuration of negative domination seek to challenge the DC outcome at which the game equilibrium has arrived. They employ several counter-hegemonic tactics of reciprocity, based on the means available. The following are five examples of such tactics: issue linkage, hydrogeography, creating clusters, capturing the right moment, and taking a hostage.[7]

The Syrian support for the Kurdish rebels was linked to the water issues of the Tigris-Euphrates Basin. The Syrian commitment to ban Kurdish rebels from its land, through the 1998 Adana Agreement, was considered a breakthrough that paved the way for the signature of the 2001 Cooperation Protocol, which attempted to coordinate the Turkish GAP scheme with the corresponding Syrian developmental projects (Wolf and Newton 2009: 243–246). Another example of issue linkage is the British recognition of Ethiopia's independence in return for the Ethiopian commitment in the 1902 Treaty not to construct any hydrologic project on the Blue Nile without taking the prior approval of the British and Sudanese governments (Sect. 3.2.3).

There are other examples in which hydrogeography plays a vital role in shaping counter-hegemonic moves, particularly when the hegemonic and counter-hegemonic states share more than one transboundary water resource and their geographic positions are reversed on each. In 2001, Turkey used the Euphrates to flood downstream lands in Syria, and the latter reciprocated by using its upstream dams on the Asi River to flood lands in the Iskandaron, a politically important province in Turkey (Zawahri 2008: 286). The same constellation is obvious between Israel and Jordan. When Israel, in several occasions, slowed down the water flowing in the pipeline that supplies Jordan with its share of the Jordan River based on the 1994 Treaty, the latter threatened to do the same with the Wadi Araba aquifers that partially feed agricultural communities in Israel (Zawahri 2008: 286–287).

Another tactic employed by the counter-hegemonic states is creating clusters through building coalitions in order to be effective in challenging the equilibrium maintained by the hegemon. The Nile upstream riparians have been moving in one cluster along the CFA negotiations and the NBI activities since launching the multilateral path in the 1990s in the attempt to institutionalize basin-wide cooperation with the support of international developmental partners (Sect. 5.2.2). The upstream cluster signed individually the CFA disputed draft in May 2010 despite the objection of the downstream riparians to it for not recognizing their historical water rights based on the existing agreements (Sect. 5.4.3). In April 2011, when Ethiopia announced unilaterally the commencement of the GERD, it was aware that forming a new cluster with Sudan is crucial to proceed with the dam construction in spite of the Egyptian expected opposition. The direct economic benefits of the GERD to Sudan were enough to assure Khartoum's consent for the dam, in that the regulated flow and the generated hydropower will help attract foreign agricultural investments and meet urgent energy needs in Sudan respectively (Sect. 6.2.3).

---

[7] Raub and Weesie (2000), cited in Axelrod (2000), introduced the concept of "hostages," where the "trustor" takes a hostage from the "trustee" to assure the latter's compliance and reduce its incentive to defect.

It is also crucial that the counter-hegemonic state captures the right moment to initiate its unilateral action. The Ethiopian decision to build the GERD without prior consultation with its downstream co-riparians, violating the Articles 11–19 in the UNWC, was taken at a critical moment in the history of Egypt and Sudan. In early 2011, Cairo was extremely busy with domestic affairs after the spark of the January 25 Revolution and Khartoum was facing a severe shortage in its energy supplies after the separation of South Sudan, the land where most of its oil fields exist (Sects. 6.2.1 and 6.2.2). In the meantime, Ethiopia had been witnessing relative economic and political stability that allows for financing the dam domestically, being sure that the Egyptian opposition will cut off financial support from international institutions.

Furthermore, the counter-hegemonic state may resort to take a hostage from the hegemon as a threat of retaliation. In the same example, the location of the GERD on the Ethiopian-Sudanese border puts the downstream communities in Sudan and Egypt under the threat of disastrous floods if any thought of destroying the dam through a military action (Whittington et al. 2014: 598). Such a threat is associated with grave risks to the settlements and lands lower the dam in case of its failure for any geological or structural reasons (IPoE 2013).

Turning to positive leadership, the virtue of this configuration is the awareness of the hegemonic state about the vitality of sharing benefits with the other riparians while achieving its own developmental goals in the basin. The Orange-Senqu River Basin, which involves Lethoso, South Africa, Botswana, and Namibia, is considered the most developed and institutionalized basin in the Southern African Development Community (SADC) region (Jacobs 2012). With the aim of securing its water needs, South Africa has led the implementation of numerous large dams' projects and inter-basin water transfer schemes (Heyns 2003), besides the signature of several water agreements (Turton et al. 2004). Yet, such plans have been beneficial to a large extent for the other riparians (Jacobs 2012). As described by Zeitoun and Warner, the hegemon here employs a "utilitarian" strategy that combines "compliance-producing sticks" with "compliance-encouraging carrots" through providing the others with considerable incentives to accept its integrative basin-wide developmental approach (Turton 2003: 12). The previously mentioned LHWP example (Sect. 2.2.3) shows explicitly how South Africa was able to get the needed amounts of water transferred from Lethoso, rewarding the latter with a satisfying payoff from the generated hydropower (R) and thus generating a sustainable cooperative scenario (CC).

The Ganges-Brahmaputra-Meghna river system is another example in which the hegemon has used the strategy of breaking down the game and setting agendas to safeguard its position in the basin. As briefed by Mirumachi et al. (2013), India has been able to manage transboundary water interactions with Nepal and Bangladesh over the river system through isolated bilateral agreements, avoiding the establishment of an inclusive basin-wide institution. Moreover, because it is a midstream country, India has adopted different approaches of shared water management in negotiating with Nepal and Bangladesh, its upstream and downstream co-riparians.

With Nepal, negotiations were based on benefit sharing of hydropower, whereas, with Bangladesh, the focus was on water resources allocation. The hegemon in this example defined the form of cooperation payoffs (R) obtained by others. However, Nepal and Bangladesh, together with Pakistan, have been carrying out unofficial talks for building a multilateral platform on the non-governmental level, involving Indian representatives, to open more opportunities for basin-wide resources management. In addition, Nepal has renegotiated bilateral agreements with India for a more equitable share of its economic benefits.

It is worth noting here that those examples of positive leadership align with the theory of "hegemonic stability" by Keohane (1984/2005), which argues that cooperation led by a hegemonic state among others is substantially capable to survive. This was confirmed by Martin (1994) who contends that heterogeneous capabilities and interests indicate that everyone has something to offer in order to initiate cooperation (e.g. through mutual trade). However, those examples of negative domination shown earlier illustrate that such argument is not always true, which points to the varying influences of the game context.

2.3.5 How Does the Context Influence the Game Outcomes?

In all the examples illustrated in this chapter, it is evident that interactions over the central issue of water is greatly affected by the game context. As stated by Axelrod and Keohane (1985: 238), the "basic concepts" of "payoff structures, iteration, and the number of players provide only a framework for analysis," but these "take on greater significance, as well as complexity, when they are viewed within the broader context of other issues, other games, and the institutions that affect the course of world politics," which leads us to "turn to the question of how the context of interaction affects political behavior and outcomes." They named issue linkage and connections between international relations and domestic politics (Axelrod and Keohane 1985: 239–241).

This question is of great importance in the realm of hydropolitics, since the interests, capabilities and beliefs of riparian states over shared waters are evidently interconnected with other various issues at the multiple levels. Allan (2001: 337) highlights the fact that managing "watersheds" cannot be addressed without considering the "problemshed," which is the term that "captures the operational context in which decision makers and their problems exist." Cascao and Zeitoun (2010: 40) points to the crucial role of the "determining political and physical context," which incorporates, for example, the timing of elections and military coups and the occurrence of droughts and floods. The comparative study of Mostert (2003: 40–41) on river basin organizations concluded, among its key messages, that "there is no single best way to manage international freshwaters," in that "what may work in one context does not necessarily work in another context."

It was shown above that signing the Indus Water Treaty (Sect. 2.2.6) was seriously challenged by the entrenched political disagreements between India and Pakistan. The World Bank, as a result, made enormous effort to focus on the water issue, resorting to the "suboptimal" solution of separating water resources management of the Eastern and Western tributaries instead of integrated basin-wide approach (Wolf and Newton 2009: 194). Since the 1950s, disputes over water rights in the Jordan Basin have been overshadowed by the Arab-Israeli conflict and the heated territorial claims adopted by its riparian countries (Wolf and Newton 2009: 198–205). The 1953–1955 Johnston Plan, for instance, was technically and hydrologically rational enough to be accepted by the riparians (Phillips et al. 2007). Even though, it was not ratified due to the rejection of the Arab riparians to recognize the legitimacy of Israel (Wishart 1990). During the 1990s, the water issue was a strong component of the comprehensive peace talks that involved multilateral and bilateral negotiations between Israel, Jordan, and the Palestinian Authority (Wolf and Newton 2009: 223–231). The concluded 1994 Israeli-Jordanian agreement defined the water rights of both countries in the Yarmouk and the Lower Jordan and comprised cooperation on water transfer and storage besides the commitment to find alternative sources of water (e.g. desalination) (Anisfeld 2010: 277). Whereas, Palestinians in the West Bank and Gazza remain under water stress, and Syria is not allowed to access the Jordan and Yarmouk Rivers from the Golan Heights since the latter was occupied by Israel in 1967 (Asser 2010).

In the Senegal River Basin, political tensions between Guinea and Senegal in 1967 led the former to suspend its participation in the multilateral talks between co-basin riparians (Niasse 2010; Wolf and Newton 2009: 239–243). Consequently, Guinea was not a full member in the OMVS during its establishment in 1972. Nevertheless, Guinea neither participated nor boycotted, but was involved as an observer. In the last decade, Guinea eventually joined the OMVS (World Bank 2013b). In contrast, the context may play a positive role, particularly when cooperation over other political issues helps riparians cooperate over water. The EU broader perspective of common interests contributes positively to the sustainability of shared water resources between its members. Moreover, the advanced standards of the EU Water Framework Directive have been boosting cooperation in the Danube, Rhine, and other European rivers.

Concerning multilevel influences, it is obvious that the intervention of international donors, regional/international powers, and domestic actors have evidently influenced the progress of interactions and consequently the game outcomes. In the case of the Gabcikovo-Nagymaros Project, the environmental groups in Hungary played a role in its decision to halt the construction work. The United States and Arab countries, among others, were effective actors along the different phases of the Jordan's complex negotiations. The World Bank crucially contributed to the signature of the 1960 Indus Water Treaty and 1986 LHWP Treaty. International developmental partners also facilitated cooperation on the Senegal River by providing its riparians with required funds to undertake their water projects.

2.4 Concluding Remarks

Based on the illustration shown above, this study finds the game theoretic model of Axelrod (1984, 2000) and Axelrod and Keohane (1985), in addition to the concept of heterogeneities introduced by Keohane and Ostrom (1995), comprehensive enough to analyze the complex interactions over transboundary river basins. Throughout the empirical analysis, the Nile riparians are modeled as rational actors playing a PD iterated game. The outcome of every move along the game is defined by whether each side (an individual riparian or a cluster of riparians) utilizes the river unilaterally or in collaboration with the other side. Four outcomes, then, may be generated: sole utilization by either, cooperation, and contention. What is substantially significant in Axelrod's work is that even if utilizing the river solely is more favorable than cooperatively, it is still possible that players find cooperation more profitable than sole utilization on the long term on condition that the game is iterated. This is because cooperation will help players avoid the pitfalls of defection in which they enter into a close circle of retaliation and counter retaliation.

In addition, this approach allows for investigating the inherited heterogeneities among the riparian states over their interests, capabilities and beliefs. To what extent is each in need of utilizing the river? What is the capacity of each to utilize the river and fulfill its needs? How does each perceive its rights in utilizing the river and trust the utilization of the others? The level of heterogeneities over these three aspects defines the point at which the game comes to equilibrium, which indicates the status quo created in the basin. At the same time, the approach considers the influences of the changing contexts on the game progression, both through different issues linked to the main issue of water and through the intervention of various actors at multiple levels. In the following four chapters, the Nile case will be further analyzed against the iterated PD framework employing the process tracing method as discussed in the introductory chapter.

Chapter 3
The Colonial Round: The Opening Scene of Utilizing the Nile Waters Outside Egypt

> *No one can hold Egypt securely unless he holds also the whole valley of the Nile. The sources of the river in hostile, or even in indifferent, lands must always be a grave cause of danger, or, at the best, anxiety. ... Everything depended on the Nile. The more Egypt was developed, the greater grew the need for the regulation of the water. The rulers of Egypt need have troubled little about the fate of countries divided from them by so many leagues of rainless desert, but for the link of the all-important river ...*
>
> *Sir Sidney Peel (1904: 112, 135)*

3.1 The Contexts of the Colonial Round

The colonial round in the Nile game started with the longstanding configuration shaped by the natural and climatic characteristics (Appendix A) that made Egypt the only actual user of the Nile waters with no considerable competition. However, such a situation, which had persisted for many centuries, was changed gradually by the British occupation of Egypt in 1882. During the 1890s, France's desire to threaten British interests in Egypt, though unsuccessful, led to the first real attempt to build a dam to alter the water flow to Egypt. Britain then sought to control the entire Nile Basin, either under its direct administration or through concluding agreements with other colonial powers and with Ethiopia. By the beginning of the twentieth century, Britain's dominating position on the Nile led to the comprehensive development of cotton cultivation in its colonies across the basin. This was associated with the commencement of the first large irrigation scheme and regulator dams in Sudan, the attempt to build a storage reservoir at Lake Tana, and the generation of hydropower at the outlet of Lake Victoria in Uganda. As a result, the colonial round comprised different scenes of conflict and cooperation over the Nile waters, ranging from the exchange of threats to the conclusion of agreements.

M. M. Samaan, *The Nile Development Game*,
https://doi.org/10.1007/978-3-030-02665-3_3

3.2 The Second Half of the Nineteenth Century

3.2.1 The Nile Before the British Occupation of Egypt

During the nineteenth century, Egypt upgraded its water infrastructure on the Nile under the Muhammad Ali dynasty who exerted tremendous effort to centralize and modernize the country. During the reign of Muhammad Ali Pasha (1805–1848) and his successors up to the British occupation in 1882, barrages and intricate networks of distribution and irrigation canals were constructed to regulate the Nile flow in Egypt, expand the cultivable land, and promote the productivity of irrigated agriculture (Dumont 2009: 15). For instance, the Delta Barrage, which was constructed from 1845 to 1861 at the apex of the Delta north of Cairo (Fig. 3.1), was the first diversion dam in its magnitude across the Nile and contributed significantly to the expansion of the Delta's irrigated agriculture (AUB 1961: 41; Awulachew et al. 2012: 11). The main aim of these developmental projects was to transform the agriculture system in Egypt from the traditional basin irrigation

Fig. 3.1 Map of Egypt showing the location of the Nile Delta and the Suez Canal. (Source: Contributed by Bartholomew 1885, available by the Library of Congress at: https://www.loc.gov/item/2009580108/ (accessed on August 19, 2016))

(once per year) into the perennial irrigation (twice or more per year), allowing water supply during the Nile low season, which led to the growth of cotton agriculture as a summer crop (Tignor 2011: 215). In addition to infrastructure projects, the Muhammad Ali Dynasty built Egypt's educational and scientific capacity by establishing the first specialized schools and transferring knowledge from Europe to Egypt in the fields of agriculture, industry and construction (Awulachew et al. 2012: 10). Simultaneously, the American Civil War of the 1860s, and the consequent shortage and sharp rise of cotton prices, contributed to the expansion of cultivation in Egypt (Degefu 2003: 21).

Despite its formal association with the Ottoman Empire, Egypt sought to act as an independent regional power. Yet, Egypt was under the inherent perception and continuous threat that the Nile River, representing its lifeline, could be altered by the Ethiopian rulers. A clear example in this respect is the 1706 letter[1] sent from the Abyssinian (Ethiopian) emperor Tekle Haymanot I to "the Ottoman Pasha, princes, and lords commanders in the city of Grand Cairo," which was written in an angry language over a severe diplomatic dispute, as follows (Lobo 1789: 184–185):

> From the august king, powerful arbiter of nations, the shadow of God upon earth, … the most mighty of Christian kings, he who keeps peace between the Mussulmans and Christians,[2] protector of the boundaries of Alexandria,[3] … inheriting from his father a mighty kingdom, issue of the lineage of David and Solomon[4] … Since the ambassadors of kings ought to be at liberty, and to pass where they will, and to be treated with owner, and not hindered or molested; neither ought any dues or tribute to be exacted from them. It is in our power to return the injury, if we pleased to revenge the insult offered by you to our messenger. The Nile might be made the instrument of our vengeance, God having placed in our hands its fountain, its passage, and its increase, and put it in our power to make it do good or harm. At present, we require and exhort you to cease from offering any injury to our ambassadors, and from disquieting us by stopping those who are on their way to us …

Using the Nile as an "instrument of vengeance" by Ethiopia took place several times over different political and religious tensions with Egypt. Yet, this remained a mere threat and never entered into force, obviously due to the lack of technological

[1] The letter is recorded by the Portuguese missionary Father Jerome Lobo (1789: 184–185) during his "voyage to Abyssinia," and available at: https://archive.org/details/avoyagetoabyssi00johngoog (accessed on August 16, 2016).

[2] At the heart of the relations between Egyptian and Ethiopian rulers for centuries was the safety of the Christian minority in Egypt and the Muslim minority in Ethiopia.

[3] Since Saint Athanasius of Alexandria consecrated Frumentius, the first Primate of Abyssinia, at the beginning of the fourth century, every Orthodox Primate of Abyssinia had been consecrated by the Coptic Patriarch of the Church of Alexandria (Hozier 1869: 4). Only in the 1950s, the Coptic and Abyssinian Churches were separated in light of severe hydropolitical as well as geopolitical disputes (Sect. 4.2.3).

[4] Legendary chronicles tell, that "from this district [Abyssinia] the Queen of Sheba travelled to visit the capital of the Jewish Solomon, returning thence to the seat of her power at Axum, in Tigre, with the ark of the covenant and a Hebrew colony, and there raised a temple to the true God," and that "the royal house of the country has directly descended from the Queen of the South, and the Negoos, or ruler of modern Ethiopia, claims descent from Menilek, an asserted child of Solomon by the Queen of Axum" (Hozier 1869: 3).

capacity needed to control the sweeping waters flowing through deep steep canyons on the Ethiopian Highlands. Nevertheless, the Muhammad Ali Dynasty sought to secure Egypt's water needs by mobilizing forces beyond the borders with the aim of exploring and capturing the sources of the Nile. It extended its domain militarily along the Nile Basin, including Sudan and the territories around the Equatorial lakes (Said 1993: 107), except over Ethiopia and Uganda.

Ethiopia had suffered from long periods of disintegration before the reign of King Theodore II (1855–1868) and then King Johannes IV (1871–1889). Shukry (1938: 258) described both as "two ambitious kings, who were notoriously hostile to Egypt." King Theodore II was able to establish a central government in Ethiopia. In his correspondence to the Earl of Clarendon on April 7, 1855, published in the British and Foreign State Papers, Vol. LII[5] (1868: 840), Consul Walter Plowden in Massawa noted that the newly-crowned king was "persuaded … that he is this destined Monarch," referred to in the "ancient prophecy that a king of this name [Theodorus] should reform Abyssinia, restore the Christian faith, and become master of the world." Plowden also recorded that King Theodore II "dreams even of turning the waters of the Nile into another channel, and thus to ruin or subdue Egypt."

However, King Theodore II committed suicide after the successful military intervention of Britain in 1867–1868 to release the European hostages whom he detained (Fueter 1922: 342). Before Britain sent the expedition to Magdala, where the hostages were imprisoned, "measures were taken to obtain the permission of His Highness the Viceroy of Egypt to land on Egyptian soil" (Great Britain War Office 1870: 40). Cairo responded to the British request stating that "permission had been granted for the passage of the expedition through Egyptian territory, and that Egyptian officers would proceed immediately to Massowah to ensure every facility being afforded, and, further, that Viceroy's ships of war in the Red Sea were placed at the disposal of Her Majesty's Government in case of need" (Great Britain War Office 1870: 41).

Few years later, King Johannes IV safeguarded the centralization of Ethiopia and attempted to expand in East Africa, colliding with Egypt and later with Italy and Sudan (El-Atawy 1996: 8). An armed dispute broke out between Egypt and Ethiopia in 1875 and 1876, which ended with the failure of Egyptians to defeat the Ethiopians, on the one hand, and with the full control of the Egyptians over the African coast of the Red Sea, on the other (Budge 1907/2010: 234; Shukry 1938: 258, 259, 268, 269). In 1889, King Johannes IV was killed and his forces were routed in the confrontation against the Sudanese (Collins 1971: 73). On the other side, the Egyptian forces reached the borders of Buganda (now Uganda) and offered its Kabaka (King) to be placed under Egypt's protectorate. He refused the offer and requested Egypt's military support as from an equal ally against Bunyoro, his traditional rivalry. However, such a demand was refused by the Egyptians (Low and Smith 1976: 341–342). Thus, the Egyptian control over the Nile did not exceed the borders of Ethiopia and Uganda. Yet, none of them was capable of threatening Egypt's water security by utilizing the Nile waters.

[5] Available at: https://archive.org/details/britishandforei00offigoog (accessed on August 16, 2016).

3.2.2 *Opening the Suez Canal and the British Occupation of Egypt*

Since the Napoleonic campaigns (1798–1801), the great European powers gave special attention to Egypt throughout the nineteenth century, particularly Britain and France. The policy of Muhammad Ali Pasha concerning foreign affairs was mainly centered on knowledge transfer and adopting features of Western civilization in order to build the modern Egypt that would be self-sufficient in terms of food and industrial products. Muhammad Ali also relied on foreign expertise in constructing barrages, railroads, as well as other public works and projects. At the same time, he did not welcome the idea of digging the Suez Canal by Europeans, an international waterway linking between the Red Sea and the Mediterranean Sea (Fig. 3.1), as he feared that this project would aggravate foreign intervention in Egypt. In addition, he sought to expand beyond Egypt's borders, not only towards the south to secure the Nile sources, but also towards the east and northern east fronts to challenge the domination of the Ottoman Empire over the entire region. Muhammad Ali's expansionist policy satisfied neither the Ottoman Empire nor the European powers, especially Britain, leading to collusion with him in several occasions in different ways. In the end, Muhammad Ali was forced to sign the 1838 Anglo-Turkish free trade agreement followed by the 1841 London Treaty, which limited his rule to Egypt and Sudan, and maintained the Ottoman ultimate control over him and his successors (Abul-Magd 2008: 109).

Years after, the Suez Canal project was implemented by a French company and Egyptian labor under the reign of Khedive Said (1854–1863) and was opened in 1869 during the reign of Khedive Ismail (1863–1879). As a result, Egypt's position significantly increased in importance with respect to international navigation, which ultimately changed the Nile game entirely, as illustrated in Ward and White (1972: 28, 37), Collins (1990: 30–33), El-Atawy (1996: 14–15), Degefu (2003: 20), and Kitissou et al. (2007: 26). The Suez Canal placed Egypt at the heart of Britain's focus, and it sought to gain a foothold on this strategic sea-route serving its colonies in India (Fig. 3.2). Moreover, the substantial debts that Khedive Ismail owed to Europe for the modernization of Egypt and the opening of the Suez Canal resulted in the financial collapse of the Egyptian economy, leading therefore to direct foreign intervention in Egypt, mainly by France and Britain. By the end of 1870s, under European pressure, the Ottoman Sultan deposed Khedive Ismail due to his opposition to foreign intervention. These events led to a period of internal disturbance and gave birth to the nationalist movement led by Colonel Ahmed Orabi. Such developments aroused Britain's anxiety over its interests in Egypt, and consequently led to its decision to occupy the latter in 1882. This unilateral action by Britain antagonized France, which wanted to keep Egypt in its sphere of influence.

Britain and France were not the only global players at the time as there had been other European powers playing important roles in the region, namely Germany, Belgium and Italy (Fig. 3.3). This configuration is demonstrated in Fueter (1922: 326–327), Wright (1972: 30, 64), Collins (1990: 33–34), El-Atawy (1996: 15–17),

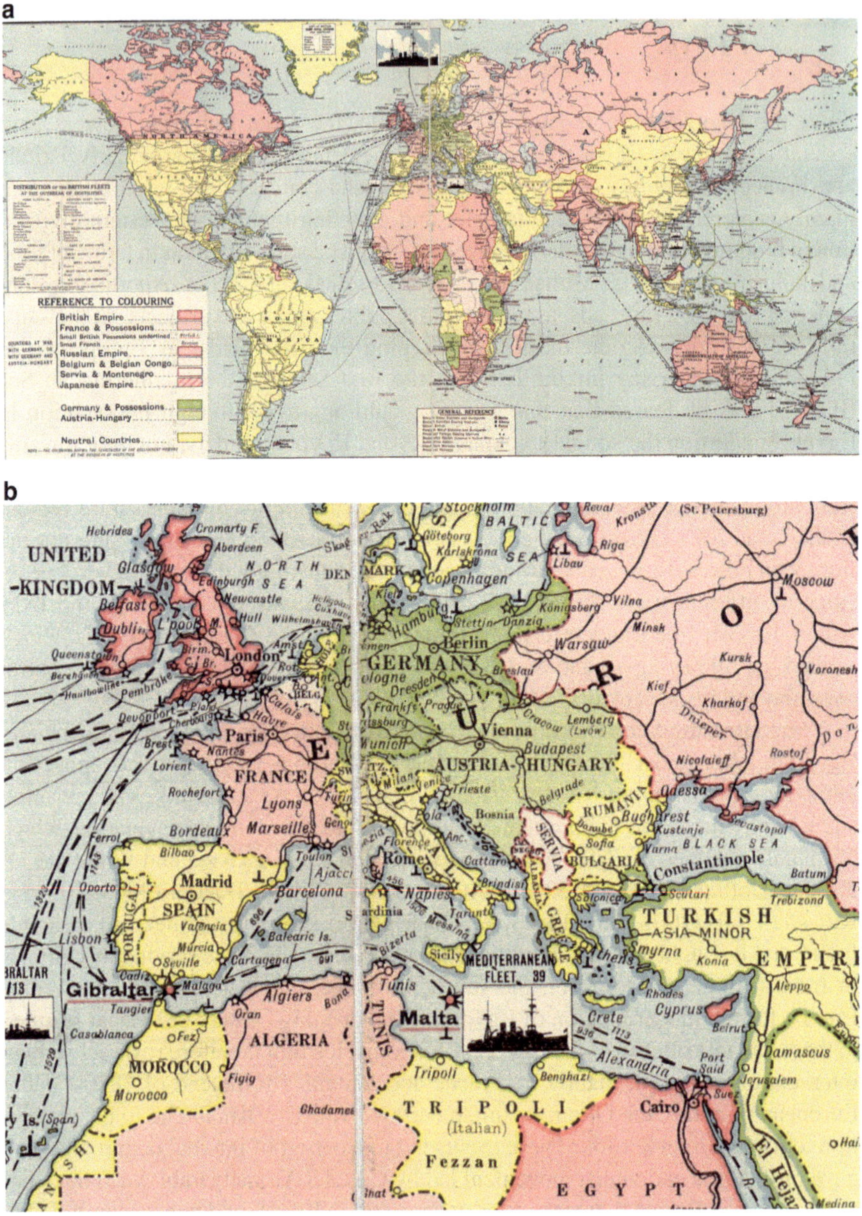

Fig. 3.2 The world during the colonial era. (**a**) The 1917 Daily Mail World Map of War and Commerce. (**b**) Closeup of the British sea route through the Suez Canal. (**c**) Comparing the military resources of the European powers. (Source: Contributed by George Philip & Son, London Geographic Institute, Daily Mail world map of war and commerce (1917), available by Library of Congress at: https://www.loc.gov/item/2013593057/ (accessed on August 19, 2016))

c

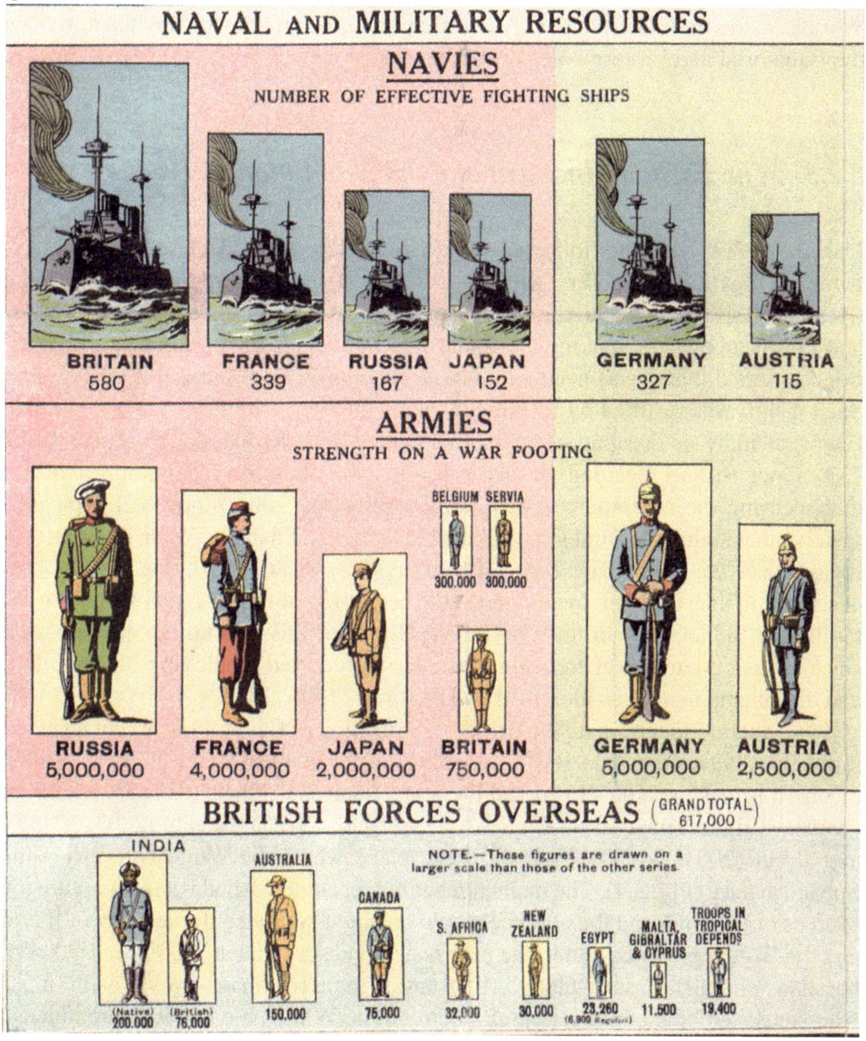

Fig. 3.2 (continued)

and Degefu (2003: 37–41, 53–58). Although Germany, in the meantime, was not seeking to establish a worldwide empire, giving priority to the consolidation of its power in Europe, it decided to expand its presence to East Africa after the British occupation of Egypt. The territories targeted by Germany comprised one of the White Nile tributaries in Uganda. Germany's main purpose of maintaining Britain's occupation of Egypt was to strengthen its position in world politics by ensuring British dependence on its support to control East Africa and by preventing Britain to ally with France; its major rival at the time. By the early twentieth century, Germany further expanded its presence in the Middle East by launching the Baghdad Railway project, which aimed at connecting Berlin to the Arab Gulf through Istanbul and Baghdad, creating a new international route linking the Eastern and the Western Worlds. Aside from Germany, Italy and Belgium attempted to occupy territories on the Equatorial and Eastern Nile sub-basins respectively.

3.2.3 *The Fashoda Incident and the Nile Colonial Treaties*

During the few years that followed 1882, Britain repeatedly declared that its occupation of Egypt is temporary with the purpose of securing free navigation in the Suez Canal. But even when the 1888 Suez Convention was signed between Turkey and European powers, making the Canal "free and open whether in peace or war," London found that it was necessary not to give up Egypt (Mahajan 2003: 95–96). Such a shift was justified by its policy makers on the grounds that "if we evacuate Egypt, anarchy or occupation by another power must be the result" (Dilke 1891: 732). Once Britain regarded its stay in Egypt as a long-term condition, it realized that securing the Nile sources was necessary for the stabilization of Egypt, ultimately envisioning the strategic goal that led to the establishment of an international regime over the entire Nile Basin (Collins 1971: 79, 80, 1990: 36). This is best described in the statement mentioned at the beginning of this chapter, which is written by Sir Sidney Peel in his 1904 book. However, this was not an easy mission since British control over the entire Nile Basin required employing both military and diplomatic means, as identified by El-Atawy (1996: 18–27).

In an attempt to terminate the British colonization of Egypt, France sent an expedition in 1893 to the Nile Basin, resulting in the incident of Fashoda as widely recorded across the Nile's historical literature, such as Collins (1971: 83, 84, 1990: 48–59), Wright (1972: 164–189), El-Atawy (1996: 23–27), Degefu (2003: 32–37), and Tvedt (2004: 44–50). Fashoda is the place where the White Nile meets the Sobat tributary (Fig. 3.4). The main aim behind reaching Fashoda was to capture the sources of the Nile, and thus France would be able to threaten the security of Egypt and the British presence there. The plan rested on "speculative" (Collins 1990: 49), but also "simplistic" and "faulty" (El-Atawy 1996: 16), made by Victor Prompt,[6] who suggested "the construction of a dam on the White Nile below its confluence with the Sobat which could be so regulated in order to ruin Egypt by drought or flood" (Collins 1990: 49). Although it contributes only to 14% of the total Nile flow, the White Nile was of great importance to Egypt at the time, particularly in dry years in which the Blue Nile flood is low; since there had been not yet large storage dams regulating the water over the years. The French decision was facilitated in varying degrees by Belgium and Ethiopia.

In July 1898, the French expedition reached Fashoda and signed an agreement of protectorate with the local polity there. In response, Britain decided to send an

[6] Victor Prompt is a French hydrologist. He outlined his proposal in his lecture delivered before the Institute d'Egypte in Paris, 1893, under the title of "Soudan Nilotique" (Collins 1971: 83, 1990: 49–50; Wright 1972: 45; El-Atawy 1996: 16).

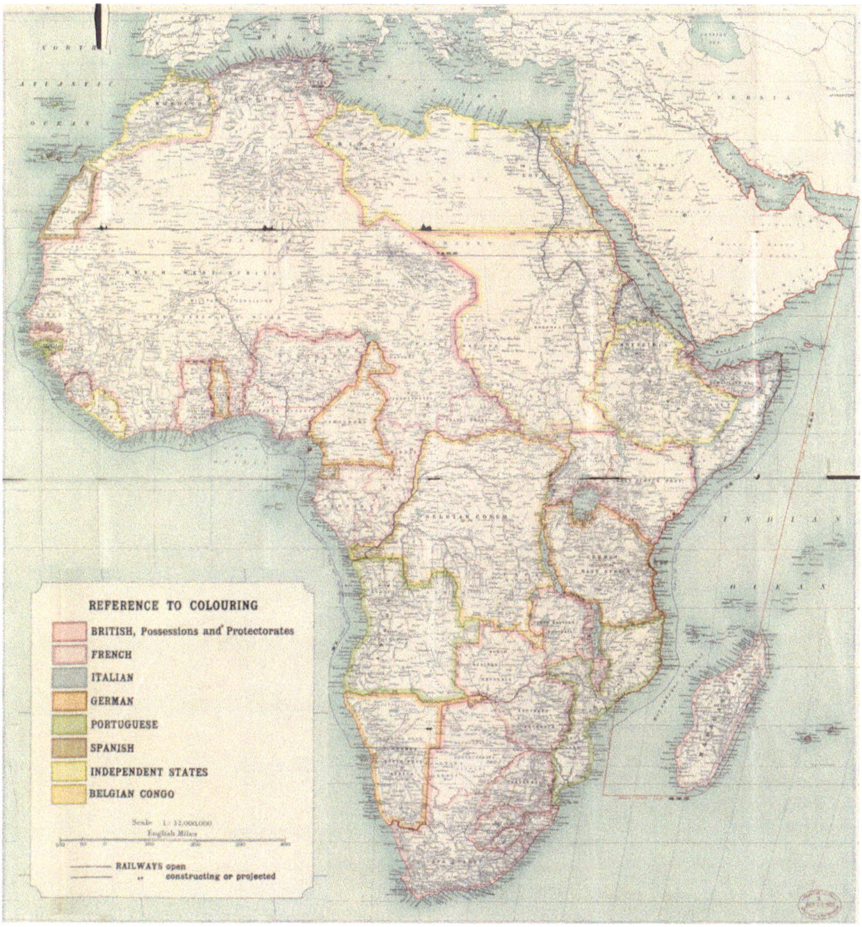

Fig. 3.3 Map of Africa showing the distribution of territories among European powers. (Source: Contributed by E. Hertslet, R. W. Brant and H. L. Sherwood, published by Harrison and Sons (1909), available by Library of Congress at: https://www.loc.gov/item/99446110/ (accessed on August 19, 2016))

Anglo-Egyptian force to re-conquer Sudan, which was at the time out of the Egyptian control due to the Mahdist movement in the 1880s. Before 1898, Britain did not find the Mahdists a severe challenge to its interests in Egypt, as they lacked the technological capacity of blocking the Nile. However, retaking Sudan under the control of Egypt became a necessity to prevent any French presence upper the Nile. The Anglo-Egyptian force reached Fashoda in September 1898, and an armed confrontation was about to take place. The leaders of the expeditions chose to avoid the outbreak of war between Britain and France. Rather, they agreed that the Egyptian, British and French flags should fly over Fashoda and that the dispute should be settled by negotiations. France claimed that Fashoda was not possessed by any

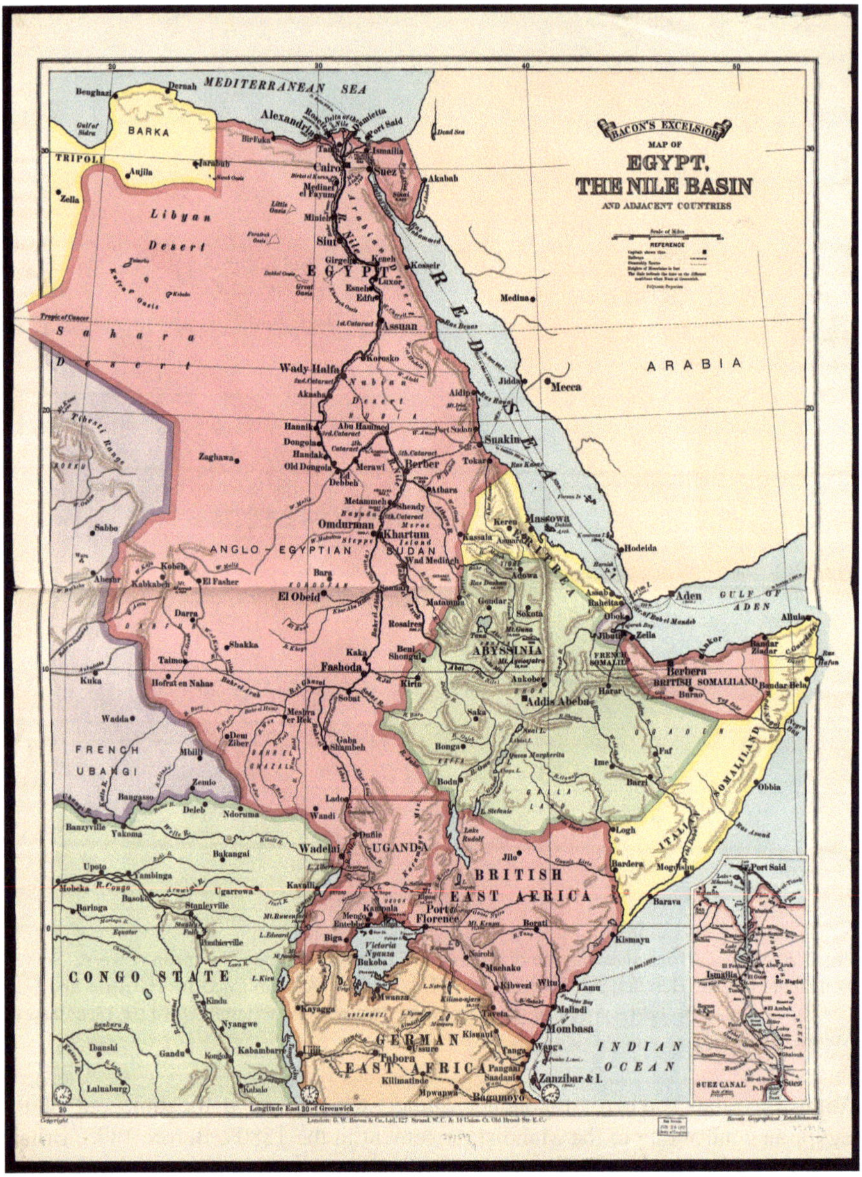

Fig. 3.4 Map of the Nile Basin showing the entities controlled by Britain and other European powers. (Source: Contributed by G. W. Bacon & Co. (1916), available by Library of Congress at: https://www.loc.gov/item/2009580100/ (accessed on August 19, 2016))

power at the time and that the local polity agreed to be under its protectorate, whereas Britain claimed that Egypt had the right to retake the territory that it possessed in the past. In fact, each side was seeking to extend its dominion in Africa: Britain, from Egypt in the north to its South African colonies; and France, from Djibouti in the east to its West African colonies.

Eventually, the balance of power between both and their common desire to prevent war forced France to withdraw from Fashoda. Although the French threat to alter the Nile did not come true, the Fashoda incident fueled Egypt's fears about the use of the Nile waters as a weapon in political conflicts, which is best summarized by Waterbury (1979: 64), stating, "From then on Egypt has been actually sensitive to any exogenous forces impinging upon its water supply." In March 1899, Britain and France agreed that "the sources of the Nile and the Congo rivers should mark their spheres of influence" (Blair 2015: 9). Moreover, the issue was completely resolved through the 1904 "Entente Cordiale" Agreement (Simpson and Jones 2009: 339), in which France was ultimately excluded from the Nile and Britain, in return, supported France's claims in Morocco (Figs. 3.3 and 3.4).

With respect to the other European powers, Britain was able to place the entire Nile Basin under its regime with less militarized and more diplomatic effort. Concerning the Equatorial Lakes sub-basin, Germany accepted by 1890 to evacuate territories up to the first degree of South Latitude for Britain, including Uganda and Kenya (Figs. 3.3 and 3.4), while Tanganyika (now Tanzania), Burundi and Rwanda remained under the German control (Collins 1990: 48; Awulachew et al. 2012: 11). Additionally, Britain signed a series of agreements with Belgium in 1908, 1911 and 1912, limiting the Belgium presence upper the Nile (Fig. 3.3) and securing the flows of two tributaries feeding Lake Albert.

As for the Eastern Nile sub-basin, Britain recognized Ethiopia as an Italian protectorate in 1891, and in return Italy agreed to keep the Blue Nile beyond its control and to secure the flows of the Atbara tributary (El-Atawy 1996: 20–21). After the defeat of the Italian forces at Adwa in 1896, Ethiopia regained its sovereignty, while Eritrea remained an Italian colony until 1941 (Waterbury 2002: 62). On May 15, 1902, Britain signed a treaty[7] with Emperor Menelik II (1889–1913), recognizing the Ethiopian independence, defining the borderlines between Sudan and Ethiopia (Figs. 3.3 and 3.4), and securing the Nile flow to Egypt and Sudan. Article III, which relates to the Nile, reads as follows:

> His Majesty's the Emperor Menelek II, King of Kings of Ethiopia, engages himself towards the Government of His Britannic Majesty not to construct, or allow to be constructed, any work across the Blue Nile, Lake Tsana, or the Sobat which would arrest the flow of their waters into the Nile except in agreement with His Britannic Majesty's Government and the Government of the Soudan.

As noted in Waterbury (2002: 62), the Ethiopian officials had continuously referred to the differences in the English and Amharic versions of this article, in that the latter states the obligation of asking prior approval concerning the works on the Nile

[7] Available at: http://gis.nacse.org/tfdd/treaties.php?page=full&origin=river&tn=36 (accessed on August 16, 2016).

from the British government alone, not from both the British and Sudanese governments. Furthermore, in 1906, Britain concluded the Tripartite Treaty with France and Italy, obtaining their acknowledgement of the British interests in Ethiopia (McCann 1981: 670).

3.3 The First Half of the Twentieth Century

3.3.1 The British Vision on the Nile

Britain's interest in the Nile was not limited to the security of Egypt in general and the Suez Canal in particular but it also included the development of the cotton production in its colonies. This twofold objective made the regulation of the waters along the entire Nile a priority for Britain to maximize cotton cultivation. The essential concept of the British vision over the Nile is again perfectly described in the following lines written by the British scholar, Sir Sidney Peel under the thirteenth chapter of his 1904 book, "The unity of the Nile" (Peel 1904: 134–135).[8]

> It is no longer possible to think of our occupation of Egypt as merely a stepping-stone on the road to India … Egypt is only a portion of the great country of the Nile. Looking southward from Alexandria or Suez, the horizon is only bounded by the sources of the Nile, and these do not well up at Assouan … The story of the binding of the Nile, incomplete as that story is at present, makes one thing, at least, perfect clear, and that is, that all Nileland is one country. No divided sovereignty is possible; there must be one firm hand over all … the occupation of the whole valley of the Nile was inevitable … It sounds as far cry from the snows of Ruwenzori, the lakes and swamps of Equatorial Africa, or the rain-swept hills of Abyssinia, to the cotton-mills of Lancashire … We have our own direct commercial interest in holding the Valley of the Nile, and Egypt is still on the road to India. Apart, therefore, from the duties which rest upon us as a civilized Power, we are doubly responsible for the welfare of the people of the Nileland.

Such a developmental vision was primarily adopted by the British administration in Egypt and was translated into the implementation of barrages and dams along the Nile Valley and Delta in addition to the preparation of advanced studies and comprehensive plans for regulating the flows upper the Nile.

The water projects implemented in Egypt under this vision, included the Assiut Barrage (1902), Zifta Barrage (1903), and Esna Barrage (1909). The main aim of these barrages was raising water for summer irrigation (Kliot 1994: 32–33). In addition, the Aswan Low Dam (1902) was constructed below the first cataract in southern Egypt (Fagan 2013: 108). At the time of its construction, it was the largest masonry dam around the world (Novokshshenov 1993). It aimed initially at storing water volume of 1 BCM (Starr and Stoll 1988) to secure water for irrigation and drinking during the summer months, while in the flood season, silts with rich nutrients ran naturally downstream (Willis et al. 2011: 19). The dam was heightened twice, from 1907 to 1912 and from 1929 to 1933, increasing the volume of its water

[8] Available at: https://archive.org/details/bindingofnilenew00peel

reservoir to about 6 BCM. By 1946, the reservoir was overflowed, which opened the idea of building a larger structure upper the dam's site instead of heightening it for the third time (Zunes and Tucker 2010: 166). This larger structure was then the Aswan High Dam (AHD), which is discussed thoroughly in the next chapter.

With respect to the British studies and plans on the Nile, hydrologists and engineers carried out multiple expeditions to advance the knowledge about the entire basin and its potentials. One of the most important contributors in this regard is Sir William Garstin, the British adviser to the Egyptian Ministry of Public Works (1892–1907), after which he was appointed as the British Government director of the Suez Canal Company. His published reports during the 1900s put the foundations of many of the Nile projects and plans afterwards (Collins 1990: 156–157; Kliot 1994: 35, 39; Waterbury 2002: 59). In particular, his 1904 study, "Report upon the basin of the Upper Nile: With proposals for the improvement of that river"[9] was produced after 5 years of intensive field surveys and proposed numerous large schemes to be built on the Blue and White Nile sub-basins (Garstin 1904). The main aim of these proposals was to boost irrigated agriculture in Egypt and Sudan by regulating the Nile waters along different seasons and years of floods and droughts. The initial conclusion of Garstin's report indicated that all projects aiming at the prosperity of agriculture in both Egypt and Sudan are "closely connected" and "well-nigh inseparable",[10] since both depend on the same river and every action by either affects the other. Concerning the proposed schemes, the report regarded the Equatorial Lakes, on one side, and Lake Tana, on the other (Fig. 3.4), as "natural reservoirs" that could supply waters lower the basin both regularly and sufficiently (Garstin 1904: 193–195). To exploit these potentials, Garstin recommended constructing a series of regulating dams at the outlets of those lakes and on the Blue Nile and Atbara tributaries. In addition, he proposed digging a channel to bypass the vast swamps of the Sudd; where most of the runoffs coming from the Equatorial Lakes to Egypt and Sudan are lost due to evaporation (Fig. A.3). Garstin also indicated the great potential of implementing a large scheme for cotton cultivation in the Gezira region (Fig. 3.5), south of the confluence where the Blue Nile meets the White Nile.

As shown in the following subsections, some of the Garstin's proposals have materialized, such as the Sennar Dam (1925) and Owen Falls Dam (1954). In addition, the other proposed projects were the subject of further investigations and negotiations for long periods, such as the Tana Project and the Equatorial Nile Project (Collins 1990: 156–157; Kliot 1994: 35, 39; Waterbury 2002: 59). Garstin ended his report by the following lines reflecting his deep-seated belief in the feasibility of these schemes (1904: 195):

[9] Available at: https://archive.org/details/reportuponbasin00dupugoog (accessed on August 16, 2016).

[10] The exact wording of Garstin is worth reading, which stated, "In all projects connected with Nile regulation, the interests of Egypt are so closely linked with those of the Sudan, as to be well-nigh inseparable. Both countries must derive their water supply from the same sources, and the agricultural prosperity of both is mainly dependent upon the one river. It is, therefore, impossible to consider any important irrigation scheme, projected for the one country, without touching upon its possible effects as regards the other" (Garstin 1904: 193).

> Fortunately, in the Sudan, equally as in Egypt, there can be no doubt of the eventual return to be obtained from any well-considered, and sound irrigation project. Given a sufficient population, the combination of the sun, the soil and the water renders its success a certainty. An attempt has been made, in these pages, to explain the utility of the different schemes. Should all, or even a portion of them ever become accomplished facts, it is difficult to estimate the extent of the benefits that will have been secured to a not inconsiderable area of the continent of Africa. The limits of cultivation in Egypt are far from having been as yet reached. With a Nile under control throughout its entire length, and with the power of adding to its volume, by drawing upon the almost inexhaustible resources of the natural reservoirs which supply it, the agricultural wealth of that country should increase, to an extent beyond the dreams of the most sanguine reformer.

One significant remark should be highlighted here is Garstin's wording in the above passage, "Given a sufficient population," in which he underpinned the fact that the major obstacle to the expansion of irrigation in Sudan is the lack of human resources (Garstin 1904: 193). He referred to what he stated in his 1901 report, "The Sudan is scarcely ready yet for the introduction of irrigation works upon a large scale. Were an increased supply of water to be granted at this moment, the country is not in a position to make effective use of the boon. Its chief want, for many years to come, must be population" as quoted in Garstin (1904: 193). He indicated that other opinions suggested attracting immigrants from China, India or the U.S. to be prospect "colonists" in Sudan. He commented on this and recommended giving the priority

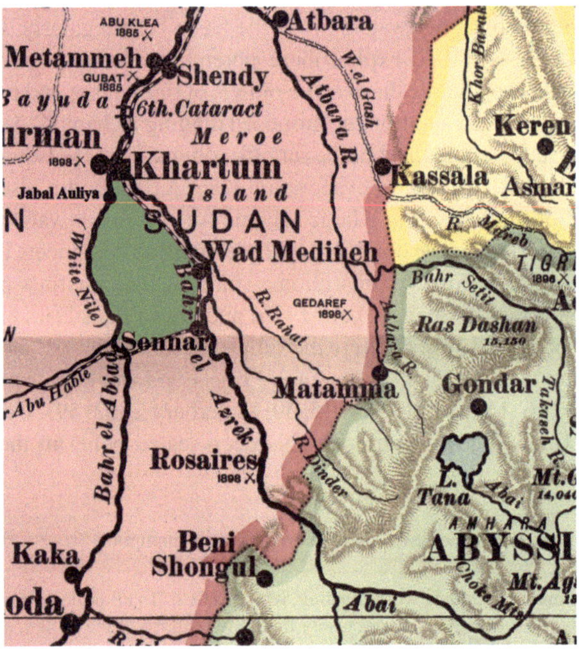

Fig. 3.5 Map of Sudan showing the locations of the Gezira, Sennar Dam and Jabal Auliya Dam (Source: Contributed by G. W. Bacon & Co. (1916), available by Library of Congress at: https://www.loc.gov/item/2009580100/ (accessed on August 19, 2016))

to coordinate with the government of Abyssinia to bring labor from the Gallas, who were skilled in agriculture and whose climatic conditions were similar to that of Sudan. However, he advised not to delay the implementation of Sudan's proposed schemes due to the population insufficiency, contending that these projects would be the proper solution for developing the country and establishing communities.

The above-mentioned study and its proposed schemes represent a turning point in the progress of the Nile regime. Since ancient times, Egypt was the only country that possessed the capacity of utilizing the Nile waters. The unprecedented development of the concept of executing large schemes on the Nile outside Egypt was crystallized in Garstin's (1904) study, and from that time onwards, plans for large-scale upstream schemes have been rapidly evolving on both official and scientific fronts. Nevertheless, this concept came at a time in which the Nile Basin, albeit under the British control, was not an "integral unit" but conversely a group of "separate entities" (El-Atawy 1996: 29). In general, the British administrations in the colonies, together with the native administrators, advocated their own interests on the Nile. This was evident, for instance, in Uganda and Kenya, and later in Tanganyika that was placed under Britain's administration instead of Germany as a consequence of the First World War (McIntire 2009: 457). In Egypt, nationalists, led by Saad Zaghloul, were striving for independence, which ultimately resulted in the 1919 revolution and the nominal independence of Egypt in 1922. Even in Sudan, the British governor-general, despite the fact that he subordinated the Anglo-Egyptian Condominium since 1899, sought a degree of interdependence from Cairo, particularly in expanding the country's cotton cultivation. In addition, Ethiopia was already an independent country, as it was able to maintain its independence except from 1936 to 1941, when it was temporarily occupied by Italy.

Accordingly, these factors led to several interactions amongst the Nile riparians over Garstin's proposed schemes to run bilaterally with the essential role of Britain, not as a "master," but as a "moderator," as described by El-Atway (1996: 14, 28). The following subsections discuss in detail the cases of the Gezira scheme in Sudan, the Tana concession in Ethiopia, and the Owen Falls Dam in Uganda.

3.3.2 *The Gezira Scheme and the 1929 Nile Waters Agreement*

The Gezira region, located some 125 km south of Khartoum (Fig. 3.5), is characterized by its fertile soil and gentle slope with accessible waters from the Blue Nile, which made it an ideal place for long-staple cotton cultivation (McCann 1981: 671). Under an international condition of relative shortage and rising cotton prices, the promising British studies, undertaken at the beginning of the twentieth century, made the British Cotton Growers Association push for the commencement of the Gezira scheme as an alternative to the Egyptian monopoly of cotton (Waterbury 1979: 45). At the same time, the British administration of Sudan saw the project as a driver for its economic viability and consequently for boosting the country's independence from Egypt (El-Atawy 1996: 32). The double benefit of the project for Britain and Sudan materialized the first Nile irrigation scheme implemented outside Egypt.

In 1904, the Egyptian Ministry of Public Works allowed the pumping of waters during the "timely season" (from mid-January to mid-July) by the Sudanese government in order to irrigate 10,000 feddans in the Gezira region (Waterbury 1979: 66). Four years later, the construction of a water-storage dam in Sennar on the Blue Nile (Fig. 3.5) was formally suggested in order to increase the irrigated land to 500,000 feddans (Waterbury 1979: 66). Shortly thereafter, Egypt raised the limit of the irrigated land to 20,000 feddans, while on the ground, the actual irrigated area reached only 16,416 feddans by 1919 (Waterbury 1979: 66). In 1910, Britain completed the railroad connecting the Gezira scheme to the Khartoum-Red Sea line operating since 1906, which raised the value of the scheme by linking it to the cotton mills in Manchester and Lancashire (McCann 1981: 671). In 1914, Egypt accepted, under British pressure, the construction of the Sennar Dam on the Blue Nile in return for the implementation of Jabal Auliya Dam on the White Nile (Fig. 3.5). The latter aimed at storing water to cover Egypt's increasing needs when the Blue Nile is exhausted (Collins 1990: 117–118). Figure 3.6 shows that the Blue Nile's rainfall pattern divides the year into two main seasons; before mid-July (the low season), and after mid-July (the high season). The Egyptian acceptance therefore was conditioned by limiting the utilization of the Nile water in the Gezira scheme in a way that would not affect meeting the country's minimum requirements during the low season and also in dry years of low flood. The implementation of the Sennar Dam started immediately in the same year. However, the construction works and the search for an acceptable way to expand the Gezira scheme were postponed due to the First World War (El-Atawy 1996: 35, 36).

Two years after the outbreak of the First World War, a heated and publicized technical debate reopened the issue, which is explained precisely in Collins (1990: 119–142). The debate was centered on the claim of the British irrigation adviser in Egypt, Sir Murdoch McDonald,[11] who argued that, according to the Nilometer records in Sennar, even the extraordinarily low Blue Nile flood of 1914 was hypothetically sufficient to fulfill Egypt's water needs while irrigating up to 300,000 feddans in the Sudanese Gezira scheme. This claim was acutely refused by other British experts in both Sudan and Egypt, mainly Kennedy Pasha and Sir Willcocks respectively, who accused Sir McDonald of falsifying the field results and destructing the public documents upon which he based his claim. Sir Willcocks raised several questions, involving the extent to which the proposed dams on the Blue and White Nile would lead to "prejudicing Egypt's well established water rights," as quoted in Collins (1990: 123). A commission of six members was formed in 1918 by Britain to judge upon this dispute. Sir Willcocks questioned the neutrality and credibility of its members, while the Egyptian press raised a campaign supporting his general doubts.

As a result, Britain formed another international commission in 1920, including representatives from Britain, India and the U.S.; the Nile Projects Commission. To defend his reputation, in front of the newly formed commission, but also the

[11] Sir Murdoch McDonald was an adviser to the Egyptian Ministry of Public Works after the retirement of Sir William Garstin and up until 1921 (Kliot 1994: 35).

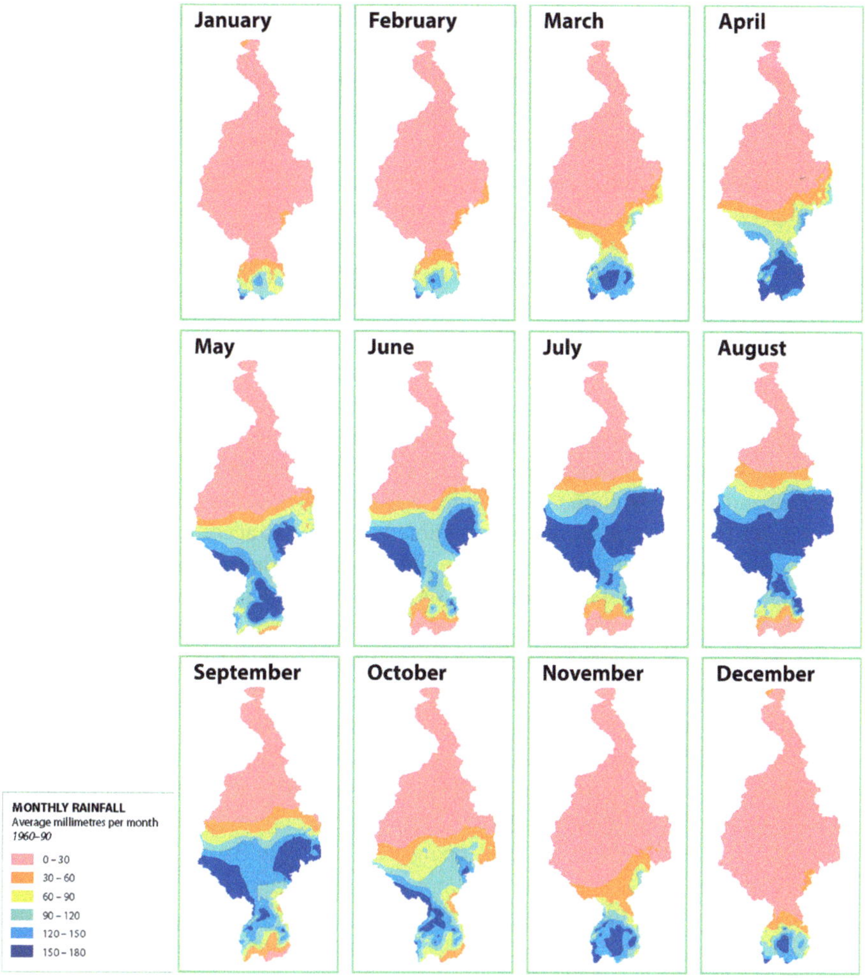

Fig. 3.6 Monthly rainfall precipitation in the Nile Basin. (Source: Prepared by NBI, State of Basin Report (2012a: 31) after the data obtained from the Climatic Research Unit)

Egyptian public opinion, McDonald published his 1920 Nile Control report. He introduced in that official publication an elaborated vision upon Garstin's work for achieving the "over-year storage" required for promoting agriculture to cope with Egypt's rapid population growth (Collins 1990: 129). Although McDonald's storage plan was founded on the principle of securing Egypt's historic water rights, the Egyptian public opinion could not "accept the means of achieving it" because of the concept of building most of the proposed dams upstream, which in turn aroused their anxiety over the external control of Egypt's fate beyond its borders (Collins 1990: 130).

After months of intensive field visits, careful investigations, public sessions and hearings of witnesses, the commission cleared Sir McDonald from any intentional falsification or destruction of documents. Vindicated by the commission, Sir McDonald charged both Sir Kennedy and Sir Willcocks with "defamatory libel" (Collins 1990: 138–139). The Nile Projects Commission also endorsed McDonald's Nile Control plan, which comprised the Sennar Dam, Jabal Auliya Dam, Nag Hammadi Barrage, a regulator dam at Lake Tana outlet, another regulator at the Lake Albert outlet, and a canal bypassing the Sudd swamps. In addition, the two British members of the commission stated the historic rights of Egypt, given the country's existing irrigated lands estimated to be 5,400,000 feddans, while acknowledging Sudan's right to irrigate 300,000 feddans in the Gezira scheme and additional 100,000 feddans in other pumping schemes (Collins 1990: 141–142). Nevertheless, the commission neither translated these figures into volumes of water nor specified the criteria of distributing the potential benefits brought by McDonald's plan. This outcome called for the necessity to reach an accord setting the allocations of Egypt and Sudan from the Nile waters. Egyptian nationalists, in their turn, heavily criticized Britain's position on the whole issue, since, according to their views, it underestimated Egypt's water needs and supported the idea of building dams beyond its sovereignty, which implied a threaten to the country's security and economy (Waterbury 1979: 66; Collins 1990: 120–122; El-Atawy 1996: 35).

The tension between Egypt and Britain over the Gezira scheme reached its peak in 1924, after the assassination of Sir Lee Stack, the Sirdar and British governor-general of Sudan, in Cairo on November 20. As recorded in the Annual Register, Vol. 166[12] (1925: 124), the British government, through Lord Allenby, the High Commissioner in Cairo, immediately dispatched an "ultimatum" to Saad Zaghloul, Egypt's Primer Minister since January of the same year, considering that the "murder" was "the natural outcome of a campaign of hostility to British rights and British subjects in Egypt." The ultimatum comprised seven requirements: (1) presenting "ample apology"; (2) bringing the "criminals" to justice; (3) suppressing "popular political demonstrations"; (4) paying a "fine of £E500,000"; (5) ordering "within 24 hours" the withdrawal of the Egyptian army from Sudan; (6) accepting that "the Sudan Government will increase the area to be irrigated in the Gezira from 300,000 feddans to an unlimited figure as need may arise"; and (7) complying with new measures for the "protection of foreigners" (1925: 124). Zaghloul's answer was that he would accept "the first, second and fourth, and also the third," but he emphatically rejected the last three requirements since they "affected the sovereignty of Egypt" (1925: 124). Lord Allenby responded firmly that he sent instructions to Khartoum to put in effect the withdrawal of the Egyptian army and ensure "liberty to irrigate the Gezira area to an unlimited extent" (1925: 124). As "no further move was made by Zaghloul, … ships of the [British] Mediterranean fleet were sent to Egyptian waters," which led to the resignation of Zaghloul's cabinet (1925: 124).

The British action stirred up the anger of Egyptian nationalists and threatened the instability of the country (Waterbury 1979: 65; El-Atawy 1996: 36). It was also "the

[12] Available at: https://archive.org/details/in.ernet.dli.2015.79540 (accessed on August 16, 2016).

subject of much cynical comments in the press of foreign countries," which led the opposition Labour Party to issue a "strongly worded protest against making a political assassination the occasion for asserting new imperialistic claims" (The Annual Register 1925: 124–125). Mr. Ramsay Macdonald, the then leader of the opposition, stressed that "the Government's handling of the affair had done the country great damage in the eyes of the world" and that "it was a mistake to mix up just and proper indignation against murder with certain political matters that could only be settled by agreement" (1925: 125). London retreated in order to avoid any probable escalation that might disrupt its interests in Egypt, which were specified in four elements since the Egyptian nominal independence in 1922: Egypt's defense, the Suez Canal, the foreign citizens in Egypt, and Sudan (Collins 1990: 147; El-Atawy 1996: 30). Mr. Austen Chamberlain, the then Secretary of State for Foreign Affairs, commented on "the demand for unlimited water from the Nile, which was the point in the British Note most generally and severely criticized." He "admitted that the Government had gone too far" and "pleaded in excuse that when one was obliged to act in a great hurry it was not always possible to choose the happiest phraseology." He "promised to invite the Egyptian Government to join with them in an inquiry under a neutral chairman as to the amount of water available for the Sudan after making allowance for Egypt" (The Annual Register 1925: 134). This was followed by sending a formal letter (fn. 22) from Lord Allenby on January 26, 1925 to Zewar Pasha, the new Egyptian prime minister, stating:

> I need not remind your Excellency that for forty years the British Government watched over the development of the agriculture well-being of Egypt, and I would assure your Excellency at once that the British Government, however solicitous for the prosperity of the Sudan, have no intention of trespassing upon the natural and historic rights of Egypt in the waters of the Nile, which they recognize today no less than in the past … Moved by these considerations and in proof of their intentions, His Majesty's Government are disposed to direct the Government of the Sudan not to give effect to the previous instructions in regard to the unlimited development of the Sudan Gezira mentioned in the note of the 23rd November, on the understanding that an expert committee composed of Mr. J. J. Canter Cremers, Chairman, who has been chosen by agreement between the two Governments, Mr. R. M. MacGregor, British Delegate, and Abdul Hamid Soliman Pasha, Egyptian Delegate, who has been selected by the Egyptian Government, shall meet not later than the 15th February, 1925, for the purpose of examining and proposing the basis on which irrigation can be carried out with full consideration of the interests of Egypt and without detriment to her natural and historic rights.

In effect, the events of 1924 had a great influence on the hydropolitics of the Nile. On the one hand, the evacuation of the Egyptian army from Sudan in 1925 strengthened the independent decision-making of the Sudanese government. On the other hand, Britain sought keenly to reach a compromise that would satisfy both Egypt and Sudan. The committee mentioned above in the letter of Lord Allenby submitted its report in March 1926, stating the measures recommended to "reserve absolutely to Egypt the natural flow of the river during the low season [from January 20 to July 15]," along which Sudan would rely mainly on the stored water behind the Sennar Dam (fn. 22). This two-man report, together with that of the Nile Projects

Commission, constituted the broader Anglo-Egyptian negotiations that led to the 1929 Nile Waters Agreement,[13] which was signed between Egypt and Britain acting on behalf of Sudan, Kenya, Tanzania and Uganda.

Based on the agreement, Sudan got an annual share of 4 BCM to irrigate the Gezira scheme and assured Egypt's financial support for the project, while Egypt obtained a recognition of its "natural and historic rights", estimated as 48 BCM per year. In addition, the agreement stated that "no irrigation or power works or measures are to be constructed or taken on the River Nile … in Sudan or in countries under the British administration" without the "previous agreement of the Egyptian government." Egypt also reserved the right of collecting hydrological data and constructing required facilities for this purpose.

By 1926, the Sennar Dam had been already completed with a storage capacity of 0.8 BCM (Waterbury 1979: 66; Collins 1990: 154–157; Kliot 1994: 39; El-Atawy 1996: 37). The Jabal Auliya Dam was completed in 1937, and its water reservoir was fully filled in 6 years with a storage capacity of 3.5 BCM, while the Nag Hammadi Barrage was completed in 1930 (Collins 1990: 158; Kliot 1994: 33, 39). In addition, the Aswan Low Dam was heightened in 1933, the Assiut Barrage was strengthened in 1938, and the new Delta Barrage was completed in 1940 (Collins 1990: 158; Kliot 1994: 33, 39).

In terms of water allocations, the 1929 Agreement had prevailed until the signature of the 1959 Egyptian-Sudanese Agreement. However, the principle of which Sudan was obligated to take the prior approval of every project related to the Nile had strained the relation between the Sudanese government and the Egyptian Ministry of Public Works (Collins 1990: 209). Moreover, in 1938, when Khartoum requested increasing its water share defined by the agreement, Cairo did not respond, and the outbreak of the World War II completely halted the Sudanese request (Collins 1990: 161).

3.3.3 *The Tana Concession*

Building a dam at the outlet of Lake Tana was first suggested by Sir William Garstin in his comprehensive plan for utilizing the Nile lakes as natural reservoirs so as to release water lower the basin in the low-flow months. Garstin detached C. Dupuis in an expedition in 1902 to survey the project's location. The latter asserted that the geological characteristics and the low evaporation losses at the site (Fig. A.3) confirmed the feasibility of the project. This conclusion is clear through the following statement written in Garstin's report (1904: 184):

---

[13] The 1929 Exchange of Notes between Britain and Egypt in regard to the use of the waters of the River Nile for irrigation purposes, in addition to the 1925 Nile Commission Report and the letters exchanged between Lord Allenby and Zewar Pasha, are available by the Transboundary Freshwater Dispute Database at: http://gis.nacse.org/tfdd/treaties.php?page=full&origin=river&tn=92 (accessed on August 24, 2016).

> It seems most unlikely that any site on the Blue Nile, or Dudessa, can be as well situated as is Lake Tsana, for the purpose required and, if a reservoir has to be constructed within foreign territory at all, that spot should certainly be selected for its construction.

Along the first three decades of the twentieth century, the expansion of the Gezira cotton scheme was a priority for Britain, which put the Lake Tana project on its agenda. Ethiopia, at the time, was an independent country. However, it was subject to the direct intervention of Britain, France and Italy, based on the 1902 Anglo-Ethiopian Treaty and the 1906 Tripartite Treaty. What made this particularly influential was the succession of internal crises that weakened the government. The level of intervention and the importance of Lake Tana were evident in the deposition of Emperor Iyasu V (1913–1916), the successor of Emperor Menelik, for his "failure to cooperate with European legations, on issues such as Tana," as cited in McCann (1981: 673). Ethiopia was then ruled by Empress Zauditu (1916–1930), the daughter of Emperor Menelik, and the regent-plenipotentiary Ras Tafari Mekonnen (later Emperor Haile Selassie). The latter, who aimed at strengthening the central government and modernizing the state, found Lake Tana project and its potential revenues substantial for achieving his objectives. In spite of the British interest and Tafari's enthusiasm, the negotiations over the Tana concession, from 1920 to 1935, were unsuccessful. This is mainly due to the feudal relations between Addis Ababa and the powerful local governors, the xenophobes of conservative nobility, the 1929/1930 world economic depression, and the Italian colonial ambitions in Ethiopia. The different phases of negotiations and the multiple attempts of Britain and Tafari to carry out the project, discussed in this subsection, are illustrated after in McCann (1981: 675–697), Waterbury (2002: 63–65), Collins (1990: 159–161), El-Atawy (1996: 39, 40), and Tvedt (2004: 160–183).

Lake Tana was located within the borders of Begemder and Gojjam provinces (Fig. 3.7), which were governed by Ras Gugsa and Ras Hailu respectively. Both were rivals to Ras Tafari, and they exercised effective control over their areas. Thus, since he was appointed as a regent in 1917, Tafari sought to negotiate the Tana concession with Britain through the central government to consolidate the rule of Abyssinia in Addis Ababa. The message received by the British officials from their talks with Tafari was summarized by Consul Dodds as follows. "If he were emperor … many questions would by now be settled, but he explained how his hands were tied by the empress and her supporters who were mainly obstructionists to the welfare of Abyssinia" (McCann 1981: 679).

Britain chose to adhere to diplomatic means in getting the Tana concession, as it considered the Italian defeat of 1896 at Adwa a learned lesson not to use force in capturing the Ethiopian Highlands. After several rounds of talks, Britain, in 1923, gave Tafari its offer to get the concession in exchange for generous cash payments and facilitating control over territories on the Abyssinian borders. However, London aroused the anxiety of Addis Ababa by opposing Ethiopia's admission to the League of Nations and by planning to construct a direct road linking Khartoum with Tana through which Haillu would gain additional power. Addis Ababa feared that developing the Lake Tana project would further weaken the central government and strengthen its rivals. As a result, the negotiations were broken off, and Britain decided to proceed with its alternative plan by coordinating with Italy.

Fig. 3.7 Map of Abyssinia focusing on Lake Tana showing the provinces of Gujjam and Begemder. (Source: Contributed by Bartholomew in 1885, available by Library of Congress at: https://www.loc.gov/item/2009580108/ (accessed on August 19, 2016))

In 1924 and 1925, the British governor of Sudan exchanged notes with the Italian governor of Eretria in which the latter agreed to regulate the flows of the Gash River for the sake of the Gezira scheme in return for annual rent and a share of agricultural yields. In addition, the British and Italian governments exchanged notes[14] in December 1925 "to concert for common action with the Abyssinian Government" in order to approve the construction of a railroad from Eretria to the Italian Somaliland, and to grant Britain the Tana concession. The two countries also pledged that if one government secured its desired concession, the other "would not relax their whole-hearted efforts to secure a corresponding satisfaction for the other Government concerned." Moreover, Britain promised to "recognize an exclusive Italian economic influence in the west of Abyssinia and in the whole of the territory to be crossed by the above-mentioned railway," and this was "subject to the proviso that the Italian Government on their sides recognizing the prior hydraulic rights of Egypt and the Sudan, will engage not to construct on the head waters of the Blue or White Niles or their tributaries or effluents any work which might sensibly modify their flow into the main river." Italy accepted this and in return requested from Britain "respecting the existing water rights of the populations of the neighboring territories which enter into the sphere of exclusive Italian economic influence," and understanding "that, in so far as is possible and is compatible with the paramount

[14] Available at: http://gis.nacse.org/tfdd/treaties.php?page=full&origin=river&tn=80 (accessed on August 24, 2016).

interests of Egypt and the Sudan, the scheme in contemplation should be so framed and executed as to afford appropriate satisfaction to the economic need of these populations."

Unsurprisingly, this Anglo-Italian Agreement increased the Ethiopian worries about the European imperial intensions. Tafari immediately sent official letters to the League of Nations protesting against the Anglo-Italian Agreement. Even when Britain sought to reopen direct negotiations with Tafari in the mid-1927, he completely refused due to his fears of any Italian involvement. In late 1927, Tafari made a surprising move sending his confident emissary to the U.S. to negotiate with the American engineering company, J. G. White, to build the Tana dam for Ethiopia with a draft contract of $20 million (McCann 1981: 687). By doing so, Tafari transferred the nature of the project from granting the concession to a specific country into contracting with a company to build the dam for Ethiopia. He also attempted to bring the U.S. into the scene as a counterweight power that would halt the colonial plans against Ethiopia. Waterbury (2002: 64) deduces the message sent to Britain by Tafari's decision to privately finance the project as follows, "We will carry out the project with our own resources, and it will be under the sovereign control of Ethiopia." The move both drove Britain to resume talks with Tafari and encouraged other companies from other European countries to submit offers of lower costs. Meanwhile, Tafari worked on consolidating his control over the entire country to assure that the development of the lake would benefit the central government the most, not the local governorates. His efforts led him eventually to be crowned as negus (king) in October 1928. The American officials in Addis Ababa, Cairo and Washington D.C. dealt seriously with the issue in support of the J. G. White company, and they were fully aware about the different positions of the involved countries. This is evident in the correspondence published by the U.S. Department of State (1943: 786–803) in the "Papers relating to the foreign relations of the United States" of 1928, some of which are as follows:

> Within the past few days, His Imperial Highness has received a note on the Tsana dam subject from the British Foreign Office. This note, which I have read, states in effect that the British Government is prepared to consent to the building of the dam as an Ethiopian enterprise provided the work is given into the hands of competent engineers and sufficient guarantees are given that the water will be made amply available on appropriate terms to the Sudan and Egypt. This note also mentions British understanding that the Ethiopian Government proposes giving the concession to the J. G. White Engineering Corporation.
>
> Sent from Mr. Addison E. Southard, the U.S. Minister Resident in Ethiopia, to the J. G. White Engineering Co. on April 26, 1928. (U.S. Department of State 1943: 787)

> I have the honor to report that I had an opportunity today of discussing at length with Ibrahim Fahmy Bey, the [Egyptian] Minister of Public Works, the general problem of Nile control … I was surprised to learn from the Minister that, in his opinion, the Lake Tsana project should take precedence over Gebel Aulia, Lake Albert, and the diversion of the Sudd channel. He seemed to think that the three above mentioned projects could wait even for twenty-five years. The only Nile engineering plan to which he gives precedence over Lake Tsana is that for the heightening of the Assuan dam and the subsequent development of hydro-electric power at that spot.

Sent from Mr. Franklin M. Gunther, the U.S. Minister Resident in Egypt, to the U.S. Secretary of State on October 2, 1928. (U.S. Department of State 1943: 788)

> … I have twice and at length discussed with His Majesty the J. G. White Corporation's proposals for a conference … I am convinced that he must first definitely make up his mind that he wants the dam constructed and that, second, he must overcome his apparent present reluctance to oppose British desires and decline the profitable concessions or payments which they are willing to make towards acquiring the Tsana Dam. I have endeavored in my various despatches to place before the Department my opinion that the King originally had the White Corporation approached in the Tsana matter not because he had any clear or definite idea of building a dam, but because he hoped by such action, not only to uncover British attitude in certain matters of political interest to him, but to provide further apparent reasons for the opening here of the American Legation which he has for many years sought … In illustration, it might be remarked that the British have been working on the Tsana Dam project for many years during the last fifteen or twenty of which they have on various occasions been almost near enough to grasp it … Final decision may be had in the next week, the next month, the next year, or the next decade.
>
> Sent from Mr. Addison E. Southard, the U.S. Minister Resident in Ethiopia, to the U.S. Secretary of State on November 28, 1928. (U.S. Department of State 1943: 795–796)

In April 1929, Tafari signed a "letter awarding the construction concession to the American company," and what remained was only the "approval by the British and an agreement specifying the amount to be paid annually for the water" (McCann 1981: 691). For this purpose, he launched a series of conferences with the U.S. and the British administration of Sudan. Moreover, by the death of Empress Zauditu, Tafari became Emperor Haile Selassie and was able to crush the power of his rivals, which strengthened his control over the entire country in general and Lake Tana in particular.

Nevertheless, the worldwide depression of 1929/1930 changed the situation entirely. As the international cotton prices dropped significantly, to the extent that the Gezira scheme lost more than half of its annual revenues in 6 years (McCann 1981: 695), the Tana project diminished in importance for Britain. Besides, by the conclusion of the 1929 Nile Waters Agreement, Sudan assured its water share that would cover the required expansion of the Gezira scheme, which reduced the urgency of implementing the project. At the same time, Britain recognized Egypt's right of prior approval concerning any upstream project, which indicated the complication of further negotiations over the project.

Britain, as a result, modified its strategy. Tvedt (2004: 164–165) specifies three pillars of the new strategy, after the British representative in Cairo, Sir Loraine: putting the project into "cold storage" until the Gezira scheme would need it; playing the "Egyptian card"; and insisting on "the plea of further studies." In 1931, London asked the Egyptian government, addressing it as a principal actor in the Tana project according to the 1929 Agreement, to finance the implementation of required studies. The Egyptian opposition parties and nationalist press raised a campaign to reject the British request, viewing the proposal as a "wily scheme," which Egypt would cover its costs, while Britain and Sudan would reap its benefits (Tvedt 2004: 165). Khartoum also did not have the capacity to finance the project, and in the 1933 Tana conference, the Sudanese representatives informed both the Ethiopian government and the American company that the project needed further investigation and that neither of them “could expect any financial support from the Sudan” (Collins 1990: 160). However, Egypt and Sudan continued to negotiate the basis on which the two countries would share the potential benefits and costs of the project whenever it would be implemented, and they reached a technical agreement in this regard in 1935 (Collins 1990: 160–161; Howell and Allan 1994: 120).

Although this agreement was not executed, it was considered one of the rare occasions, after 1929, in which both Egypt and Sudan had technically coincided. This is clearly shown in the comment of MacGregor, the irrigation adviser to the Sudanese government, on the agreement as cited in Collins (1990: 161).

> The only tangible outcome of six or seven years of endeavour was the preparation in full detail and proper form of a complete Lake Tsana Reservoir Project, and an agreement between Egypt and the Sudan as to its operations as a joint enterprise. A less tangible, and perhaps more evanescent gain, is to be seen in the fact that a mutually beneficial partnership had been entered into between the two Governments in a field where acute discord had once prevailed.

On the other hand, the Tana project, at that critical moment, was the only chance for Haile Selassie to obtain immediate cash that would fuel his ambitious developmental plans, but also, and more important, it was the only way to prevent the increasing danger of an Italian invasion. Haile Selassie applied for a loan from an American bank, but both Ethiopia’s poor credit rating and the declined economic value of the project due to the worldwide depression were enough reasons to reject its request (McCann 1981: 693). Haile Selassie also “unearthed” the British promise to Menelik II in 1907 to pay £10,000 per year in return for the Ethiopian 1902 engagement not to alter the Blue Nile flow, and he asked Britain to pay the “arrears”, which amounted to £230,000, but his request was rejected (McCann 1981: 693). In a final attempt, Haile Selassie held a conference in May 1935 inviting the relevant parties to reach an agreement. Yet, the Anglo-Egyptian delegation did not attend the conference. Britain was certain that an Italian invasion was soon, and as quoted in McCann (1981: 696), it preferred not to “aggravate the present unfortunate controversy between Italy and Ethiopia.” In fact, Italy mobilized its forces in October 1935, and completely captured Ethiopia in May 1936. The Emperor was exiled, whereas years later, the World War II broke out, and he returned from the exile in 1941 after the defeat of Italy in East Africa. Despite the fact that Egypt was interested in renegotiating the Tana project after the end of the World War II, Haile Selassie was suspicious of the British and Egyptian proposals (Collins 1990: 161). Furthermore, it is important to note that the talks over Tana during the late 1940s were burdened by other territorial issues, such as the Ethiopian claims over Eritrea and the Egyptian claims over Sudan. On the one hand, Ethiopia sought to tie the Tana question to the “Eritrea question” in which they put their acceptance on the former issue conditioned by the Egyptian support in the latter (Tvedt 2004: 247). On the other hand, the “question of representation” of Sudan in the talks was a great obstacle, whether to be considered by the Egyptian side or to be represented independently by the British administration (Tvedt 2004: 242). All these subsequent and interconnected events, since the 1920s, made the Tana project no more than a proposal that did not find its way to implementation until the Cold War age (see next chapter).

3.3.4 The Century Storage Scheme and the Owen Falls Dam

As mentioned earlier, there had been a progressive effort made by the British experts working at the Egyptian Ministry of Public Works to develop comprehensive plans for the full control of the Nile, the most important of which were the 1904 and 1920 reports of Sir Garstin and Sir McDonald respectively. Dr. Harold E. Hurst was the one who elaborated those plans afterwards. Hurst had been working on the Nil hydrology since his arrival at Cairo in 1906 from a lectureship in physics at Oxford, and he was appointed as the director-general of the physical department at the Egyptian Ministry of Public Works in 1915 (Collins 1990: 118, 163). Only by the defeat of the German Afrika Korps at El-Alamein in 1942, the threat to the British presence in the Middle East in general and Egypt in particular was almost over, which led the Egyptian government to press Hurst for an elaborated plan on the development of the Nile Basin (Collins 1990: 200). His intensive work until retirement in 1946 produced six volumes of the Nile Basin's half-century accumulated data, in addition to the seventh volume of the Future Conservation of the Nile in which he and his colleagues at the physical department introduced the well-known concept of the "Century Storage" scheme (Hurst et al. 1946). As a mathematician, Hurst was able to advance Garstin's holistic perspective on the Nile and McDonald's innovative concept of "over-year storage," setting the method by which water could be stored to regulate the Nile fluctuations over the years (Collins 1990: 199). Hurst's concept suggested building dams and reservoirs along the Equatorial Lakes and the Eastern Nile sub-basins with a total capacity sufficient to release annual volume of water equivalent to the Nile average flow over a century (Collins 1990: 199). Implementing the Century Storage Scheme would then regulate the Nile waters along the different high and low flood years.

As shown in Fig. 3.8, the projects proposed in the Eastern Nile sub-basin were located at Wadi Rayan in Egypt, the Fourth Cataract and Roseires in Sudan, and Lake Tana in Ethiopia (Howell et al. 1988: 36). However, the other part of the scheme, which was thereafter known as the Equatorial Nile Project or Victoria–Albert–Jonglei scheme (Fig. 3.8), was the most substantial in Hurst's plan (Collins 1990: 201, 208–210). He chose to develop the Equatorial Lakes to carry the required century storage, estimated as 155 BCM. Two reasons were behind this: the balance between the annual rates of precipitation and evaporation losses there (Figs. A.2 and A.3) and the applicability of raising the lakes' levels. He specified three major elements in the Equatorial Nile Project to attain this objective (Fig. 3.8). First, Lake Albert would be converted into a large reservoir by building a dam to raise its capacity to 140 BCM. Second, another dam would be built at the outlet of Lake Victoria to regulate the flows reaching Lake Albert and also generate hydroelectric power. Third, the Jonglei Canal would be dug from Bahr al-Jabal to the mouth of Sobat, bypassing the Sudd swamps, with a capacity of 55 Mm^3/day, which was significantly larger than those suggested before.

Few months after the publication of the Future Conservation of the Nile in 1946, the Egyptian government decided to adopt the projects proposed in it, particularly the Equatorial Nile Project, since this would increase the irrigation water in Egypt during the timely season in which the Blue Nile flood is exhausted (Collins 1990: 208). However, in order to implement such a large scheme, Egypt had to negotiate the issue with Sudan, the East African colonies, and Britain, which was co-ruling the former and administrating the latter. The advantages and disadvantages of the project were seen from different points of view by the involved parties, which extremely complicated the negotiations. This is thoroughly demonstrated in Collins (1990: 208–223) and Woldetsadik (2013: 41–42), as follows.

On the one hand, the project would negatively affect the indigenous people at the proposed sites. In fact, increasing the capacity of Lake Albert, as required, would raise the lake's level 22 m and would flood lands of 3500 km^2 in Uganda, which was already suffering land shortage, in addition to flooding lands of smaller areas in Uganda, Kenya and Tanganyika around Lake Victoria. Moreover, digging the Jonglei Canal would alter the livelihoods of the Nilotes in the Sudd. Similar to the rainfall pattern of the Ethiopian Highlands as indicated in Sect. 3.3.2, the natural cycle of the flow in the Sudd comprises the timely season (from January to July) and the untimely season (from July to January) (Fig. 3.6). During the former, the flow is low and allows the Nilotes' cattle graze on the floodplain, while during the latter, the Sudd is flooded and the Nilotes retreat. Carrying out the project would fragment such a natural regime, which was the concern investigated by the British experts in Sudan along the preceding decades. In spite of these adverse impacts, the Equatorial Nile Project would also benefit Uganda and Sudan. Uganda was in urgent need for the hydroelectric power in order to industrialize the country and meet the needs of the growing population, while Sudan would benefit, albeit limitedly, from the increased and regulated waters of the White Nile.

At first, Uganda suggested to store the bulk of required waters in the reservoir of Lake Victoria instead of Lake Albert, which would flood smaller areas of land and provide its present and future needs of hydropower at that time. Sudan opposed this suggestion, because it would lower the level of Lake Albert, and in years of high water, the dam there would be opened and the Sudd would be flooded, causing greater harm to the Nilotes than that of the original proposal. The extraordinary flood that occurred in 1946 justified the Sudanese fears. Uganda then suggested constructing a small dam at the outlet of Lake Victoria that would generate the electricity it urgently needed, but Egypt also opposed the proposal because this small dam would absolutely cancel the Equatorial Nile Project. Each party firmly adhered to its own interests: Egypt prioritized achieving the maximum capacity of the Century Storage; Sudan, assuring proper flood control and no harm to the Nilotes; and Uganda, generating required hydroelectric power as soon as possible with the smallest area of flooded land and with adequate financial support. The atmosphere in which the negotiations were taking place further complicated the process, specifically due to the strained Anglo-Egyptian relations, Egypt's distraction with the Palestinian problem, and the history of suspicions between the Sudanese government and the Egyptian Ministry of Public Works.

After numerous meetings and proposal modifications along the years of 1947 and 1948, the three parties were not able to conclude a formal multilateral agreement. Technically, the representatives of Egypt, Sudan and East African colonies

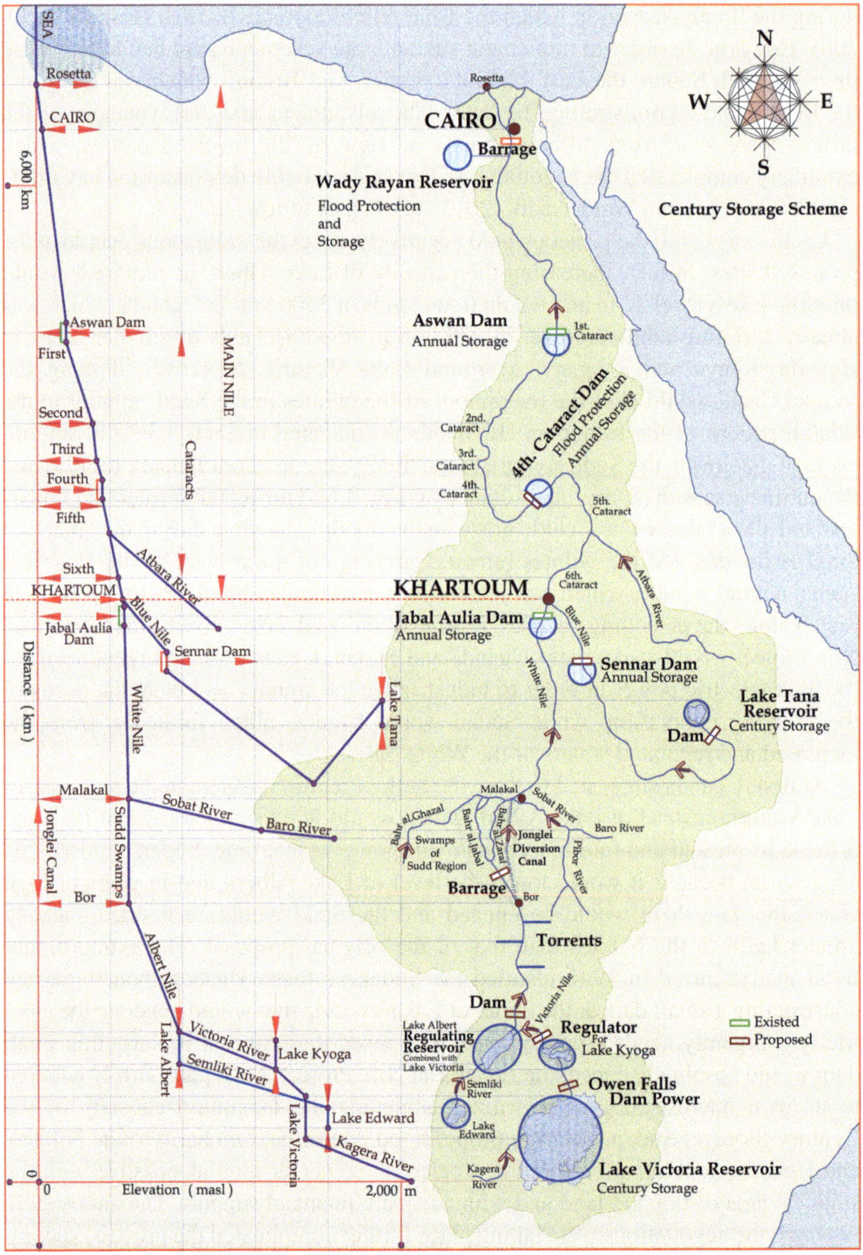

Fig. 3.8 Hurst's proposed century storage scheme including the equatorial Nile project. (Source: The map is reproduced by Michel W. Samaan after the schematic diagram of Hurst's plan for century storage as included in Collins (1990: 202), while the flipped longitudinal section is reproduced after Howell et al. (1988: 29))

reached a compromise in which they agreed upon raising Lake Albert 18.5 m, not 22 m as in the original Egyptian proposal or 14 m as the East African representative initially requested (Collins 1990: 219). It was agreed also that Egypt would finance the Lake Albert Dam completely and the dam at the outlet of Lake Victoria (the Owen Falls Dam) partially. However, this was not enough to sign the Equatorial Nile Project Agreement at the level of governments, since there were other technical and political issues that required further negotiations and longer time, while Uganda wanted to construct its hydropower station immediately.

Nevertheless, Uganda still needed the Egyptian financial support to build the Owen Falls Dam, located at Jinja (Fig. 3.8), where it planned to install its hydropower turbines. Egypt also was still interested in storing waters at Lake Victoria. This common interest led Egypt and Uganda to a separate negotiation table discussing a deal concerning the Owen Falls Dam apart from Lake Albert and Jonglei projects. In 1949, Egypt and Britain acting on behalf of Uganda agreed, in an exchange of notes,[15] upon the technical and financial details of constructing the Owen Falls Dam. According to Collins (1990: 221–222), the agreement stated that Egypt would provide the requested financial support (£E4.5 million) to build the dam in addition to compensating Uganda for the flooded lands around Lake Victoria, while Uganda financed the installation of the power station. The dam would store and release waters for Egypt during the timely season, and an Egyptian engineer would permanently stay at the dam's site to assure full coordination between the two countries. In accordance, Egypt compensated Uganda by £E980 thousand for the hydropower losses caused by releasing waters downstream to meet the Egyptian needs. In 1952 and 1953, complementary agreements were signed dealing with the financial and maintenance related issues of the dam respectively, and in 1954 the construction work was completed. Through these agreements, the "Agreed Curve" was stated, which correlated the required water releases from the dam with the levels of Lake Victoria, and this was the main task of the Egyptian engineer to follow the implementation of the statement (Mulira 2010: 133).

Although the entire Equatorial Nile Project did not see the light of day, it played a principal role in the succeeding decades, namely as it relates to the construction of the Aswan High Dam and the Roseires Dam during the 1950s–1960s, in addition to the incomplete attempt to construct the Jonglei Canal in the 1970s–1980s. However, these projects were carried out in the Cold War age in a completely different context, as will be discussed in the next chapter.

[15] Available at: http://gis.nacse.org/tfdd/treaties.php?page=full&origin=river&tn=142 (accessed on August 24, 2016).

3.4 Analysis and Discussion

While the previous sections have traced the major events along the colonial round of the Nile game, this section is dedicated to discussing the role of the contexts in inducing the game equilibrium. Based on the iterated PD framework adopted by the research, the game equilibrium is defined by the resultant of the outcomes generated in every round. These are categorized into four possible scenarios: one side utilizes the river solely, reaping the tempting payoff (T-S) and leaving the other with the sucker's payoff (S-T); both sides utilize the river jointly to yield the rewarding payoffs of cooperation (R-R); and both sides utilize the river unilaterally, getting the punishing payoff of contention (P-P). What defines which of these outcomes is generated in every move is the distribution of benefits and costs among the riparians, considering that the tempting payoff is the highest and the sucker's payoff is the lowest, while the rewarding payoff remains higher than the punishing payoff. In the following subsections, it is shown how every outcome is affected by the political and environmental contexts at the global (G), regional (R) and domestic (D) levels. The main aim is to examine the process and mechanisms through which these contexts have driven the heterogeneities among riparians over interests (I), capabilities (C) and beliefs (B) to generate different outcomes and move the game equilibrium from one point to another.

3.4.1 *Sole Utilization (T-S and S-T)*

The colonial round started with the situation that had prevailed in the Nile since ancient times. Egypt had been getting solely the greatest payoff (T) of the utilization of the Nile waters through irrigating and cultivating its valley and delta. Obviously, this longstanding outcome (T_{Eg}-S_{Ot}) is mainly shaped by the basin natural and demographic characteristics (Appendix A). Life in Egypt, this deserted and rainless country (Figs. A.2 and A.4), was founded on agriculture in the Nile Valley and Delta that supplied Egyptians with the rich fertile soil (C_{Eg}). This outcome had persisted for long times because Egypt is the only riparian completely dependent on the Nile, having no other sources of freshwaters (I_{Eg}), and because it was the only riparian that possessed the capacity to utilize the Nile waters in extensive irrigated agriculture (C_{Eg}). The need for an overarching authority to regulate the water, manage its use, promote the productivity, and facilitate the trade of yields led to the establishment of the centralized system of governance in Egypt (C_{Eg}). As a result, the Nile, from Menes the first ancient Egyptian king (thirtieth century BC) to Muhammad Ali Dynasty (nineteenth century AD), had been a central concern of every powerful government seeking the development of Egypt. However, agriculture had incomparably developed in the reign of Muhammad Ali Pasha and his successors, who transformed it from the traditional basin irrigation into the perennial irrigation and introduced new crops in Egypt as the long-staple cotton (I_{Eg}). Cotton cultivation flourished significantly in Egypt due to the American Civil War in the 1860s and the consequent rise of international cotton prices (G). Moreover, similar to the attempts made by several ancient Egyptian kings, Muhammad Ali Dynasty sought to secure the sources of the river upper the Nile, which created a sort of antagonism between Egypt and other riparians (B_{Ot}).

On the other hand, all other riparians were dependent on different sources of livelihoods rather than irrigated agriculture, such as groundwater wells, rainfed agriculture, and cattle grazing (I_{Ot}). For long ages, the polities upstream of Egypt, regardless of their levels of centralization, did not possess the technological capacity to utilize the waters effectively and alter the flow to Egypt in any considerable way (C_{Ot}). By the second half of the nineteenth century, this was still the case across the Nile, where Abyssinia was ruled by King Theodore II and thereafter by King Johannes IV, Buganda by its Kabaka, and Sudan by the Egyptians and then by the Mahdists, while the other parts of the basin were under the Egyptian control. Even in the occasions when other riparians declared that it would act upon the Nile, this was threatening Egypt to cut off the flow in response of other political issues, such as the threat of King Tekle Haymanot I and the ambition of King Theodore II. Although this was never fulfilled, the potential danger of using the waters upper the Nile as a weapon against Egypt fueled suspicions of the latter against the other riparians (B_{Eg}-B_{Ot}).

The opening of the Suez Canal (G) brought the European powers into the Nile game (G). Britain ultimately wanted to secure the Suez Canal, a strategic sea-route to its Indian colonies. France considered Britain cheated it by occupying Egypt unilaterally. Germany decided to hold a foot in East Africa and wedge between both, while Belgium and Italy already had colonial ambitions in East Africa. Consequently, the Nile game was evidently internationalized (G).

Britain's interests in Egypt were initially established on the security of the Suez Canal, but the latter required the stability of the whole Egypt, which led inevitably to controlling the entire Nile Basin (I_{Br}), as the security of Egypt meant the security of its lifeline. To attain this ultimate goal, Britain signed treaties (C_{Br}) with Germany, Belgium and Italy, and later with Ethiopia, in which it provided some territorial concessions in exchange for securing the Nile sources. Moreover, Britain occupied Sudan, in an Anglo-Egyptian expedition, and collided with France in the Fashoda incident (1898) in order to prevent any potential attempt to alter the Nile flow (C_{Br}). Although using the Nile waters as a weapon in settling other political issues was again unsuccessful, Egypt became more suspicious about the intentions of those ruling upper the basin (B_{Eg}).

Thus, the British occupation of Egypt led to the maintenance of the historic and natural status of the basin (T_{Eg}-S_{Ot}). Furthermore, the British advisers to the Egyptian government worked on developing the water infrastructure in Egypt by building new dams and barrages and improving the old ones (I_{Eg}). The Aswan Low Dam, in particular, aimed at storing waters in the reservoir behind the dam, which was a substantial advancement to the barrages system of raising the water level. Nevertheless, those British advisers, mainly Sir Garstin, noted that the prosperity of agriculture in Egypt is closely linked with the development in Sudan (I_{Eg} and I_{Su})

and the implementation of large storage schemes upper the Nile (I_{Up}). Britain then adopted this vision, as it promoted cotton cultivation along the Nile in its favor, generating for the first time the outcome of cooperation with Sudan (R_{Eg}-R_{Su}) and with Uganda (R_{Eg}-R_{Ug}), and it was going to take place with Ethiopia ($R_{Eg\&Su}$-R_{Et}), but it failed. The following subsection shows how the political context played a crucial role in the achievement and failure of those cooperation outcomes. Yet, overall, it can be said that Egypt maintained the upper hand in all those interactions, in that the acceptance of the Egyptian government upon any project on the Nile was essential for its implementation.

3.4.2 *Cooperation (R-R)*

By the turn of the century, Britain consolidated its control over Egypt, Sudan, Uganda and Kenya. As discussed earlier, although the motivation of controlling the Nile was initially the security of the Suez Canal and the stability of Egypt, Britain's growing interests to increase the cotton yields across its empire made it keen to carry out development projects in its other Nile colonies, not only in Egypt (I_{Br}). This new motivation was confirmed by Garstin's plan where he indicated the adequacy of cultivating cotton in the Gezira region in northern Sudan, where rainfalls are scarce (I_{Su}) unlike other parts as the Sudd, Uganda and Kenya. In addition, Garstin's plan shed light on the major problems that were challenging the feasibility of implementing a large-scale irrigation scheme in the Gezira region, particularly the lack of required human resources (C_{Su}), whereas he proposed attracting immigrants from neighboring entities to work in the project. Thus, for the first time, Egypt faced the situation in which large irrigation schemes were seriously proposed upstream of Aswan.

The subsequent moves between Egypt and Sudan, with Britain playing the role of a moderator, had run on completely different basis since 1904 when the Gezira scheme started on a small scale until 1929 when the Nile Waters Agreement was signed. Egypt, at first, was cautious about the project, while the ambitions of the British administration of Sudan were increasing over time. Given its interests in both countries, Britain sought to reach a compromise that would satisfy Egypt's insistence to safeguard its historic and natural water rights (I_{Eg}) and at the same time would fulfill the needs of Sudan (I_{Su}). In 1914, Britain pushed for the acceptance of the Egyptian government concerning the construction of two dams in Sudan, one for the sake of the Gezira scheme, the Sennar Dam, and another for storing and supplying waters to Egypt, the Jabal Auliya Dam.

The outbreak of the First World War (G) temporarily halted all these plans. However, the heated technical debate between Sir McDonald and Sir Willcocks reopened the issue and led to the formation of two successive commissions, the second of which prepared a decisive report in 1920 that contributed later to the agreement. Nevertheless, the political context, during the period between 1922 and 1925, was crucial in driving the process to the outcome of cooperation. The nominal independence of Egypt in 1922 (D), followed by the assassination of the British governor-general of Sudan in Cairo in 1924 (R), worsened the relations between Britain and Egypt whereby the former threatened the latter to cultivate the Gezira region without limits. The consequent resignation of Zaghloul, followed by the anger of Egyptian nationalists, prompted Britain's anxiety about potential disturbances in Egypt, which led Lord Allenby to retreat and act speedily to conclude an agreement that would solve all Nile water-related issues. A third commission was formed in 1926, which based its report on its predecessor, stating the methodology upon which the Gezira scheme could be irrigated without prejudicing the natural and historic rights of Egypt (I_{Eg}-I_{Su}).

In 1929, Britain acting on behalf of Sudan and its other East African colonies signed the Nile Waters Agreement with Egypt. The benefits and costs of the 1929 Agreement distributed among its principal actors, Egypt and Sudan, confirm the assumption that the rewarding payoffs of the cooperation outcome need be neither equivalent nor comparable (Sect. 2.2.3). Egypt gained legal recognition of its water rights estimated as an annual share of 48 BCM (I_{Eg}). It also gained the power to veto any hydrological project upstream (C_{Eg}), to collect data (B_{Eg}), and to construct storage projects upper the Nile (I_{Eg}). Sudan was allowed to store waters in the reservoir of the Sennar Dam after the timely season (during the flood months) and to use it in irrigation up to 4 BCM annually (I_{Su}). It also received financial support from Egypt (C_{Su}) to expand the Gezira scheme within its water share. The agreement was based on the current needs of Sudan, not on the requirements for irrigating its cultivable lands on the long term. Nevertheless, it was a step towards sharing benefits of the Nile between Egypt and Sudan and setting the foundation of the downstream cluster that persisted thereafter (R_{Eg}-R_{Su}).

There are two other situations where cooperation outcomes were negotiated on the Eastern Nile sub-basin, of which one was successfully generated (R_{Su}-R_{Er}) and the other failed ($R_{Eg\&Su}$-R_{Et}). The first was the agreement between the British governor of Sudan and the Italian governor of Eritrea, in 1924/1925, upon which Sudan regulated the waters of the Gash River for its Gezira scheme (I_{Su}), and Eritrea received an annual rent besides a share of the scheme's agricultural yields (I_{Er}). The second was the negotiations over the Tana concession between Ras Tafari (later Haile Selassie) and Britain. The proposed project presented different potential benefits for both. Britain needed to store the waters at Lake Tana to expand the Gezira scheme (I_{Su}). Tafari's benefits (I_{Et}) were not related to the use of waters, but to strengthening the position of the central government in Addis Ababa, blocking the colonial ambitions of Italy in Ethiopia, and receiving financial resources needed for implementing other national development plans.

The British and Ethiopian objectives of the Tana project explain why negotiations had failed several times because of the changing context of the period from 1920 to 1935. Tafari rejected the British offer when he realized that the project would benefit the local governors and weaken the central government (D). He thereafter resorted to implement the project privately with the American J. G. White company, in response to the British deal with Italians to get the Tana concession from him under pressure. The world economic depression of 1929/1930 (G), which diminished the international cotton prices, changed the situation entirely because the expansion of the Gezira scheme became no longer a priority for Britain. Moreover, the option of requesting foreign loans from American banks was rejected due to the poor credit rating of Ethiopia (C_{Et}). Even when Egypt and Sudan reached a technical agreement stating how to share the benefits and costs of the project, the growing tension between Ethiopia and Italy, until the Italian invasion in 1935–1936 (R), terminated the negotiations. Eventually, when the negotiations were reopened after the end of the World War II (G), the Ethiopian suspicions against the British and Egyptian intentions, in addition to the linkage of the Tana question to the Ethiopian claims over Eritrea and the Egyptian claims over Sudan (R), led again to the failure of the talks.

During the late 1940s, there had been another attempt to generate a cooperation outcome among Egypt, Sudan and the Equatorial colonies of Britain (R_{Do}-R_{Eq}). The negotiations over the project reflected the conflicting interests of the involved parties and the complexity of reaching a multilateral compromise. Egypt wanted the storage capacity of the project to be increased to the maximum so as to fulfill its growing needs (I_{Eg}). In contrast, Sudan would limitedly benefit from regulating the White Nile flow, and the British advisers to the Sudanese government were aware of the negative impacts of the proposed Jonglei Canal on the local inhabitants at the Sudd region (I_{Su}). Whereas, Uganda was in urgent need for hydropower to fuel its industrial projects that would accommodate its growing population (I_{Ug}). However, the project would flood large areas around Lake Albert, which was an expensive cost for a country that was already suffering from land shortage.

The technical representatives were making progress along the negotiation path, but achieving a final comprehensive solution required longer time to conduct further investigation, particularly because Sudan wanted to assure that the local people in the Sudd would not be harmed after fragmenting the natural regime of the region. Given that Uganda had no time to lose and its need for energy was urgent, besides its inadequate financial capacity (C_{Ug}), it preferred to negotiate with Egypt over one part of the large project, which is the Owen Falls Dam. The offer was to store waters at Lake Victoria for the sake of Egypt in return for approving the project based on the 1929 Agreement and providing the required financial support to build the dam. The payoffs this time satisfied both sides. They agreed in 1949 on the technical and financial details, and the dam was completed by 1954 (R_{Eg}-R_{Ug}).

During the different attempts discussed above, either to maintain the sole utilization outcome (T-S) or to generate cooperation outcomes (R-R), the threat of utilizing the waters upstream (P-P) was employed several times as shown in the following subsection.

3.4.3 Contention (P-P)

During the colonial round, the Nile water was used as a weapon in political disputes in three specific occasions. The first was the ambition of King Theodore II to divert the Nile and "ruin" Egypt (P_{Et}-P_{Eg}). The second was the attempt of France to capture Fashoda, aiming at controlling the White Nile flow and consequently terminating the British occupation of Egypt (P_{Fr}-P_{Br}). The third was the ultimatum of Lord Allenby sent to Egypt, threatening to irrigate the Gezira scheme without limits following the assassination of the British governor of Sudan in Cairo (P_{Su}-P_{Eg}). None of them came true, since, in the first, the Ethiopian King did not possess the capacity to do so (C_{Et}), while in the second, Britain disrupted the French plan (C_{Fr}), and in the third, the risk of destabilizing Egypt pushed Britain to retreat (I_{Br}). However, the three occasions deepened Egypt's mistrust concerning the potential water uses upper the basin and the presence of any foreign power over the sources of the Nile (B_{Eg}).

In two other occasions, the upstream riparians declared that they would privately construct a dam upper the Nile. The first was the contract that Haile Selassie signed with the American engineering company to build a dam at Lake Tana (P_{Et}-P_{Eg}), while the second was the threat made by Uganda to build a small hydropower station at the outlet of Lake Victoria that would mean the complete cancelation of the Equatorial Nile Project (P_{Ug}-P_{Eg}). Both also did not come true, because of the world economic depression (G), the consequent drop of the international cotton prices (G), and the inability to get private loans (D) in the first, and because of the low financial capacity of Uganda and its need for the Egyptian support to build the Owen Falls Dam (R_{Eg}-R_{Ug}) in the second. Nevertheless, there is a common feature in both occasions, whereby these potential outcomes of contention were preceded and succeeded with water negotiations to establish cooperation outcomes. In the first, negotiations between Britain and Haile Selassie had taken place from 1920 to 1935, but they all failed due to the broader political context shown earlier. Whereas, in the second, the threat made by Uganda took place during the negotiations over the Equatorial Nile Project (1946–1949) and those over the Owen Falls Dam (1949).

The above illustration indicates that the potential outcome of contention in the latter two occasions represented bargaining cards that were used to push forward water negotiations. In addition, the card of unilateral action played by Haile Selassie aimed at bringing the U.S. into the game in order to prevent the Italian invasion of Ethiopia (G). In contrast, the strategy employed in the former three occasions of King Theodore's threat, the Fashoda incident, and the Lord Allenby's ultimatum entailed the use of the Nile waters merely as a weapon in other political disputes.

3.5 Concluding Remarks

The game equilibrium in the colonial round had moved from national sole utilization of Egypt to the formation of bilateral clusters between Egypt and Sudan and between Egypt and Uganda, creating another form of sole utilization in the Nile Basin (Table 3.1). Obviously, the equilibrium in the round had been driven by the changing contexts that surrounded this period, particularly the global games of international navigation (opening the Suez Canal), the European rivalry in Africa, the international cotton market, and the First and Second World Wars. These contexts had shaped the progression of the outcomes along the round, defining the path from the point where the British domination over the Nile started in the 1880s to the point where it waned by the early 1950s. Between the two points, the balance of interests, capabilities and beliefs among the riparians had witnessed profound changes.

Under the British occupation, Egypt upgraded its water infrastructure on the Nile. At the same time, the studies of the British adviser to the Egyptian Ministry of Public Works, Sir William Garstin, opened the way for implementing water development schemes upstream, explaining how these would be to Egypt's benefit. As a result, dams were built on the Nile for the first time ever outside of Egypt. Through the 1929 Agreement, Sudan secured a share in the Nile waters and expanded its large-scale cotton cultivation scheme in the Gezira region. Uganda also began reaping benefits from the Nile after signing the 1949 Agreement through hydropower generation. Britain's interests on the Nile led it to play the role of moderator in achieving these outcomes through making the interests of those riparians more compatible and their capabilities more dependent on one another. This is evident in the construction of the Sennar and Jabal Auliya Dams in Sudan and the Owen Falls Dam in Uganda, whereby Egypt entered into negotiations to share benefits and costs with both. Nevertheless, not all the contextual events were pushing for cooperation. For instance, achieving cooperation between Egypt and Sudan was temporarily hindered several times in the context of the low-flood years in the 1910s, the First World War, and the assassination of the British general of Sudan in Cairo in 1924. Yet, the resultant of the different contextual influences led ultimately to the generation of the cooperation outcome of the 1929 Agreement.

The case was different with Ethiopia. Beliefs of Egypt and Ethiopia were already heterogeneous due to the former's past attempts to control over the Nile sources and the latter's threats to alter the flow downstream. At first, through the 1902 Agreement, Britain assured Ethiopia's pledge not to construct any water project on the Blue Nile without the prior approval of the British government and the British administration of Sudan. This was in return for the British recognition of Ethiopia's sovereignty over its territory. Later, Britain was interested in getting the Tana concession to expand the Gezira scheme in Sudan. This time, the deal comprised potential direct cash revenues to Addis Ababa. However, generation of cooperation was hindered because of the weak administrative system in Addis Ababa, the Italian imperial ambitions in Ethiopia, the world economic depression, and the World War II. Even

Table 3.1 The outcomes of the Nile colonial round according to the iterated PD framework

		Does player Y utilize the river unilaterally?	
Scenarios		No	Yes
Does player X utilize the river unilaterally?	No	Cooperation (R-R)	Sole utilization (S-T)
		1925: Sudan and Eritrea shared benefits on the Gash River	Before 1882: Except Egypt, none of the other riparians possessed the capacity to utilize the Nile waters or alter the flow (S_{Ot}-T_{Eg})
		1929: Egypt and Sudan agreed upon their water rights (R_{Eg}-R_{Su})	After 1882: Except for the Gezira scheme in Sudan and the Owen Falls Dam in Uganda, other upstream riparians did not compete Egypt at all in utilizing the Nile waters (S_{Up}-T_{Eg})
		1920–1935: Ethiopia attempted to cooperate with Britain over Lake Tana project (R_{Et}-R_{Br})	
		1949: Egypt and Uganda built the Owen Falls Dam to store water for the former and generate hydropower for the latter (R_{Eg}-R_{Ug})	
	Yes	Sole utilization (T-S)	Contention (P-P)
		Before 1882: Egypt was the sole user of the Nile waters due to the natural and demographic characteristics of the basin, and it extended its domain over the others (T_{Eg}-S_{Ot})	1853–1869: King Theodor II in Ethiopia threatened Egypt to divert the Nile flow (P_{Et}-P_{Eg})
		After 1882: Britain maintained the status quo to stabilize Egypt and secure the Suez Canal, and it concluded several treaties to control the Nile (T_{Eg}-S_{Ot})	1898: France attempted to control the Nile flow at Fashoda to terminate the British occupation of Egypt (P_{Fr}-P_{Ar})
			1925: Britain threatened Egypt to let Sudan utilize the Nile waters unlimitedly in the Gezira scheme in response to the assassination of the British governor of Sudan in Cairo (P_{Su}-P_{Eg})
			1929: Haile Selassie negotiated with an American construction company to build the Tana dam privately out of the British and Italian influence (P_{Et}-P_{Ar})
			1947–1949: Uganda threatened to build unilaterally a small hydropower dam, which meant for Egypt the absolute cancelation of the Equatorial Nile Projects (P_{Ug}-P_{Eg})

when negotiations over the project were revived in the second half of the 1940s, Ethiopia's suspicions against the British and Egyptian plans remained the major obstacle to cooperation.

Along the round, it was obvious how the contexts influenced the reciprocal tactics (Tit for Tat) played on the Nile. When Britain took its unilateral decision to occupy Egypt and dominate over the Nile, France responded by mobilizing its forces in the 1890s to capture Fashoda in an attempt to control the White Nile flow to Egypt. Yet, the British superiority at the global level abolished the French plan. In the mid-1920s, Britain responded to the assassination of the British general of Sudan in Cairo by threatening to expand cotton cultivation in the Gezira scheme without limits, but the fear that the step would trigger popular anger in Egypt led Britain to retreat. During the same period, Haile Selassie responded to Britain's agreement with Italy over the Tana concession by contracting an American construction company to privately implement the project. However, Ethiopia's limited financial capacity and the world economic depression halted the move. Thus, the success of the reciprocal tactic is evidently restricted by the contextual influences on the balance of interests, capabilities and beliefs among the riparian countries.

Chapter 4
The Cold War Round: Large-Scale National Water Schemes in a Polarized World

After the dust of Soviet Dam dedication settled at Aswan, I travelled about a thousand miles up the Nile to a spot in Ethiopia where, 40 years ago, an American company had been asked to build a dam on the Blue Nile, a river which, if diverted into Ethiopian irrigation ditches, could dry Mr. Nasser's lake into a swampy mudhole ... I told his imperial majesty that I had first started writing about Ethiopia in 1927, when he had asked the J. G. White Company of New York to build a dam on Lake Tana, headwaters of the Blue Nile, in order to keep this key control of the Nile out of the hands of the British. 'Yes, that was almost 40 years ago,' replied the Emperor, 'and the Americans never did build that dam. We also asked the Americans to build a dam on the Finchaa River, a tributary to the Blue Nile. It took them five years to make a survey [by the Bureau of Reclamation] and the dam is not even started. Meanwhile, you have doubtless heard of the big fuss further down the Nile where the Russians dedicated a much bigger dam [the Aswan High Dam] which they built not in 40 years but in about five years.' ... said the Emperor, who has been a consistent friend of the United States and has given us a base [at Kagnew] for our most important foreign tracking station.

Article by Drew Pearson (1964a) interviewing Emperor Haile Selassie on June 8, 1964, in Sarasota Journal

4.1 The Contexts of the Cold War Round

The Cold War round in the Nile game refers to the period from early the 1950s to early the 1990s. It starts with the independence of the Nile riparian states one after another and runs under the global rivalry between the Western and Eastern blocks that prevailed until the fall of the Soviet Union in 1991. The map of alignments in the Nile was not static along the round. When Egypt was firmly aligned with the Soviet Union by the late 1950s, Ethiopia was a pro-Western country, while by the mid-1970s both reversed orientations (compare Fig. 4.1 against Fig. 4.2). The

M. M. Samaan, *The Nile Development Game*,
https://doi.org/10.1007/978-3-030-02665-3_4

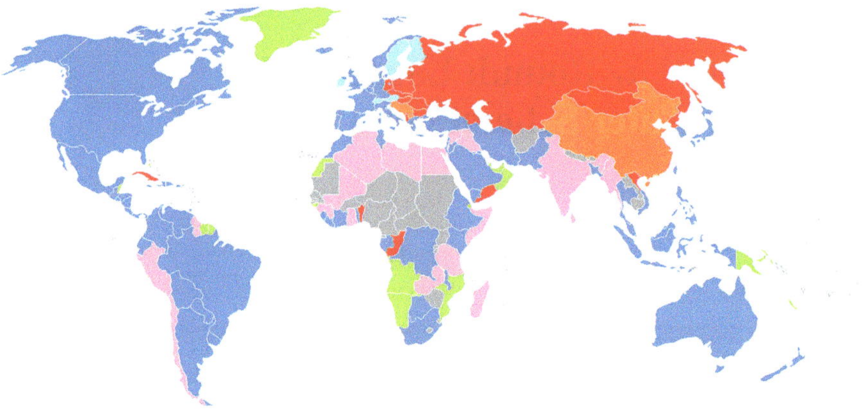

Fig. 4.1 A color-coded world map showing the alignments of countries with the global superpowers in 1970, of which Egypt was a non-communist ally with the Soviet Union while Ethiopia was an ally with the U.S. Key: Green – Non self governing possessions of US allies; Blue – US and US allies; Red – Soviet Union and other communist allies; Orange – Communist countries not aligned with the USSR; Pink – Non Communist allies of USSR; Light Blue – Non NATO members of EFTA and OECD; Gray – Unknown or non aligned. (Source: Intela (2012), available by Wikimedia Commons at: https://upload.wikimedia.org/wikipedia/commons/7/7b/Cold_War-1970.png (accessed on October 13, 2016))

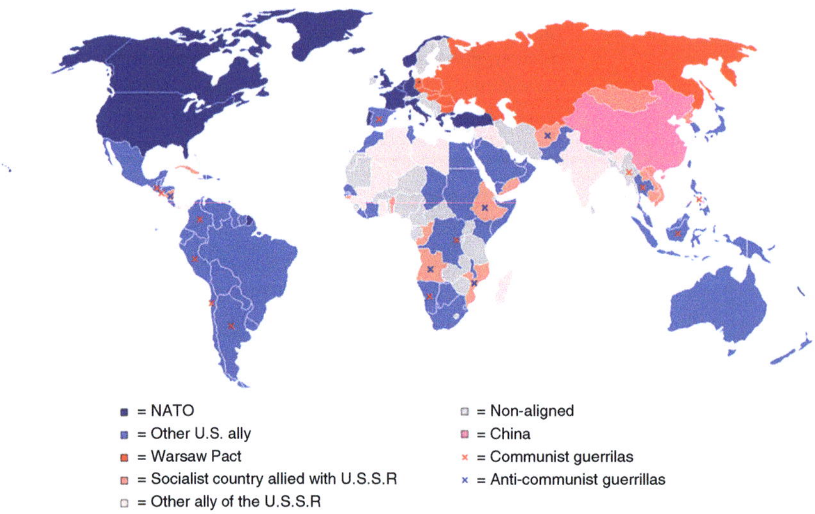

Fig. 4.2 A color-coded world map showing the alignments of countries with the global superpowers in 1980, of which Egypt was an ally with the U.S. and Ethiopia a socialist ally with the Soviet Union. (Source: Minnesotan Confederacy (2007), available by Wikimedia Commons at: https://upload.wikimedia.org/wikipedia/en/8/87/New_Cold_War_Map_1980.png (accessed on October 13, 2016))

Western-Eastern competition was obviously reflected on the development schemes in the Nile Basin. The Soviet Union assisted Egypt to build the Aswan High Dam (AHD) from 1960 to 1971, while the U.S. conducted an extensive survey for Ethiopia on the Blue Nile from 1958 to 1964, and it supported the construction of the Finchaa Dam in the late 1960s. By shifting camps of the Cold War, the Soviets progressed the surveys on Lake Tana in the late 1970s, and Ethiopia proceeded in the 1980s with the construction of the Tana-Beles project, however not completed. The U.S., on the other hand, carried out the maintenance work required for the AHD after its full operation in the 1970s, and Egypt launched the Peace Canal project in 1979 to reclaim lands in the Sinai Peninsula. Concerning Sudan, the bilateral relations with Egypt witnessed serious changes, beginning with tensions until 1958, followed by a period of close alliance based on which the 1959 Agreement was signed and the Roseires Dam was constructed. Yet, by the 1980s, relations between the two countries were disturbed. Moreover, a severe civil war broke out in Southern Sudan with a temporal relief from 1972 to 1983, during which Egypt and Sudan started constructing the Jonglei Canal in 1978, however the project was halted only 5 years later. On the other front upstream the basin, the Equatorial riparian states had been suffering from political instability. At the regional level, the Suez Crisis (or Tripartite Aggression) and the Arab-Israeli Conflict, on the one hand, and the Ethiopian-Eritrean civil war and the Ethiopian-Somali war, on the other, significantly affected the Nile game. Furthermore, the successive environmental crises that took place in the Nile, particularly the 1961–1964 great rains in the Equatorial Lakes, the 1983–1985 famine in Ethiopia, and the 1988 flood in Sudan, played an important role in initiating collaborative programs for information sharing and technical cooperation, such as the Hydromet survey and the Undugu group.

In this chapter, all of these topics will be discussed in detail, focusing on the interconnections between such dynamic context and the Nile game, and how these different events drove the inherited heterogeneities among upstream and downstream states to trigger conflictive and cooperative outcomes along the round.

4.2 From the 1950s to the 1960s

4.2.1 The Transition from the Colonial Age to the Cold War Age

Following the World War II, Britain started losing its imperial presence worldwide. It lost its control over India and Palestine in 1948 and over the Iranian oil fields in 1951. In 1956, it withdrew from Egypt and Sudan. As a result, its presence in East Africa became no longer of great importance as before, and its priority was to balance between the interests of the white settlers, the indigenous people, and the Asian immigrants in its colonies (El-Atawy 1996: 41). By the early 1960s, the Equatorial riparians that were colonized by Britain gained their full independence; Tanganyika in 1961 (unified with Zanzibar to form Tanzania in the following years), Uganda in 1962, and Kenya in 1963. The other riparians also gained their independence from Belgium; Congo in 1960 (named Zaire in 1965 then the D.R. Congo in 1997), and Rwanda and Burundi in 1962 (Fitzpatrick et al. 2006: 31).

However, the geopolitical weight of the Red Sea, which previously motivated the Europeans to compete for controlling East Africa, triggered again the competition between the U.S. and the Soviet Union. This time, the Red Sea represented a vital route of the international oil trade. The tools of dominance were also different. Instead of mobilizing forces and occupying territories, the Western and Eastern blocks worked on establishing monitoring stations and military bases, wherever possible, in addition to expanding their markets and trade zones. Thus, the bipolar rivalry that covered the world, from the Cuban Missile Crisis and the use of sugar crops sanctions in the 1960s to the Afghanistan War and the employment of religious extremist groups in the 1980s, was no less apparent in the Nile. Nevertheless, here the competition ran over water development schemes, as shown in the following subsections.

4.2.2 The Aswan High Dam, the Roseires Dam, and the 1959 Agreement

Few months after Colonel Gamal Abdel Nasser and the Free Officers threw King Farouk off his throne on July 23, 1952, the Revolutionary Command Council (RCC) decided to build a single "mighty" dam south of the Aswan Low Dam instead of waiting to implement the multiple upstream projects of the Century Storage Scheme (Collins 1990: 247). Since the overflow of the Aswan Low Dam in 1946, Egypt was contemplating the construction of larger dams upstream for water conservation and regulation, which was crystallized in Hurst's Century Storage Scheme. However, the failure to reach a multilateral agreement with Sudan and East African riparians upon the Equatorial Nile Project and with Ethiopia upon the Lake Tana project (Sects. 3.3.3 and 3.3.4) encouraged the RCC to build an extra large reservoir within the Egyptian borders.

The potential benefits of the AHD were numerous; saving the waters discharged in the Mediterranean Sea (neither stored behind existing dams nor used in irrigation schemes), preventing damages caused by floods, regulating water over seasons and years, expanding perennial irrigation along the Nile Valley and Delta, generating hydropower, developing fishing industry in the dam's reservoir, and creating job opportunities (Zunes and Tucker 2010: 166–167). Water savings were estimated to be 22 BCM annually on average. This estimation was based on subtracting the 1929 established water rights for Egypt and Sudan (52 BCM in total) in addition to the reservoir's losses due to surface evaporation and seepage (10 BCM) from the total annual runoffs as measured at Aswan (84 BCM) (Waterbury 2002: 73).

Against these benefits for which the growing Egyptian population was in a great need, the AHD's costs were enormous, not only economically, but also environmentally, socially and politically. By blocking the flow of sediments and nutrients, the agricultural land around the Nile Valley and Delta would lose its natural source of fertility. The blocked flows and sediments would cause the erosion of the delta and coasts, the intrusion of seawater and contamination of groundwater, and the deterioration of water quality due to the overuse of artificial fertilizers (French 2007: 249–251; Schwartz 2005: 358). In addition, filling the reservoir required the submergence of fertile lands and displacement of some 100,000 local inhabitants, after flooding Nubian villages in Egypt and Sudan, including Wadi Halfa in the latter (Fahim 1981: 59). Moreover, constructing the AHD meant that Egypt had to bargain with Sudan over the management of the dam's benefits and costs and to face the objection of the other Nile riparians. In addition, the lack of financial and technical capacity pushed Egypt into the global game of the Cold War, as discussed below in this section.

While Egypt was studying the construction of the AHD, Sudan was investigating the expansion of its Gezira irrigation scheme, and in 1954 the latter took its decision to build the Roseires Dam on the Blue Nile (Collins 1990: 251). Both countries sought separately to get funds from the International Bank for Reconstruction and Development (IBRD), which is the original institution of the World Bank created in 1944 to help Europe rebuild after the World War II (World Bank 2014). Given that the Roseires Dam would affect the Nile flow to Egypt on the one hand, and the AHD would flood fertile lands and displace local inhabitants in Sudan on the other (Fig. 4.3), the IBRD conditioned its approval of the required funds upon reaching an agreement between the two countries (El-Atawy 1996: 48). Negotiations between Egypt and Sudan were not smooth, passing through four rounds: from September to December 1954; from April 1955 to November 1955; from December 1957 to January 1958, and from March to November 1959 (Collins 1990: 251–258, 267–270). The first three rounds were unsuccessful, while the fourth resulted in the 1959 Agreement on the Full Utilization of the Nile Waters.

Along the unsuccessful rounds of negotiations, there was a wide gap between the Egyptian and Sudanese positions, as shown in detail in Table 4.1. Sudan refused to discuss each project individually. Rather, it insisted on negotiating a comprehensive accord that would replace the 1929 Agreement and reallocate the Nile waters in a more equitable manner, recognizing not only the existing needs of both sides, but also the potential ones. From this standpoint, Sudan argued that the Nile total runoffs (84 BCM/year) should be divided between both countries based on the available cultivable land in each, deducting the evaporation and seepage losses from Egypt's share. According to this calculation, Sudan, demanded 35 BCM/year since the first round of negotiations, and affirmed that it would not accept less than 25 BCM/year (Collins 1990: 252). Egypt refused such criteria and conversely rested its calculation on setting aside the established rights and water losses while dividing the remainder (22 BCM/year) based on each country's population, at that time estimated to be 8 million and 22 million for Sudan and Egypt respectively (Waterbury

Fig. 4.3 The map of the Eastern Nile Basin showing the Lake Nasser and the reservoirs of the Roseires Dam, Sennar Dam and Khashm al-Girba Dam. (Source: Eastern Nile Technical Regional Office official website, available by the Nile Basin Initiative at: http://entroportal.nilebasin.org/OSIKit/images/JMPMainNile.png (accessed on October 13, 2016))

Table 4.1 Comparing Egypt's and Sudan's positions during the 1954–1957 negotiations with the 1959 Agreement

Negotiation issue	Egyptian position	Sudanese position	Final agreement
What would be built?	One large dam at Aswan is more efficient in cost and water saving than building the whole Century Storage Scheme	It is more efficient to build the Century Storage Scheme elaborated by Morrice and Allan, among which the Roseires Dam is proposed	It is agreed to build the Aswan High Dam and the Roseires Dam, and that Sudan can build its water infrastructure that enables it to utilize its water share
How to share the costs?	Because the AHD would assure the annual water shares of Egypt and Sudan, both should share its costs	Sudan would not bear any costs of the AHD, since it would have to build its own projects to utilize its share. Egypt also should compensate Sudan for the loss of property and displacement of people	It was agreed that Sudan would not bear any costs of the AHD and would be compensated for flooded lands and displaced people
How much is the compensation for flooding lands and displacing people in Sudan?	Egypt started with £E9 million and £E10 million	Sudan started with £E36 million and then lowered it to £E20 million	Nasser offered to reach a compromise at the middle on £E15 million, and Sudan accepted
Which waters would be shared?	The established rights plus the water losses should be subtracted first. Then, the remainder would be divided based on the population	The total amount of waters should be divided based on the available cultivable land, and Egypt should bear the water losses from its share, since it is the one that insisted on building one large dam at Aswan	Established rights of 1929 plus the water losses were subtracted, and the net benefits were allocated as one-third to Egypt and two-thirds to Sudan
What would be the water quota of each?	Egypt offered Sudan to get 8 BCM, and then the offer was increased twice to 9.5 BCM and then to 13 BCM. The calculation was based on the populations in the meantime: 8 million in Sudan and 22 million in Egypt	Sudan demanded 35 BCM to cultivate 4.5 million feddan (8000 m^3/feddan) at first and then declared that the bottom-line of its needs to 25 BCM	The remainder (22 BCM) were divided into 7.5 BCM for Egypt and 14.5 BCM for Sudan. In total, shares are 55.5 and 18.5 BCM for Egypt and Sudan respectively

Source: Waterbury (1979: 68, 69) and Collins (1990: 251–258, 267–270)

1979: 68). Accordingly, the Egyptian offer to Sudan was an annual share of 8 BCM, a figure that was raised twice to 9.5 and 13 BCM during the first and second rounds respectively (Collins 1990: 252, 254). Sudan, in turn, rejected the Egyptian calculation, and the negotiations over the water quota were deadlocked.

For Egypt, the effort to secure the financial resources for the AHD was not limited to obtaining IBRD's funding, but also subject to exploring other options by approaching the Eastern block, which eventually changed the game. The subsequent events in this regard are summarized after Zunes and Tucker (2010: 166–168) citing Shibl (1971), Lytle (1977), and Parks (2003) (see also Kliot (1994: 40, 41) and Withana (2008: 139)).

In 1954, when Nasser became the president of the RCC and later elected as Egypt's president in 1956, he presented himself as a pan-Arab nationalist and as a neutralist and anti-imperialist concerning foreign affairs. Adopting such positions, Egypt signed two important treaties in 1954, the first of which recognized Sudan's self-governance, while the second stated the gradual evacuation of British troops from occupying the Suez Canal zone until mid-1956. Domestically, Nasser's main objective was to develop Egypt through the construction of the AHD and the modernization of the military. At first, he favored to achieve this orientation with the support of the Western block. However, when the latter refused to supply him with the demanded modern weapons given the context of the Arab-Israeli conflict, he turned to the Eastern block and realized what he needed from Czechoslovakia in return for cotton.

In December 1955, Nasser received a package offer concerning the AHD with aid funds of $56 million from the U.S. and $14 million from Britain in addition to a $200 million loan from the IBR (Zunes and Tucker 2010: 166). He did not respond immediately to this offer as he waited for another convincing offer from the Soviet Union, which did not happen, leading to his acceptance of the Western block's offer announced, on July 17, 1956. Two days later, the U.S. and Britain withdrew their offer, justified by the failure of the Egyptian-Sudanese negotiations to reach an agreement regarding the AHD. However, the opposition to such a deal in the U.S. Congress reflected other political and economic reasons, primarily associated with the hostility between Nasser and Israel, the Egyptian recognition of the People's Republic of China (PRC), and the competition between American and Egyptian cotton production, which would further expand after constructing the AHD. Nasser's response was immediate. On July 26, in front of a crowd of some 250,000 people in Alexandria, Nasser announced his decision to nationalize the Suez Canal, which had been administrated by an international company since its opening in 1869 owned by French and British capitals (Sect. 3.2.2), in order to possess the financial capacity needed for constructing the AHD. The Suez Canal's revenues in the preceding year were $100 million of which Egypt got about $2 million (Zunes and Tucker 2010: 166). Nasser in the speech said the following words, as published by the U.S. Department of State (1956: 25–28):

> The Suez Canal Company was a state within a state. It was an Egyptian Joint Stock Company, relying on imperialism and its stooges. The Suez Canal was built for the sake of Egypt and for its benefit. But it was a source of exploitation and the draining of wealth … It is no shame to be poor and work for the building of my country. But it is shameful to suck blood. They used to suck our blood, our rights and take them. The Canal was dug by the Egypt's sons and 120,000 of them died while working. The Suez Canal Company in Paris is an imposter company … But history will never repeat itself. On the contrary, we shall

> build the High Dam. We shall restore our usurped rights. We shall build the High Dam as we want it. We are determined to do it … Thus, today, citizens, when we build the High Dam we are actually building the dam to defend our dignity, freedom, and pride, and to eradicate humiliation and submission. Egypt … announces that it will fight to the last drop of its blood … for the sake of Egypt. We shall not let warmongers, imperialists, or those who trade in human beings dominate us. We shall depend on our hands and on our blood … We shall build a strong and dignified Egypt, the Arab Egypt … Today citizens, the Suez Canal Company has been nationalized … Today, our wealth has been restored to us.

As expected by Nasser, his decision antagonized Britain, France and Israel to the extent that the three attacked Egypt in late October 1956. However, the Egyptian resistance, the Soviet Union's threats, and the pressure exerted by the U.S. and the UN stopped the attack days later. Once more, the use of the water weapon was contemplated against Egypt. Three weeks following Nasser's speech, the idea of involving the Nile in the dispute was first suggested by Sir Ivone Kirkpatrick, the Permanent Secretary at the Foreign Office and a personal advisor to Sir Anthony Eden, the then British Prime Minister. The initial assumption behind this idea was articulated by the Foreign Office indicating that, "Ministers have recently shown much interest in the possibility that, by stopping or reducing the flow of the White Nile at Owen Falls, severe damage may be caused to the Egyptian economy" (Tvedt 2004: 308). The suggestion was intensively studied, from that date to late December, through a series of "top secret" reports produced by the British Foreign Office and Colonial Office in addition to consultations provided by technical and legal advisors to the British government and the British administrations of Uganda and Sudan as elaborated by Tvedt (2004: 300–311).

The main questions investigated were to what extent such a measure would be effective, and what negative consequences will affect Britain. The conclusion of these studies was summarized in a Colonial Office memorandum issued on October 9, indicating that it was "not a simple matter of turning the tap on and off". Due to the complex hydrology of the Nile Basin, Egypt would not suffer a considerable damage before the timely season of the year after next year, provided the next Blue Nile flood would be significantly low. While Egypt would not be harmed immediately, the adverse impacts on Britain would be large and immediate. On the one hand, impounding water behind the Owen Falls Dam for a while would diminish the generation of hydropower, halt navigation downstream of the dam, and flood lands around Lake Victoria, all of which would hinder the manufacture and shipping of cotton products in Uganda. On the other hand, this step would damage the "British prestige" and would anger "vocal Africans" against Britain, besides the possibility that Egypt might, as noted by Sir Kirkpatrick, "persuade the other Arab countries to deny us our oil" (Tvedt 2004: 304, 309). Such fear was justified by Nasser's charismatic leadership in Africa and the Arab World, as he was inspiring and supportive of different liberation movements in the Third World countries. Another option was also discussed, to push Sudan for acting unilaterally on the Nile through the construction of the Roseires Dam and the implementation of large-scale irrigation schemes with the aim of diverting the Nile waters from Egypt. Nevertheless, legal experts confirmed that this would violate the 1929 Agreement and it would take several years before Egypt would be affected. In conclusion, the Colonial Office stated on December 20 that after 5 months of thorough investigation under "absolute secrecy," it planned to undertake no further studies since the subject was no longer "a live issue" (Tvedt 2004: 311).

By the termination of the British-French-Israeli attack (named the Suez Crisis or Tripartite Aggression), the new superpowers of the Cold War, the U.S. and the Soviet Union, completely ended the imperial influences of Britain and France on the Suez Canal. It is also worth noting here how the two international waterways, the Nile River and the Suez Canal, were closely connected, in that the Suez Canal was the initial reason for Britain's colonization of Egypt in 1882 and its control over the entire Nile by the turn of the century. In addition, the attempt to fund the AHD domestically motivated Nasser to nationalize the Suez Canal, whereby Britain contemplated the use of the Nile weapon to press Egypt to retreat.

In the following year, Egypt received an offer from the Soviet Union to assist in constructing the AHD both financially and technically. In contrast to the other offers, this did not require reaching an agreement with Sudan, at the time when the bilateral relations between both countries were generally deteriorating. By 1956, both countries restored their full sovereignty, whereby the unity project between them failed (Massoud 2012: 196). In 1957, Sudan endorsed the "Eisenhower Doctrine" that provided military support for the Middle Eastern countries against the "communist aggression" (Waterbury 1979: 70). In February 1957, the tension between Egypt and Sudan reached its peak over the Hala'ib Triangle, a disputed territory on their common border (Dzurek et al. 2001: 5). Moreover, the Sudanese government had already commenced unilaterally, in 1956, the Managil extension of the Gezira scheme in order to irrigate additional 800,000 feddans in three stages, and by July 1958, it accomplished the first stage of the extension and started heightening the Sennar Dam by which 221,000 feddans were then ready for irrigation (Collins 1990: 265). Although Khartoum kept Cairo informed with its water projects, it did not ask the Egyptian prior approval, which was perceived as "openly repudiating" the 1929 Agreement (Waterbury 1979: 71).

Furthermore, in June 1958, some months after the failure of the third round of negotiations, Sudan announced its own master plan on the entire Nile, published in Humphrey Morrice and Nimmo Allan's Report on the Nile Valley Plan. The plan principally elaborated Garstin's and Hurst's work. However, as described by Collins (1990: 259), this was the first time, not only in the Nile, but in any other basin, to use computer in modeling the hydrologic system of the river and simulating the operation of proposed reservoirs. The plan added new projects to those studied in the previous ones, including the Roseires Dam, Khashm al-Girba dam (see below) and Tana-Beles dam (see Sects. 4.2.4 and 4.3.6). The AHD was also included, however significantly smaller than the one proposed by Egypt (Collins 1990: 264). Thus, the plan did not adopt Egypt's perspective to implement the concept of Century Storage exclusively in the AHD, but went for constructing regulators and hydropower plants along the basin. Actually, the Nile Valley Plan represented a large step toward the extensive use of waters upstream of Egypt.

The acute tension between both countries was broken through the bloodless military takeover in Sudan by General Ibrahim Abboud in November 1958. This move was induced by the civil war sparked in South Sudan since 1955, the breakdown of the Sudanese parliament formed in 1956, and the deteriorated Sudanese-Egyptian relations (Waterbury 1979: 71; Collins 1990: 255). Egypt welcomed the new leadership in Sudan, and in few months, the fourth round of negotiations started, headed this time by military figures from both countries (Waterbury 1979: 72). There was a political will in Egypt and Sudan to reach a comprehensive agreement that would resolve all the suspended issues. In fact, the Soviet assistance needed no agreement with Sudan, and the public opinion in Egypt would no longer understand any delay in implementing the great national project for which they had fought. However, Nasser was aware that Egyptians would not accept flooding lands and displacing people in their southern neighbor without concluding a satisfying deal with the "Sudanese brothers" (Collins 1990: 267–268). At the same time, Abboud was in great need for the agreement to consolidate his control over Sudan with the support of Egypt, and no less important he still lacked the financial capacity to construct the Roseires Dam (Collins 1990: 266). In addition, both countries wanted to forestall the proposal discussed by the British parliament in May 1956 to establish an International Nile Waters Authority that would govern the Nile Basin at the international level and in which Britain would be a full member (Collins 1990: 266). In August 1959, Britain, on behalf of Kenya, Tanganyika and Uganda, sent official notes to Egypt, Ethiopia, Sudan, and Belgium that was administrating Congo, reserving the water rights of its East African riparians, while also calling for an international conference to discuss the establishment of its proposed Nile authority (Waterbury 1979: 72).

On November 8, Egypt and Sudan signed the 1959 Agreement on the Full Utilization of the Nile Waters[1] through which the major issues were settled. In terms of the water shares, the division was applied on the remainder amount, setting aside the 1929 shares in addition to evaporation and seepage losses. The net additional water was estimated to be 22 BCM/year in average, from which Egypt received 7.5 BCM/year while Sudan received 14.5 BCM/year. As a result, the established rights of Egypt and Sudan increased respectively from 48 to 55.5 BCM/year and from 4 BCM/year to 18.5 BCM annually. It was also agreed that Egypt would compensate Sudan with an amount of £E15 million in relation to the flooded lands and displaced communities as shown in Table 4.1, which provides a comparison between the provisions stated in the final agreement and the different positions adopted by the two countries during negotiations. Moreover, unlike the 1929 Agreement, it was defined how to share on equal basis both the costs of future water conservancy projects and the new waters added to the system in case of implementing such projects. In specific, the agreement named the Jonglei Canal as a prioritized water conservancy project. Furthermore, Article 5.2 discussed the possibility that the other riparians seek shares in the Nile waters, as follows:

[1] Available at: http://gis.nacse.org/tfdd/treaties.php?page=full&origin=river&tn=230 (accessed on October 13, 2016).

> As the riparian states, other than the two Republics, claim a share in the Nile waters, the two Republics have agreed that they shall jointly consider and reach one unified view regarding the said claims. And if the said consideration results in the acceptance of allotting an amount of the Nile water to one or the other of the said states, the accepted amount shall be deducted from the shares of the two Republics in equal parts, as calculated at Aswan.

Based on the agreement, the Permanent Joint Technical Commission (PJTC) was established in January 1960, and headquartered in Khartoum, with the purpose of planning, supervising and monitoring water projects along the Nile (Collins 1990: 274). In addition to improving their deteriorated bilateral relations, the 1959 Agreement opened the way for the construction of the AHD in Egypt and development projects in Sudan, which have brought various benefits and costs to both countries. With the Soviet assistance (a credit of 400 million rubles) (Collins 1990: 266), construction of the AHD started in 1960, whereas the first stage was completed in 1964 by launching the reservoir's first filling. The dam structure was completed by 1970, and the reservoir reached its full capacity by 1976 (Zunes and Tucker 2010: 167). The AHD is a rock-fill dam, 3820 m long and 110.7 m high above the riverbed, which resulted in the creation of Lake Nasser behind it, with a total capacity of 160 BCM, out of which 31 BCM is for dead storage and 90 BCM for live storage, while the remaining capacity is for high-flood waters (Kliot 1994: 41). To fill the reservoir, local Nubian people were displaced to New Nubia (newly reclaimed lands near Kom Ombo in Upper Egypt), however the conditions there did not satisfy them, and they kept the dream of returning back to their original homes (Fahim 1981: 93). The dam generates hydropower with a capacity of 2200 MW, which, as shown in Fig. 4.4, covered 63% of Egypt's total electricity production when it came

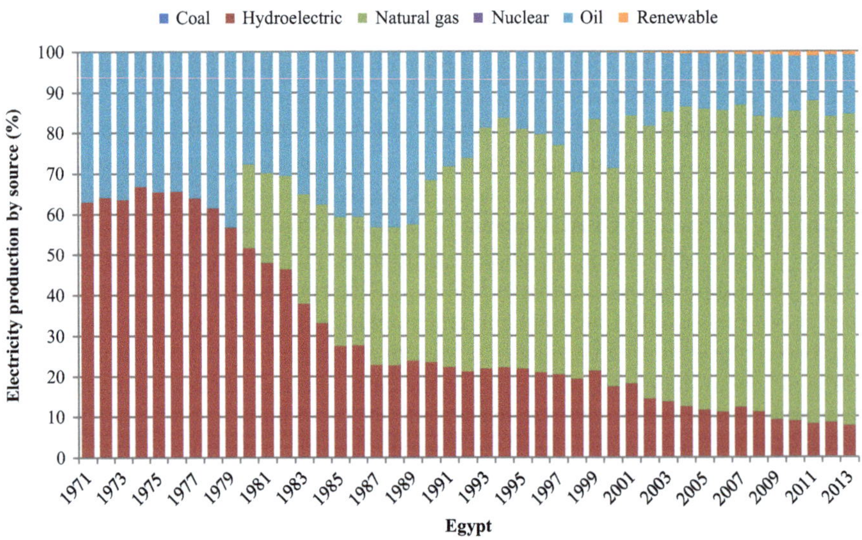

Fig. 4.4 Electricity production by source in Egypt. (Source: International Energy Agency Statistics (2014), available by World Development Indicators at: http://data.worldbank.org/data-catalog/world-development-indicators (accessed on November 2, 2016))

into operation in 1971; however this percentage has steadily decreased due to rapid population growth until it reached 7.7% in 2013 (OECD/IEA 2014). In addition to these factors, most of the annual flowing sediments (124 million tons) have been blocked behind the dam at Lake Nasser, out of which only 10 million tons was deposited on the floodplain along the Nile Valley and Delta. As a result, the soil lost its natural source of fertility and the erosion rates in the delta's coasts increased by three times (Schwartz 2005: 358).

Sudan, on the other hand, began the construction of the Roseires Dam on the Blue Nile (Fig. 4.3) after receiving financial assistance from the IBRD ($32.5 million) and West Germany ($19 million) (Collins 1990: 274). The Roseires Dam was completed in 1966, impounding a reservoir of an initial capacity of 3 BCM (which decreased over years due to siltation), with the possibility of 10-m heightening to increase the storage capacity to 7.6 BCM (Taha 2010: 196) (see Sect. 5.3.4). In addition, the Managil extension was implemented by 1964 over five stages (Taha 2010: 193). Sudan also constructed the Khashm al-Girba Dam on the Atbara tributary (Fig. 4.3) with a capacity of 1.3 BCM to host the displaced Nubian inhabitants in the New Halfa irrigation project, creating a new environment comparable to their old one in which they had previously settled (Taha 2010: 194). However, the Egyptian compensation of £E15 million was not enough, and Sudan had to provide additional budget to complete the displacement program. The process of displacement did not run smoothly and was characterized by "violence" (Collins 1990: 272). Moreover, since its completion in 1964, the Khashm al-Girba project faced serious problems, mainly bringing together "two different and conflicting societies" (the Halfa's settled cultivators with the local nomads), in addition to the high rates of the reservoir's siltation,[2] which significantly lowered its water storage capacity (Taha 2010: 194, 195; Fahim 1981: 97).

Concerning the water conservancy projects, the focus since the beginning of the twentieth century had been mainly to save the waters coming from the Equatorial Lakes and lost in the vast swamps of the Sudd. Furthermore, the ideas, which were previously depicted in Nile plans of Garstin, McDonald and Hurst (Chap. 3), underpinned the high potential of constructing regulators at the Equatorial Lakes and a channel bypassing the Sudd swamps in order to increase the runoffs of the White Nile. Yet, these projects were not undertaken (Sect. 3.3.4), whereby the many changes that took place in the Nile Basin between the late 1940s and the early 1960s made the implementation of the project more difficult. As briefed by Waterbury (1979: 75; 2002: 72) and Collins (1990: 276), these changes entailed several factors. First, the construction of the AHD substituted the major objective of the Equatorial Nile Project; to provide Egypt with the timely waters by regulating the flow across different seasons. Consequently, the PJTC concentrated on two principal projects: the Jonglei Canal to be implemented jointly by Egypt and Sudan; and a storage dam at Lake Albert to be built in coordination with the Equatorial Nile riparians. Second,

[2]The Khashm Al-Girba reservoir lost about 40% of its water capacity by late 1970s, and the Sudanese government, with the support of the World Bank and other international donors, carried out a large rehabilitation plan in 1980 (Taha 2010: 195).

the independence of the six Equatorial Nile states by the early 1960s and the spark of the civil war in Southern Sudan since 1955 complicated the implementation of the Albert Dam and Jonglei Canal respectively. Third, both Egypt and Sudan were completely busy, in terms of time and financial resources, with the construction of the AHD, Roseires and Khashm al-Girba projects. Fourth, the June 1967 Arab-Israeli war, followed by Nasser's death in 1970, shifted Egypt's focus onto rebuilding its military and capacity to restore the Sinai peninsula occupied by Israel. Fifth, the central government in Khartoum had witnessed successive political shifts that affected its capacity to launch new large projects. In 1964, civilian demonstrations led to the termination of Abboud's rule, and the succeeding civilian parliamentary rule lasted only until 1969, when Major-General Jaafar Nimeiri took control of Sudan. All these circumstances halted the implementation of those conservancy projects. Even when the excavation of the Jonglei Canal started in 1978, it stopped few years later, as further discussed in Sect. 4.3.4.

4.2.3 *The Adoption of Conflicting Political Agendas*

As noted above, the Nile game along the 1950s and the 1960s was greatly influenced by the broader game of the Cold War (Fig. 4.1). On the one hand, Nasser's Egypt had aligned with the Soviet Union since the 1955 arms deal with Czechoslovakia, followed by the Soviet technical and financial assistance in the AHD construction, and the explicit adoption of socialism in the 1960s. On the other hand, Ethiopia was led by Emperor Haile Selassie since his return from exile in 1941, who then employed an explicit pro-Western strategy, which involved the construction of the U.S. air base at Kagnew in Eritrea (Coker 1985: 42), in addition to hosting a U.S. military mission, marine research center, and NASA observation center (Laidi 1990: 13). Thus, Ethiopia provided the U.S. with a foothold on the Red Sea and formed a front, together with Iran, Israel and Turkey, to face the growing communist presence in the Middle East (Coker 1985: 42; El-Atawy 1996: 51; Ambrosio 2002: 146). In return, the U.S. endorsed Ethiopia's claims over Eritrea, oriented a considerable portion of its military aids for Black Africa to Ethiopia, and played a crucial role in suppressing the Imperial Guard's attempt in 1960 to overthrow Haile Selassie (Laidi 1990: 13).

Moreover, bilateral relations between Egypt and Ethiopia were strained over several other issues, mainly the Nile waters and the alliances with each other's enemies (Egypt with Eritrea and Somalia while Ethiopia with Israel). During the Egyptian-Sudanese negotiations over water allocations, the imperial government announced on February 6, 1956, in its official press, Ethiopian Herald, that Ethiopia "would reserve for its own use those Nile waters in its territory" (Collins 2000: 265, 267). In September 1957, Ethiopia sent official notes to the diplomatic missions in Cairo, stating that Ethiopia "reasserted and reserved its right to utilize the water resources of the Nile for the benefit of its people, whatever might be the major of utilization of such waters sought by riparian states" (Ibrahim 1984: 116). Subsequently,

Ethiopia did not recognize the 1959 Agreement because it was not consulted in its making (Degefu 2003: 75). The consequences of this water dispute affected even the historic institutional marriage of the Egyptian-Ethiopian Orthodox Church, which lasted for more than sixteenth centuries (fn. 12). Although there had been other organizational reasons behind the separation of the two churches, it only took place with the initiative of Haile Selassie in the late 1950s in the context of the deteriorated relations with Nasser regime (Haggai 2002: 112, 139).

At the regional level, on the one hand, Egypt supported during the early 1960s the Eritrean separatist movement (Haggai 2002: 130) and strengthened the Somali position over a border dispute with Ethiopia (Ehteshami and Murphy 2011: 34). On the other hand, the diplomatic, technological and economic cooperation between Ethiopia and Israel was manifestly growing, motivated by their common hostility with Nasser, as explained in Bishku (1994: 41) and Haggai (2002: 160), which prompted great worries in Egypt regarding the Nile development plans upstream (El-Atawy 1996: 51). For example, Shimon Peres and Yitzhak Rabin, important Israeli military figures at the time, visited Ethiopia in March 1963 and had a "first-hand look" at the Blue Nile "where the life in Egypt begins," as quoted in the press in Haggai (2002: 160). During this visit, Haile Selassie told Peres, "We must always view our dangers in a cold and realistic light and device the means to overcome them with the wisdom born of experience ... We must study the ground carefully and till it together," as quoted in Bishku (1994: 41).

Nevertheless, a temporary relief from such strained relations took place by the mid-1960s (El-Atawy 1996: 51–52). This was mainly due to the common interest of the two countries in the establishment of the Organization of African Unity (OAU) in 1963 through which Nasser sought to maintain his leading role in the continent, choosing to "moderate his position" on Ethiopia by "downgrading his support for both the Eritreans and Somalis" (Ehteshami and Murphy 2011: 34). During the same year, Egypt closed the office of the Eritrean Liberation Front (ELF) in Cairo (Haggai 2002: 147). Yet, such attempts to create friendly relations were still confused with the broader tensions of the Cold War, which is well illustrated in Haile Selassie's answers to the question of Pearson (1964b) about the Egyptian support for Somalia, as follows:

> Somalia is a football field for the players of the East and the West ... The reason for the influx of foreign aid to Somalia is to carry out the ideologies of the United States on one hand, Russia on the other. The only sufferer, however, will be Ethiopia ... Egypt pretends to be a friend of Ethiopia ... but we know that all the time it has been helping Somalia; and also getting aid from the communist countries.

Nevertheless, even after the loss of several Arab territories in the Six-Day War of 1967 with Israel, Syria increased its support for the ELF recognizing the great importance of the Eritrean coasts on the Red Sea, while Nasser preferred not to antagonize Ethiopia (Abir 1974: 97). Moreover, as a gesture of friendship, Haile Selassie visited Cairo in 1968 to participate with Nasser in inaugurating the new Coptic Orthodox Cathedral (Gabra 2009: 88). Still, the real breakthrough, albeit too brief, took place during the early 1970s when both countries were no longer firmly aligned with the two conflicting poles of the Cold War, as further discussed in Sect. 4.3.1.

4.2.4 The Blue Nile Study by the Bureau of Reclamation

The 1958–1964 Blue Nile Study was conducted at a time when the relations between Egypt and Ethiopia were extremely tense. In August 1957, the Ethiopian government signed a contract with the Bureau of Reclamation at the U.S. Department of Interior (USBR) to carry out a comprehensive survey for the exploitation of the Blue Nile potentials, including the Abay and Atbara tributaries (Collins 1990: 278). It was the first time ever that the Blue Nile would be surveyed both extensively and intensively. The survey started in 1958 and lasted for 5 years after which 17 volumes were submitted in 1964, comprising the main report and six appendices (plans and estimates, geology, hydrology, land classification, power, as well as agriculture and economics). In its introduction, the main report asserted that the principal purpose of the study was to propose various projects that would enable Ethiopia in meeting its growing needs, but more important to build Ethiopia's technical and organizational capacity in order to be capable of developing and implementing such proposals. The main conclusion of the survey can be summarized in the following paragraph stated in the study's main report (USBR 1964a: iii, 40–41):

> The land and water resources of the Blue Nile River Basin are extensive and will contribute much to the economic growth of the nation if the projects are properly planned, constructed, and operated. The report outlines plans for projects that would irrigate 433,754 hectares of land now contributing little to the national economy, and for power projects capable of producing 8,660,000 kilowatts to add to the present limited supply. … The flows now reaching the Ethiopia-Sudan border would be depleted by irrigation uses and evaporation losses from reservoir surfaces by about 5,400 million cubic meters annually (10%). In addition, a smaller quantity may be required to meet domestic, municipal, and industrial requirements in future years as the population increases and economic developments cause the now sparsely used areas of the Empire to become utilized. Regardless of the depletion to meet the needs of Ethiopia, where the stream originates, the developments as considered should have little if any adverse effect on projects of downstream nations dependent upon these flows for hydroelectric power, irrigation, navigation, and other minor uses, because of the regulation that can be provided by the upstream storage and the trapping of sediment before it reaches the reservoirs of the downstream nations. In fact, the upstream development, if operated in the most efficient manner considering the requirements of all users, would benefit downstream users by extending the useful life of the reservoirs, providing an assured water supply even in drought years, and making available a source of electrical energy that would permit the downstream nations to utilize more of the water entering their countries in the beneficial use of irrigation for the production of crops.

Regarding the hydropower plants, four large dams were proposed on the Blue Nile (Abay) downstream of Lake Tana: Karadobi, Mabil, Mandaya and Border Dams (Fig. 4.5) with 32.5 BCM, 13.6 BCM, 15.9 BCM and 11.1 BCM of stored waters and 1350 MW, 1200 MW, 1620 MW and 1400 MW respectively (USBR 1964b, Table enclosed with p. 2). These four projects, in total, would create a water storage capacity of 73.1 BCM (about 1.5 times of the Abay's total runoffs) and generate 5570-MW hydropower (2.5 times that of the AHD). Dams proposed on other tributaries were smaller, with a generation capacity ranging from 7.5 to 320 MW. The report indicated that the plan had taken into consideration the fact that

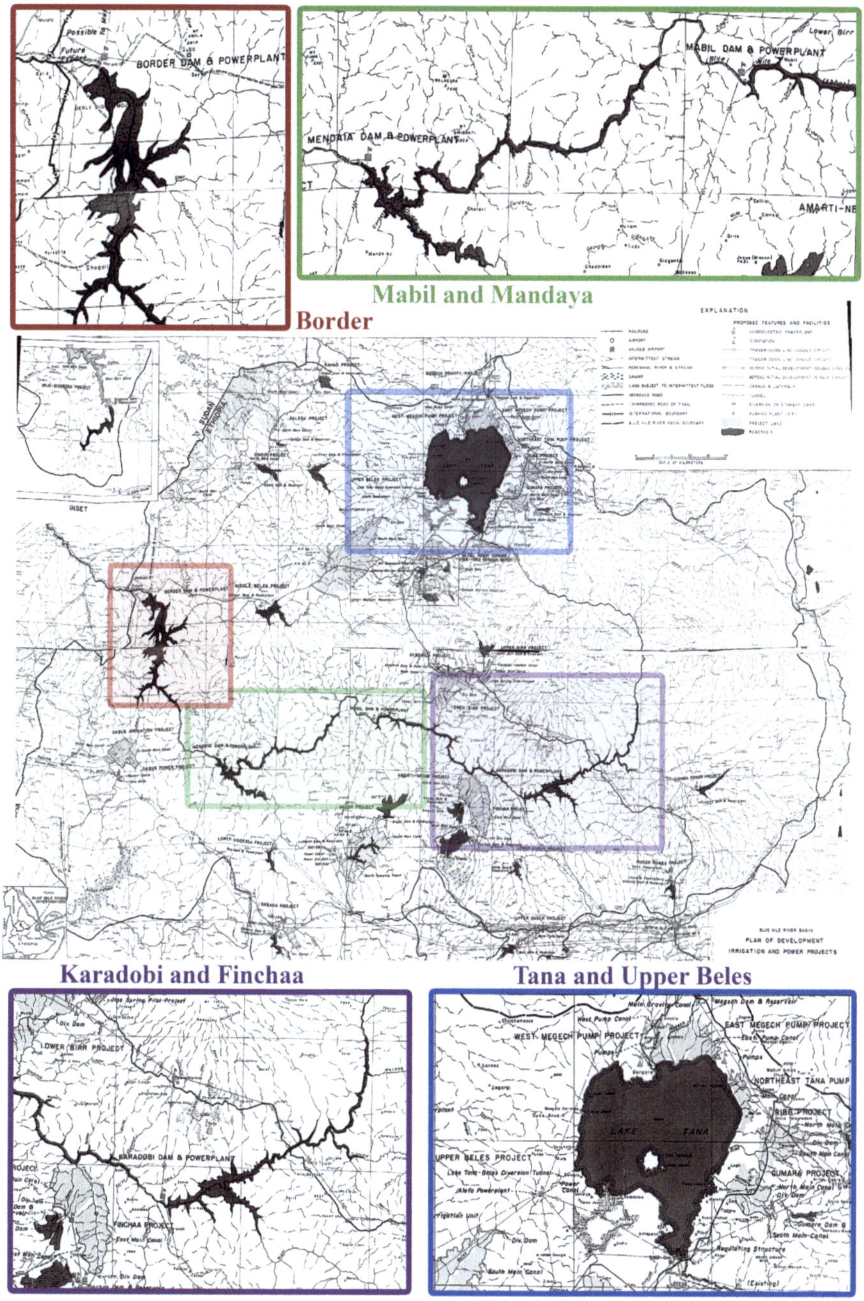

Fig. 4.5 The map of the USBR Blue Nile Study showing the proposed irrigation and power projects. (Source: Land and water resources of the Blue Nile Basin: Ethiopia (1964), the Bureau of Reclamation of the U.S. Department of the Interior (Frontispiece of App. I: Plans and estimates, Vol. I) [Indications added by the author])

sediment loads "are fairly high in streams throughout the basin, averaging about 1% in average by weight of the total runoff," whereby within 50 years, 20 BCM of sediments would be intercepted in the 26 proposed reservoirs (USBR 1964a: 40). The report also stated that the plan should be implemented on long-term basis, since neither Ethiopia's capacity nor its needs, at the time of conducting the study, would allow carrying out all the projects immediately (USBR 1964a: 127–128). Moreover, it was mentioned that the reservoirs should be constructed in an upstream-to-downstream sequence in order to control the accumulated sediments appropriately (USBR 1964a: 40).

Similar to Garstin's Nile report produced 60 years earlier, the 1964 USBR study was a turning point in the Nile history. While the former had founded the holistic vision of water resources planning in the Nile with the purpose of regulating flows to Egypt, the latter underpinned the great potential of implementing large-scale consumptive and non-consumptive water projects on the Blue Nile in Ethiopia and Sudan. Since its publication to date, the contained messages in the USBR study have profoundly influenced the development plans on the Nile. The first is a message sent by Ethiopia to Egypt, informing it implicitly that Ethiopia had previously reserved its right to use the Nile waters, and by this study, it took the first step to do so with the U.S. support. This is concisely abstracted by Waterbury (2002: 69) that Ethiopia, through inviting the U.S. to conduct the study, seemed telling Egypt, "You have the Russians and the Aswan High Dam, but we have the United States and, more important, the water." Second, Ethiopia sent another message to Sudan, by taking it into account in the study, that cooperation over upstream water development schemes would bring massive benefits for Sudan as well as Ethiopia, mainly in flood management, water regulation, irrigation expansion, and sediment control. The third is a message sent by the U.S. to Egypt and its Soviet ally, indicating that it became technically possible to harness and control the sweeping waters flowing to Egypt from the Ethiopian Highlands. While the fourth is a message perceived by Ethiopia itself, that those ambitious large-scale projects would require high technological and financial capacities in order to be implemented. In addition to these political messages, the 1964 study formed the "baseline" of all the succeeding studies conducted on the Blue Nile (Waterbury 2002: 109, 116).

Yet, none of the 26 projects were implemented during the Cold War round except the Finchaa Dam (see next subsection), mainly due to the lack of financial and organizational capacities (Collins 1990: 279; Waterbury and Whittington 1999: 155; Waterbury 2002: 69, 117–118; Degefu 2003: 212). Another factor was the protests made by Egypt at the international agencies not to alter the flows running downstream (El-Atawy 1996: 50–51; Arsano 2010: 167). These two factors are obvious in the case of the Tana-Beles project that Ethiopia attempted to implement in 1980s (Sect. 4.3.6).

4.2.5 *The Finchaa Dam*

From the Ethiopian point of view, the aid provided by the U.S. to Ethiopia with respect to the construction of dams on the Blue Nile, was insufficient, as declared by Haile Selassie in his interview with Pearson (1964b) on June 11, 1964:

> Egypt … gave great publicity recently to the help it had received from Russia in building the Aswan Dam, but gave the United States no credit for the food it received. When the dam is finished, Egypt will be independent and able to grow enough wheat for the needs of its people … I do not like to belittle the aid given to Ethiopia by the United States … It has been helping to a certain extent. But if, for instance, the United States had built the long-discussed dam at Lake Tana, it would have done for us what the Russians have done for Egypt at Aswan.

Three days earlier, Haile Selassie told Pearson (1964a) in another interview (cited at the beginning of this chapter) that he whished the U.S. had built him the Tana dam decades ago, and that even when he recently asked for the American support to build the Finchaa Dam, the USBR took 5 years conducting the Blue Nile survey whereas the dam construction did not start until the day of the interview. Sneddon (2015: 97) further comments on this point by highlighting the different approaches employed by the U.S. and Ethiopia; the former insisted on conducting a "thorough" and "long-term" analysis to exploit the potentials of the entire basin, while the latter preferred to execute a "quickie type of survey" to build the required dams in the least possible time. Figure 4.5 shows the Finchaa Dam amongst the extensive plan of the 1964 USBR study. As quoted in Sneddon (2015: 98, 104), the U.S. ambassador in Addis Ababa commented on Haile Selassie's opinion, stating that the USBR study aimed "for a comprehensive renaissance[3] of the area, not for building a dam," and that "the identification of potential dam sites was incidental to the much larger objective." The ambassador referred to mistaken information in the facts mentioned in Haile Selassie's answers to Pearson (1964a), given that the Finchaa Dam "was requested nine months ago (October 1963), not 'five years' as Pearson quotes the emperor as saying" (Sneddon 2015: 98, 104).

Nevertheless, Haile Selassie's complaints in the press prompted the attention of Johnson administration to intensify its support for the emperor in constructing the Finchaa Dam. This is obvious in the statements of the U.S. officials quoted in Sneddon (2015: 99) commenting on the dam, whereby the head of USAID stated that "the project as conceived was technically and financially sound, and that it would make a significant contribution to the economic development of Ethiopia." In addition, the director of the Bureau of the Budget declared, in a memorandum to President Johnson in 1966, that the U.S. "has a serious political stake in Ethiopia as a moderately pro-Western state with a good record of political stability and

[3] It is worth noting that the word "renaissance" that the U.S. ambassador used to describe the 1964 USBR study has been used almost five decades later in naming the Grand Ethiopian Renaissance Dam, which is located at the same site of the Border Dam included in the study but with a completely different size.

considerable influence in the African world." Hence, he summarized the U.S. view saying that the Finchaa project is in "both economic and political terms ... a worthwhile project."

As illustrated by Waterbury (2002: 118–119), the project cost was $34 million in total, and in 1969 the World Bank financed the major portion of the required expenses in foreign exchange, with an amount of $23 million, out of which USAID contributed $2.8 million. In November 1970, Haile Selassie laid the cornerstone of the project, and the dam was commissioned in 1972. The initial volume of the dam's reservoir was 900 MCM, with an installed capacity of 90 MW, which supplied Addis Ababa's needs of electricity, as the dam is located 170 km northwest of the capital. Although the USBR study proposed the project at Finchaa for hydropower generation and for irrigation, the latter did not operate during the Cold War round.

4.2.6 The 1961–1964 Great Rains and the Hydromet Survey

In 1955, whilst still under the British colonization, Kenya, Tanganyika and Uganda created the East African Nile Waters Coordinating Committee (EACC) after the proposal of Andrew Cohen, the Governor of Uganda, and Dr. Paul Howell[4] (Collins 1990: 286, 1994: 125; Howell 1994: 86). The EACC committed itself to the 1929 Agreement with Egypt and Sudan (El-Atawy 1996: 44).

Howell (1994: 98, 102) recorded the subsequent talks of the EACC with Egypt and Sudan as follows. In August 1959, few months before the conclusion of the Egyptian-Sudanese Nile Waters Agreement, Britain, on behalf of its colonies, reserved their rights to a share of 1704 MCM/year, estimated to meet their irrigation needs. This development took place "under the shadow of the Suez Crisis," after which the British-Egyptian relations were "severed" and not resumed until April 1961. In October 1961, whilst approaching full independence, these three riparians launched "technical and informal" talks with the PJTC, which represented Egypt and Sudan, contending that the figure put forward by Britain undermined their "long-term requirements of East Africa (50–75 years)," estimated to reach a figure of 5 BCM/year. The PJTC responded that it would be ready to discuss only the "immediate requirements," which would be "required for actual use during the next five years," while the "ultimate requirements" of 5 BCM were "not supported, as was the earlier statement of requirements in August 1959, by technical data."

Moreover, in 1962, Tanganyika announced the Nyerere Doctrine, named after its first president, Julius Nyerere, claiming that newly independent states should review and renegotiate the former colonial treaties including the 1929 Agreement. The claim was adopted by Kenya in the following year, and Uganda informally endorsed it, while the other Equatorial riparians (Burundi, Rwanda and Congo) were not a

[4] Paul P. Howell is a British anthropologist who worked on the Nile since 1930s in Sudan then in Uganda from 1950s to 1961 (Howell and Allan 1994: iv). He was the chairman of the Jonglei Investigation Team (see Sect. 4.3.4).

part of the agreement (UNEP 2005: 52). However, the hydropolitical interactions on the Equatorial Lakes were less disputed than those on the Blue Nile, mainly due to the following two main reasons (Waterbury 2002: 154). First, the country that would be affected the most by abstracting waters from the Lake Victoria is Uganda, since the lake's level would decrease and the generated electricity at the Owen Falls Dam and any of its future hydropower projects would be significantly reduced. Second, more than half of the water flowing from the Equatorial Lakes downstream is lost in the Sudd swamps (Howell et al. 1988: 21).

In addition, the special hydrological characteristics of the Equatorial Lakes and the Sudd as well as the endemic uncertainties of climate variability had played a crucial role in emerging cooperation with Egypt and Sudan. These factors combined led to conducting an extensive hydrometeorological survey on the Equatorial Lakes. Although technical cooperation had been first proposed by the PJTC in 1960 and the proposal was endorsed by the EACC, the actual step was taken under the initiative of the latter induced by what is known as the "great rains" of 1961–1964 (Collins 1990: 274, 285–286). The negative impacts caused by the "dramatic but unpredictable" rainfalls in those years were enormous, which can be realized in the following figures recorded in Collins (1994: 124–125). The discharge at the Owen Falls increased from 20.6 BCM in 1961 to 38.6 BCM in 1962, to 44.8 BCM in 1963, reaching 50.5 BCM in 1964. This unprecedented rising of the Equatorial Lakes by 2.5 m generated disastrous floods and huge damages around the lakes, doubling the size of the Sudd floodplain from 13,100 to 29,800 BCM. The Nilotes, who were already suffering from the miseries of civil war, lost tens of thousands of lives and over 120,000 heads of livestock.

As noted by Collins (1990: 286–289), a team from the World Meteorological Organization (WMO) conducted an initial survey on Lake Victoria, from 1962 to 1963, guided by the EACC and financed by the UNEPTA and the FAO. Those three countries then invited Egypt and Sudan to join in the "spirit of unity" as promoted by the OAU. In 1966, the UNDP funded the extensive hydrometeorological survey (Hydromet) on Lakes Victoria, Albert and Kyoga. In the following year, the signatories approved the technical program presented by the WMO to start the operation of the project and establish the headquarters in Entebbe in 1968. With an annual fund of £2 million, numerous hydrometeorological and hydrological stations were constructed and African staff were trained. The project had run initially for 5 years and then extended for another 3 years in 1976. By the continuity of the international support, the Hydromet gradually opened its focus by the 1980s from collecting data to studying and proposing water conservation projects in the entire basin. In addition, Ethiopia joined the project in 1971 and Zaire (previously Congo) in 1977, but only as observers.

4.3 From the 1970s to the 1980s

4.3.1 Shifting Camps of the Cold War

The serious changes that took place at the domestic and regional levels during the first half of the 1970s led to profound shifts in the positions of Egypt, Sudan and Ethiopia with respect to the global superpowers (Fig. 4.2).

President Anwar Sadat, the successor of Nasser, abandoned the alliance with the Soviet Union. In 1972, he decided to expel the Soviet military experts from the Egyptian army in order to act freely from the Soviet objectives (Dawisha 1979: 159) and to give a "deceptive" message to Israel (Finklestone 1996) that he would not be ready to fight in the meantime for the restoration of Sinai, which Egypt lost in 1967. On October 6, 1973, while Israelis were fasting on Yom Kippur, Sadat, together with Syria and with the support of Arab countries, launched a "surprise attack" against Israel (Kumaraswany 2006: 75). Sadat's main objective was to break the state of "no war no peace" by carrying out a military action that would pave the way for better conditions of negotiation (Sela 2013: 44). As reviewed in Sela (2013: 62), Tal and Tucker (2008: 11–12), Rubinstein (2008), El-Shazly (1980: 246, 269), and Gupta (1978: 51), the war events progressed as follows. Within 6 h after launching the attack, the Egyptian forces were able to cross to the Eastern bank of the Suez Canal and penetrate the Bar-Lev Line, mobilizing its troops 3–5 miles deep into Sinai. With the support of the U.S., Israel counterattacked on the two fronts with Egypt and Syria and was able to advance on the latter front. Sadat decided to advance beyond the effective missile-cover to mitigate the losses on the Syrian front. As a result, the Israeli forces found a gap between the two divisions of the Egyptian army through which they crossed to the Western bank and encircled the third army camping in Suez city, to be in turn surrounded by Egyptians. At this point, both the Soviet Union and the U.S. pushed for halting the war, and on October 22, both agreed to ceasefire with the approval of the UN. Sadat, thereafter, started the negotiation path with Israel under the U.S. umbrella, ultimately leading to the 1978 Camp David Accords and the 1979 Peace Agreement upon which Egypt restored all of its occupied lands.

The Britannica's "Guide to the 100 most influential world leaders of all time" summarized how Sadat emerged from the October War, as follows. "The Egyptian army achieved a tactical surprise in its attack on the Israeli-held territory, and, though Israel successfully counterattacked, Sadat emerged from the war with greatly enhanced prestige as the first Arab leader to have actually retaken some territory from Israel" (McKenna 2010: 299). In addition, some months after the October 1973, the U.S. Senate Committee on Foreign Relations (1974: 2) described concisely the war's overwhelming effect on Egyptians in general and on Sadat in particular as follows:

> The October War gave rise to a surge of self-confidence in Egypt, and this recovery of national pride seems to have been the major factor in Egypt's willingness to negotiate directly with Israel. By comparison with 1967, the Egyptians seem confident of the future

> and comparatively free of bitterness. … Egyptian officials believe that the balance is swinging in their favor—because of their comparative military success, because of the Saudi alliance, the bringing to bear of the formidable oil weapon,[5] and the new-found unity of the Arab world. That unity has been made possible largely by the personality of President Sadat.

Following the relative success achieved by Sadat, his strategy of war and peace had clear reflections on Egypt's position on several fronts, including foreign affairs and domestic policies. By 1974, the U.S. intensified its efforts to mediate between the Arab countries and Israel in order to settle peace in the Middle East. It also helped Egypt, together with Britain and France, to clear and reopen the Suez Canal for international navigation. Sadat, therefore, considered the U.S. a "full partner" in rebuilding Egypt's economy after the tough war years extending between 1967 and 1973, where he adopted an open door economic policy, seeking to establish cooperation with the Western block and to attract foreign investments (Ibrahim 1988: 26). During the 1970s and the 1980s, the U.S. support for Egypt extended to the maintenance works required for the Soviet-constructed AHD (TFDD 2008).

Sudan aligned with Egypt in such a dramatic transition with respect to the two camps of the Cold War. Major-General Jaafar Nimeiri, who took power in Sudan in 1969, followed Nasser's alignment with the east and then Sadat's rapprochement with the U.S. afterwards, especially because Sadat had supported him to restore his rule after the unsuccessful military coup of the communist party in 1971 (Ibrahim 1988: 26). During the October War, Nimeiri firmly upheld Sadat's position and declared the Sudanese resources, including men, equipment, airfields and ports, under the service of the Arab forces (Abdelwahab 2012: 132–133). Moreover, Nimeiri did not oppose Egypt's successive moves after the war, including Sadat's visit to Jerusalem in 1977 and the signature of the peace accords in 1978/1979, while most Arab countries broke their diplomatic relations with Egypt by the end of 1978 (Abdelwahab 2012: 134). Another direct consequence of the 1973 War was the policy adopted by the Gulf States, which witnessed an economic boom following the dramatic rise of international oil prices, to transform Sudan into the "bread basket of the Arab World," in that they found the Sudanese arable land a safe opportunity to meet their food security. In 1976, the Arab Fund for Economic and Social Development (AFESD) launched large investments in irrigated and rainfed agricultural sectors in Sudan, focusing on wheat, sugar, and meat products (Fig. A.5).

The changes witnessed by Ethiopia during the 1970s were no less dramatic. Between 1970 and 1973, there was a brief transition period in the relations between Ethiopia and Egypt, characterized by a more relaxed atmosphere different than the Cold War polarization and by growing Afro-Arab solidarity. Two years before Sadat expelled the Soviet experts in 1972, the importance of the American airbase at Kagnew had decreased significantly by 1970, mainly due to the technological

[5]The Arab Gulf States, together with Iraq and Iran, used their oil exports as a weapon to exert a pressure upon the countries that supported Israel. Commenting on this issue within 48 h of launching the attack, King Faisal of Saudi Arabia told two envoys from Sadat, "You have made us all proud. In the past, we could not lift our head up. Now, we can" (Vassiliev 2012).

development in satellite communications and the decision to construct a complementary station at Diego Garcia (Coker 1985: 43). Moreover, the U.S. did not support Haile Selassie in his confrontation with Somalia as the Nixon administration was already preoccupied by the Watergate crisis and the isolationist tendency induced by the Vietnam War (Metaferia 2009: 69). This is obvious in the statement of the U.S. Secretary of State regarding his country's policy in Africa, declaring that it "wants no military allies, no sphere of influence, no big power competition in Africa" (Greenville News 1970).

Under such conditions, there had been several positive signs shown by Ethiopia to Sudan and Egypt. In February 1972, Haile Selassie, with the World Council of Churches, mediated between Khartoum and Southern Sudanese insurgents to terminate the civil war, whereby both parties signed as a result the 1972 Addis Ababa Agreement (Waterbury 2002: 139–140). The events of 1973, as recorded by Haggai (2002: 163, 200), indicate the exceptional rapprochement between Egypt and Ethiopia in such a unique atmosphere. In March 1973, Haile Selassie invited an Egyptian journalist to view Ethiopia's water development plans and to visit the actual sites of the proposed projects on the Blue Nile and Lake Tana. Two months later, the emperor flew to Cairo to ask Sadat's support for preserving the OAU's headquarters in Addis Ababa and for restraining the Somali struggle with Ethiopia. Sadat endorsed these requests, and during the same month, Haile Selassie in return promised Sadat at the African summit "never" to "interfere" with the Nile flows without "consulting with the other riparian countries" (Haggai 2002: 200). Similar to most African countries, Ethiopia broke off relations with Israel on October 23, 1973 in the backdrop of the October War (Parada 2012: 115). Burundi, Zaire, Rwanda, Tanzania and Kenya had also taken the same decision on May 16, October 4, October 9, October 18, and November 29, 1973, respectively (Parada 2012: 115). While, Uganda severed its relations with Israel 1 year earlier on March 30, 1972, during the reign of General Idi Amin Dada (1971–1979), who enthusiastically supported the Arab countries in the conflict with Israel (Amin 1998: 99–100).

As mentioned above, this exceptional period was too brief, as the radical transformation in the political regime of Ethiopia had completely changed its orientation and consequently its relations with Egypt. In January 1974, a military disintegration began spreading across the country, and in June a group of junior military officers formed what became to know the "Derg" (the committee in Amharic) that ousted the emperor and abolished the monarchy in Ethiopia in September (Shinn and Ofcansky 2013: 124). Records refer to the "severe famine" of 1973 and its sharp consequences on the Ethiopians as a major contributor to the downfall of Haile Selassie (Carol 2012: 20; Metaferia 2009: 70). Major Mengistu Haile Mariam, the Chairman of the Derg at the time of its establishment, took power in November and was able within 2 years, along a "bloody path," to "liquidate" all of his opponents and to form a communist government (Haggai 2002: 164). Subsequently, the Soviet Union considered the transformation in Ethiopia an important gain, and supplied the Mengistu regime with military hardware. However, the Soviet support for Mengistu, as described by Metaferia (2009: 68), did not "make Ethiopia a showcase for other African countries by developing Ethiopia's economy, which would have been more impressive than flooding the country with armaments."

4.3.2 *The Soviet Survey on the Blue Nile and the 1977 UN Water Conference*

Furthermore, the development schemes on the Nile had been a source of conflict under the influence of the Cold War. By the mid-1970s, the map of coalitions in the region had significantly changed. Egypt aligned with the U.S. along with Iran, Jordan, Saudi Arabia and Somalia. Ethiopia, on the other hand, aligned with the Soviet Union and Cuba, together with Iraq and Syria, later joined by Iran and Afghanistan during the late 1970s. With the support of Sadat[6] to the Somali President Siad Barre against Mengistu during the Ogaden War waged in July 1977, acute tension arose between Egypt and Ethiopia (Ehteshami and Murphy 2011: 52). However, the war ended in March 1978 with the withdrawal of the Somali forces from the Ogaden province, mainly due to the military backing of both the Soviet Union and Cuba to Mengistu, however the U.S. support for Somalia continued with the aim of preventing the expansion of the Soviet presence in the region (Griffiths 2016: 80).

Yet, such conditions overshadowed the concurrent events taking place in the Nile. Sadat worried that Mengistu might use the water weapon under the pretext of the Ethiopian-Somali war (Ehteshami and Murphy 2011: 52). These worries were based on Ethiopia's water policy adopted at the time. On March 14–25, 1977, the Ethiopian delegation participating at the UN Water Conference (held in Mar del Plata, Argentina) emphasized in a country paper "the sovereign right of any riparian state, in the absence of an international agreement, to proceed unilaterally with the development of water resources within its territory" (Timberlake 1985/2013: 171). The mentioned projects aimed to cultivate 91,000 and 28,000 ha on the Blue Nile and the Baro tributaries respectively, abstracting a total volume of 4 BCM from the Nile waters (Howell et al. 1988: 78). On February 16, 1978, Mengistu challenged Egypt's "historic water rights" (Caso 2010: 319). Simultaneously, the Soviet experts started to conduct a survey on Lake Tana to harness the Blue Nile water resources (Swain 2004: 103). On May 13, 1978, Sadat declared, "We depend upon the Nile 100 percent in our life, so if anyone, at any moment thinks to deprive us of our life, we shall never hesitate [to go to war] because it is a matter of life or death," as quoted and interpreted by Waterbury (1979: 78). He also said, "Any action that would endanger the waters of the Blue Nile will be faced with a firm reaction on the part of Egypt, even if that action should lead to war" (Brunnee and Toope 2002: 106; Kendie 1999). In response, the Ethiopian Ministry of Foreign Affairs issued a statement reiterating that of 1957 (see Sect. 4.2.3), indicating that "no one in his right mind could question Ethiopia's inalienable and self-evident right to use its natural resources for the benefit of its people" (Degefu 2003: 150). The 1978 statement also

[6] Sharawi (2012: 305) refers to the secret alliance, the "Safari Club," that gathered the intelligence agencies of France, Saudia Arabia, Egypt, Iran and Morocco, with informal links to the USA and Israel. The Safari Club supported the Mobuto regime in Zaire against Angola and Barre against Ethiopia in the Ogaden War, adopting an anti-communism strategy to counter the Soviet and Cuban presence in Africa. It also facilitated the meetings between Egypt and Israel that led to the 1979 Peace Agreement.

mentioned that Ethiopia "resents" the situation of Egypt and Sudan developing water projects without "ever having felt constrained to consult Ethiopia before their plans to dam the Nile" (AIPC 1979).

4.3.3 The Peace Canal

Not only were the development plans upper the Nile conflictive, but also those ones in Egypt. Sadat's development policy after the October War was founded on creating new communities in the desert to redistribute population between the overcrowded Nile Valley and Delta and these new frontier areas (Fig. A.4), such as the Sadat city, the 6th of October city and the Ramadan 10th city. In fact, since the 1950s, Nasser had made several attempts to reclaim desert areas around the Delta. A clear example was the Tahrir Province scheme in the hinterlands west of the Nile Delta. Nasser's plan also included reclaiming lands in Sinai; however this endeavor was greatly challenged by the availability of freshwater and was completely halted by the 1967 War. Interestingly, the Egyptian government had issued a postage stamp in 1966 on the occasion of the 14th anniversary of the 1952 Revolution, titled, "Development of Sinai." The stamp illustrates a blue line coming from the Nile and feeding a water tap in the Sinai desert, which ultimately irrigate lands and create communities (Figs. 4.6 and 4.7).

More than one decade later, Sadat decided to implement a large irrigation scheme in Northern Sinai through the construction of the Peace Canal (or El-Salam Canal) to overcome the shortage of water resources in the peninsula. As shown in Fig. 4.8, the canal originates at 210 km on the Damietta branch, running 89.4 km to southeast, then crossing beneath the Suez Canal in four tunnels (phase I), and extending 175 km in Northern Sinai (phase II) (Othman et al. 2012: 100). The canal aims at providing 4.5 BCM of mixed Nile freshwater with recycled drainage water (ratio 1:1) (Othman et al. 2012: 100), which allows the reclamation of 220,000 feddans (phase I) west of the Suez Canal and 400,000 feddans in Sinai (phase II) (ESIS 2006). This was not an individual project, but part of Sadat's comprehensive plan: to lead urban expansion in Egyptian deserts, to firmly connect the Nile Valley and Delta with Sinai, and to be ready to develop the peninsula after the complete withdrawal of the Israeli forces in April 1982. This is implicitly stated through the following lines of his article in the Britannica Book of the Year (Sadat 1981):

> This year, 1981, I shall be opening the Suez Canal for the third time. The first was the original opening by the khedive Ismail in 1869. Then I reopened it in 1975 after it had been closed for eight years. Now we have the third opening. It is a completely new canal. We worked silently for five years, widening and deepening that canal. I have already opened the tunnel under the canal to Sinai after six years of work. This project is a masterpiece, one of the wonders of the world. We live most of us in this narrow Nile Valley, occupying only 4% of Egypt's total land area. We have lived on this narrow 4% when we were a population of 17 million, then 20 million, then 30 million, now 42 million.

Fig. 4.6 Postage stamp of the 14th anniversary of the 1952 Revolution showing the vision for Sinai. (Source: Egypt Post (1966), available by StampData at: http://stampdata.com/image.php?id=336202 (accessed on October 13, 2016))

Fig. 4.7 Postage Stamp of the 15th anniversary of Sinai Liberation Day showing the Peace Canal. (Source: Egypt Post (1997), available by StampBoards at: http://stampboards.com/viewtopic.php?f=13&t=57843 (accessed on October 13, 2016))

The concept of transferring waters to Sinai was originally introduced by Theodor Herzl within a broader proposal for obtaining concessions to establish Jewish settlements in Palestine and around the Mediterranean. After long negotiations from 1896 to 1903, the Ottoman Sultan Abdul Hamid refused Herzl's offer to grant the Jews an independent state in Palestine in exchange for assistance in solving the empire's financial problems, which drove him to Britain to discuss the possibility of estab-

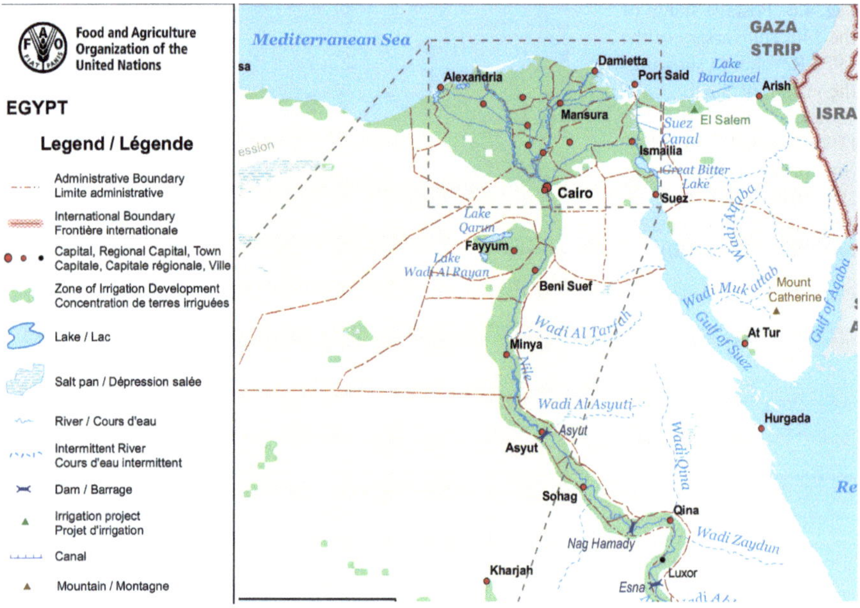

Fig. 4.8 The map of Northern Egypt showing the Peace (or el-Salam) Canal. (Source: FAO (2015), Aquastat Database: Maps by country, available at: http://www.fao.org/nr/water/aquastat/countries_regions/EGY/EGY-map_detailed.pdf (accessed on October 13, 2016))

lishing Jewish settlements in one of the British colonies (Katz 2003: 39–40). Cohen (1959: 287–301) records the story after Herzl's diary. Immediately thereafter, Herzl went to meet Joseph Chamberlain, Britain's Secretary of State for the Colonies and asked him, "Would you be satisfied if we founded a Jewish colony in the Sinai Peninsula?" Chamberlain replied, "Yes, if Lord Cromer favored it." Herzl contacted Lord Cromer, the High Commissioner of Britain in Egypt, who expressed his initial acceptance to study the matter if the project's "actual conditions were suitable." Herzl formed an expedition including J. H. Stephens, an irrigation expert, to prepare a feasibility study of the project, which concluded, "In present circumstances unsuitable for settlement; but if water is provided suitable for settlement." Lord Cromer, as a result, referred the whole issue to Sir William Garstin, who was in charge of every detail concerning the Nile waters. He studied the proposal submitted by the expedition and refused it, clarifying that "five times as much water would be needed as Stephens had estimated, and that the laying down of the pipes would interfere with the traffic in the Suez Canal for several weeks." Consequently, the project was firmly rejected for its negative impacts on the cotton cultivation in Egypt and the international navigation in the Suez Canal.

When Sadat adopted the project in 1979, more than 75 years after the refusal of Herzl's proposal, Israel was also involved. Although the Peace Canal aims primarily to reduce the population density in the Nile Valley and Delta, Sadat had the idea to link the project with the unresolved conflict between Israeli and Palestinians: Egypt extends the pipeline to Negev in exchange for "solution of the Palestinian problem and the liberation of Jerusalem" (Krishna 1988: 32; Said 1993: 273). None of the involved parties welcomed Sadat's proposal, both the Israeli officials and the nationalists in Egypt and Israel (Wolf 1995: 57). For instance, Ariel Sharon, the Israeli agricultural minister at the time, declared, "I would hate to be in a situation in which the Egyptians could close our taps whenever they wished," as quoted in Wolf (1995: 57) after Spector and Gruen (1980: 10). However, the concept of transferring waters to Israel or even to Sinai provoked another verbal tension between Mengistu and Sadat. Mengistu sharply opposed the project and threatened to reduce the Blue Nile flows in retaliation (Swain 2004: 103). As quoted in Krishna (1988: 33–34), Sadat responded in the following strong words:

> We do not need permission from Ethiopia or the Soviet Union to divert our Nile water ... If Ethiopia takes any action to block the Nile waters, there will be no alternative for us but to use force. Tampering with the rights of a nation to water is tampering with its life and a decision to go to war on this score is indisputable in the international community.

The escalation of this political tension did not last for a long time, since, on October 6, 1981, Sadat was assassinated by a group of Islamic extremists during the annual military parade of the 1973 War. Years earlier, Sadat had adopted the U.S. anti-communism strategy in the region, but also fought communism inside Egypt. As reviewed in Mannes (2004: 79–80), one of the means adopted to do so was releasing Muslim Brotherhood leaders from prison to suppress communists. Concluding peace with Israel and the pressing economic conditions in the following years after the war catalyzed the spread of radical ideas, which led to the formation of terrorist groups by the late 1970s. Ultimately, those extremists were the ones who ended Sadat's life on the day of celebrating his achievements of war and peace along his 10-year reign. After Sadat's assassination, the exchange of threats between him and Mengistu was terminated, however the construction works of the Peace Canal did not stop (Fig. 4.7), albeit progressing in slow rates because of technical and financial challenges, whereby the heated legal contest continued between both countries over the project's concept and objectives.

Meanwhile, the global debate over the general principles of international water law was growing, represented in the announcement of the 1966 Helsinki Rules and then the discussions conducted by the UN International Law Commission (ILC). For example, this was evident in the recommendations adopted by the 1977 UN Water Conference, urging the ILC to conclude as soonest its draft for the UN Watercourses Convention (Salman 2007: 3–4). However, the draft was only submitted to the UN General Assembly in 1994 and opened for signature in 1997 (see Sect. 5.4.2 for further discussion).

Waterbury thoroughly lists the points of advocacy adopted by the Egyptian and Ethiopian high-level officials during the 1970s over the Peace Canal and similarly in the 1990s over the Toshka scheme. Such debatable perspectives, which still prevail today, are explicitly found in the author's interviews with the officials in both countries concerning Egypt's development plans in Sinai and the Western Desert (M.A.S. 2016a; M.M. 2016b; T.B. 2016). In a nutshell, Ethiopia contends that the plans of reclaiming the desert in general and the Peace Canal in particular contradict with the general principles of the international water law for four main reasons. First, the principle of prior notification should be applied on both the upstream and downstream countries, and thus Egypt and Sudan should consult Ethiopia and other riparians prior to implementing their water projects. Second, the obligation not to cause significant harm is measured upon the consequences of the upstream projects on the existing uses by downstream riparians; thus constructing irrigation projects downstream would create new facts on the ground that would forego future projects upper the river. Third, the Peace Canal, in particular, stretches eastwards extending beyond the drainage area of the Nile Basin, which violates the obligation not to transfer the waters out of the basin. Fourth, fulfilling the upstream water needs should be given the priority in case there is any surplus water in the Nile shares of the downstream riparians.

From the Egyptian point of view, these four reasons are inaccurate and unfounded. First, downstream countries lower the river are not obliged to consult the upstream riparians prior to construction of water projects unless they would be significantly harmed, as in case of flooding lands upstream for instance. Second, downstream countries have the full right to manage their water shares and to plan for water projects that enable them to meet their national needs. Third, the Peace Canal would not transfer the Nile waters out of the basin, since the Nile branches in the past were different from today as there were other branches running eastwards from the Delta (Fig. 4.9). Fourth, what is actually transferred by the Peace Canal is mixed drainage water not the Nile freshwater. Moreover, and before all, Egypt is a desert country and depends heavily on the Nile waters, in contrast to the upstream riparians where precipitation rates are considerably high (Fig. A.2). However, Ethiopia responded to the last point in particular, claiming that due to the uneven distribution of rainfalls in space and time, it has to supplement its agricultural projects with irrigation schemes (Fig. A.5) so as not to be vulnerable to frequent droughts and famines.

4.3.4 The Jonglei Canal

Following the termination of the first civil war in Southern Sudan in 1972 (Sect. 4.3.1), Sadat and Nimeiri, who were seeking to develop their countries after years of instability, found no reason to delay the implementation of the Jonglei Canal, the water conservancy project to which the PJTC had been giving priority. In February 1974, the two presidents signed an accord launching the construction of the project (Waterbury 2002: 140).

The Jonglei Canal had been studied for several decades earlier, since there had been a debate among the British experts in both Egypt and Sudan over the advantage of saving waters downstream vis-à-vis the disadvantage of altering livelihoods of local people. In short, the Jonglei Canal aims at modifying the path of the flow coming from the Equatorial Lakes in order to bypass the Sudd swamps and march-

Fig. 4.9 The map of Lower Egypt drawn by W. Heather based on the 1798 reports of Napoleon Bonaparte's expedition to Egypt, depicting the Nile branches and canals during the flood season. (Source: Heather (1798), published by Heather and Williams at the Navigation Warehouse, London, available by PicClick LLC at: http://picclick.co.uk/Antique-1798-Canvas-Navigation-Map-of-Lower-Egypt-182254404514.html (accessed on October 15, 2016))

lands, the "blockage" of vegetation that can "absorb and dissipate half or more of the water it receives" (Howell et al. 1988: 21). This concept was first introduced by Sir Garstin (1904: 194–195), as follows:

> The immense advantage to be realized, by leaving the great Swamp Region all together to one side, and by conveying the water to the north, in a well-constructed, and fully-controlled channel [from Bor to the Sobat junction], are so evident, that they can be understood by all. The proposal will particularly appeal to those who have ever followed the Bahr-el-Gebel, in its long and winding course through the dreary papyrus marshes. This channel would be by far the best solution of the difficult problem of obviating the present waste of water in the Upper Nile, and the advantages of this scheme are so great, that they are well worth incurring a heavy expenditure to secure. Should the levels prove the construction of this channel to be impracticable, then the remodeling of the Bahr-el-Zaraf, so as to enable it to carry the required discharge, should be proceeded with. … Once it has been secured, by one or other of these methods, that the summer discharge of the Bahr-el-Gebel shall reach the White Nile in undiminished volume, then the work of regulating the Victoria and Albert Lakes, at there outlets, should be proceeded with.

Garstin's lines quoted above had inspired the British experts in Egypt as they set out the development of those schemes by proposing either to dig a channel from Bor to the Sobat junction or to remodel Bahr el-Zaraf, besides building regulators at Victoria and Albert Lakes. Both McDonald (1920) and Hurst et al. (1946) included the channel and the regulators in their proposed plans. In addition, Butcher (1936, 1938) focused on the design of the channel, and proposed to dig it from the upper Atem at Jonglei (north of Bor) to run northwards and join the Zaraf lower downstream (Howell et al. 1988: 34). The Egyptian government had pushed for implementing these schemes since the mid-1920s. However, only by the 1940s, it was determined that the project would be implemented in two phases: Jonglei I, building the channel without the need to wait for constructing the Victoria and Albert regulators; and Jonglei II, widening and deepening the channel to increase its capacity for carrying additional waters released after the construction of the regulators (Howell et al. 1988: 35). All these proposals had paid little attention to the negative impacts that would harm the Nilotic inhabitants at the Sudd by changing the entire hydrological characteristics of the region; thus disrupting the pastoral economy on which they relied. These impacts were first studied by Roberts (1928) and Winder (n.d.), which in general terms supported such fears. The anxiety of Sudan concerning the socio-economic impacts of the canal was one of the main factors that led to halting the negotiations over the Equatorial Nile Project (1946–1948) (Sect. 3.3.4).

The most comprehensive study was that conducted by the Jonglei Investigation Team (1948–1954) chaired by Dr. Paul Howell (JIT 1954). The team was highly concerned with the magnitude of harm and the remedial measures that should be taken to provide alternative economic systems for the Nilotes, such as irrigated pastures, irrigated agriculture and modernized fishing industry. However, the JIT (1954) report concluded that the project would directly affect about 600,000 inhabitants after altering their livelihoods in addition to another 400,000 inhabitants that might expect pressure on their resources. Moreover, the report also found that those alternatives could be economically unviable and socially unacceptable (Howell et al. 1988: 42). One year after printing this report, civil war broke out in Southern Sudan, making it impossible to undertake further required studies, since the situation there was catastrophic, best described in Waterbury (2002: 141) as: “It would be hard to find any spot on earth where life has been so miserable for so long.”

Yet, the great rains of 1961–1964 (Sect. 4.2.6) increased the permanent area of the Sudd swamps from 7000 to 20,000 km^2, reaching 40,000 km^2 in the rainy season, which motivated the PJTC to start a new survey in 1969 to touch upon such considerable hydrological changes. However, this was not effective until signing the 1972 Addis Ababa Agreement and the termination of the first civil war in Southern Sudan (Waterbury 2002: 140, 142). In 1974, Egypt and Sudan took actual political and technical steps to start constructing the Jonglei I project. Nevertheless, the benefits and costs of the project changed significantly compared to those studied by the JIT (Waterbury 2002: 144). On the one hand, the Jonglei I would “undo” the permanent submerging of large areas in the Sudd and would allow for the implementation of local infrastructural projects, year-round navigation through the channel and irrigation schemes around it, besides the primary objective of saving waters downstream. In numbers, Jonglei I aimed to save 3.85 BCM, while implementing the other parts of the conservancy scheme, Jonglei II, Bahr el-Ghazal, and Sobat-Machar would save 3.44 BCM, 5.7 BCM and 3.3 BCM respectively, with a total amount of approximately 16 BCM as measured at Aswan (Waterbury 2002: 144). On the other hand, the project would impede the "traditional transhumant routes" affecting the free movement of large numbers of people and their cattle. In addition, there might be a great difficulty for machines to work in the "vertisols" (clayey soil with little organic matter that occurs in regions of distinct wet and dry seasons) covering the area around the project. On the ground, there was a common sense among the Southern Sudanese that this project would only benefit Cairo and Khartoum and seriously harm them, and there were even rumors that Egyptian peasants would move to the place to cultivate the new irrigable lands. Hence, both countries knew that the key factor of success was the popular acceptance of the Southern Sudanese. According to the 1972 Agreement, Nimeiri provided autonomy to Southern Sudan in which the latter had its own government, but it subordinated the central authority in Khartoum. With the aim of persuading the Southern Sudanese, the Southern Regional President Abel Alier said the following words in his speech before the regional assembly, as quoted in Waterbury (1979: 77):

> It was thought that the present financial climate in the Arab World [after the 1973 War] was conducive to obtaining financing easily and at advantageous terms. It was therefore imperative to implement the project immediately while this climate existed. … Need I also assure the honorable members that no Egyptian nationals will come to settle in the area? The Egyptians are presently engaged in developing their own country and in recovering their lost lands [the Sinai peninsula]. They have the since to know the bitterness generated against those who occupy other people's land. … The people (in the South) cannot even have one full meal a day, and children of school age cannot go to school because of our underdevelopment, backwardness, and poverty. Yet we are asked to accept all this … and remain in a sort of human zoo for anthropologists, tourists, environmentalists, and adventurers from developed countries of Europe to study us, our origin, our plights, the sizes of our skulls and the shape and length of our customary scars. …. I wish to say that although this is a Central Government [i.e. Khartoum] project, the Regional Government supports it and stands for it. If we have to drive our people to paradise with sticks, we will do so for their good and the good of those who come after us.

In 1978, the excavation works started from a point on Bahr el-Ghazal (40 km west of the Sobat junction) moving southwards, and by 1983 two-third of the canal total length (267 out of 360 km) was already excavated (Fig. 4.10), but the dramatic events that took place in that year completely halted the project (Waterbury 2002: 144). In June 1983, Nimeiri abolished the self-autonomy of Southern Sudan and re-divided the south into three sub-regions. Moreover, in September, he introduced the Shari'a law across the entire country, including Southern Sudan (Carol 2012: 24). As a result, civil war erupted again for the second time.

As reviewed in Arnold and LeRiche (2013: 63–68), there were different groups of rebels due to the complex local tribal configuration in the area, all of which were competing against one another over leadership. Amongst all, Colonel John Garang, with his Southern Peoples' Liberation Army and Liberation Movement (SPLA/M), was able to convince the Mengistu regime to back him by presenting himself as a Marxist revolutionary not a separatist reactionary. In fact, Ethiopia did not want to support a separatist movement in Southern Sudan whilst combating the separation of Eritrea. The SPLA/M's manifesto of July 1983 announced its ultimate goal to spark a "socialist transformation of the whole country" and not to separate the

Fig. 4.10 The map of South Sudan showing the excavated part of the Jonglei I project. (Source: FAO (2015), Aquastat Database: Maps by countries, available at: http://www.fao.org/nr/water/aquastat/countries_regions/SSD/SSD-map_detailed.pdf (accessed on October 14, 2016))

southern people from the north. By November 1983, the Jonglei Canal was completely cancelled, after the SPLA/M attacked the excavation works, detained a number of French engineers and international workers, and destroyed the equipment installed in the site. Interestingly, Waterbury (2002: 146) notes that Garang himself had done his PhD study at Iowa State University in 1981 on the socio-economic development in the Jonglei area, in which he concluded that the project was unacceptable because its major benefits would appear downstream, while if the local people were the primary beneficiaries, the case would be completely different.

Mengistu's support for the SPLA/M had significant consequences on the intricate map of local actors in Sudan and Ethiopia. On the one hand, Khartoum retaliated by backing the Eritrean liberation movements, which was directly confronting Addis Ababa and indirectly supporting other insurgent groups in Ethiopia (Mogire 2011: 11). On the other hand, the SPLA/M, in return for his support, fought with Mengistu against those liberation movements inside Ethiopia (Arnold and LeRiche 2013: 65, 296). In general, this was the main virtue of that age: every country extended support for the rebellion movements inside its adversaries, which was aggravated by the spread of arms exported by the superpowers to the region for decades. In light of the accumulated economic and political crises, including the Southern Sudanese civil war and the pressure of refugees escaping from the famine in Western Sudan (see next subsection), Nimeiri was ousted after the outbreak of civilian demonstrations (Woodward 2011: 92). In 1986, a government was elected and headed by Sadiq al-Mahdi, after which relations with Egypt were not "particularly warm" due to the asylum granted by Cairo to Nimeiri and the sensitivity of Sudanese nationalists against Egyptian intervention in Sudan (Rubin 2014: 66).

4.3.5 The 1983–1985 Famine in Ethiopia and the 1988 Flood in Sudan

During the period of 1983–1985, up to one million people died in the horrible famine that took place in Ethiopia. This was triggered by a severe drought and exacerbated by the use of food weapon in the wars fought by the Derg regime against insurgents in Tigray and Eritrea (CQ Researcher 2011: 242). Indeed, Ethiopia's low capacity of water storage hindered its ability to mitigate the effects of the drought. However, the internal fighting in Ethiopia exaggerated the misery of the famine. Keneally (2011: 212) recorded, "The military budget in 1984, just as the famine reached its height, consumed 60 per cent of national income, and Mengistu's Ethiopian army was 300,000 strong." He added, "Wherever these rebel groups operated …, Mengistu denied food relief—in any case, the first food priority was the army. Rightly, wrongly or unknowingly, Western aid agencies supplied the Ethiopian army for their battles against insurgence" (Keneally 2011: 213). In December 1984, Mengistu's acting minister of foreign affairs at the time, Tibebu Bekele, was quoted telling the U.S. Chargé d'Affaires, David A. Korn (1986: 137), "Food is a major element in our strategy against the secessionists" (Human Rights Watch 1991: 156). Such a disastrous event led to several direct and indirect political consequences at the regional and domestic levels.

In 1986, Djibouti, Ethiopia (later Eritrea joined after its independence), Kenya, Somalia, Sudan and Uganda formed the Intergovernmental Authority on Drought and Development (IGADD). The FAO, UNDP, World Bank, African Development Bank (ADF), and International Fund supported this initiative for Agricultural Development. As reviewed in Waterbury (2002: 147–148), IGADD's primary objective was to promote food security in the Horn of Africa, which appeared highly vulnerable to the risks of droughts and desertification. Yet, IGADD's agenda combined both economic and political aspects, since it had been obvious that the ongoing warfare would halt any effort to combat severe climatic variability in the region. Such a political role motivated the U.S. to invite the Intergovernmental Authority on Development (IGAD), which was created in 1996 to supersede IGADD, concerning the participation in solving the problem, and the organization adopted the principle of self-determination for Southern Sudan. Khartoum saw the process as foreign intervention in its affairs, as further discussed in detail in the next chapter of the post-Cold War round.

In the Nile Basin, the extraordinary drought years affected Sudan and Egypt, since the extremely low flows caused significant reduction in the stored waters behind dams and in terms of generated hydropower. Lake Nasser, in particular, shrank dramatically and reached its lowest level, from 172.42 m in 1975 to 154 m in 1987 (Kliot 1994: 44), which made the revival of the Jonglei project a necessity for Egypt. President Hosny Mubarak, who succeeded Sadat, was aware that resuming the work in Jonglei would be impossible as long as the boiling condition in Southern Sudan prevailed. At the same time, his general policy, even in light of the cold relations with al-Mahdi, was prominent to the unity and stability of Egypt's southern neighbor, not only to prevent the breakup of an Arab state, but also for the security of the Nile regime established by the 1959 Agreement (Funk and Fake 2009: 204).

Realizing that the solution of the Southern Sudanese problem would be only produced in Addis Ababa, Mubarak attempted to establish relations with Mengistu and to mediate between Sudan and Ethiopia over the problem of supporting each other's rebels. Invited by Mubarak, Mengistu conducted his first formal visit to Cairo in April 1987 (Weir 1988/2005: 86). In order to "facilitate an agreement between Sudan and Ethiopia," Mubarak attended the proclamation of Ethiopia as a People's Democratic Republic held by Mengistu in September of the same year (Legum 1989: B-306). However, these meetings were not particularly effective, as Egypt had "few cards to play" (Weir 1988/2005: 86). Moreover, the high flood of 1988 increased the level of Lake Nasser (reaching 165 m) and thus mitigated Egypt's losses in the preceding years (Kliot 1994: 44). In contrast, the 1988 flood had destructive consequences on Sudan, after which Khartoum was extensively harmed. Immediately, Egypt's Prime Minister Atef Sedki flew to Khartoum to send a message that Cairo would "contribute to reconstruction efforts," and he considered his visit a "gesture of solidarity amid the hardship caused by the flood, which made about two million people homeless and caused damage worth millions of dollars" (TFDD 2008). However, problems in Sudan were escalating particularly due to economic depression, the Southern Sudanese civil war, and internal disputes over religious issues. As a result, al-Mahdi's government was overthrown by General Omar al-Bashir on June 30, 1989. Cairo initially welcomed the step hoping to improve relations with Sudan (Rubin 2014: 66). Nevertheless, Sudan's support for Iraq during the Gulf War in 1990–1991, its role in the assassination attempt against Mubarak in 1995, and the renewal of dispute over the Hala'ib Triangle were all different factors that tensioned the relations between the two countries, as will be discussed in detail in the next chapter.

4.3.6 The Attempt to Implement the Tana-Beles Project

Both the Mubarak and the Bashir regimes entered into dispute with Mengistu over the Tana-Beles project. This was the second scheme from the 1964 USBR study that Ethiopia attempted to implement after the Finchaa Dam (Fig. 4.5). Tana-beles is a multi-purpose project that requires the diversion of water from Lake Tana to the Beles River Basin through a tunnel with the aim of generating hydropower and irrigating lands along the Beles River, which runs downstream and joins the Abay River before the Ethiopian-Sudanese border. The first step of implementing the Tana Beles project was to construct a regulator at Lake Tana in order to control the water diversion process. This was completed by 1986 (Waterbury 2002: 116–117). As mentioned in the report of Human Rights Watch (1991: 372), amongst the European donations to Ethiopia after the famine, Italy assisted by $246 million in 1988, including a "generous aid to the development project of Tana-Beles." Mengistu contracted an Italian company to carry out the construction work (Waterbury 2002: 117).

Egypt, whilst witnessing Lake Nasser was at its lowest level, "protested vigorously" against the Tana-Beles (Degefu 2003: 212), which led, according to Kendie (1999), to blocking an expected ADB fund. Not only did Egypt oppose the Ethiopian steps, but also Sudan. As stated in the international water events dataset in TFDD (2008), Yaakoub Abu-Shura, the then Sudanese Irrigation Minister told the daily "Al-Sudan Al-Hadith" on January 11, 1990, that Sudan opposed "building new dams in Ethiopia on the Nile" because "this will affect revenues of Nile waters to both Sudan and Egypt." In fact, this was not the only reason for the Egyptian and Sudanese opposition but more important because of the possible Israeli involvement in the project, especially following the Ethiopian decision to restore "diplomatic relations" with Israel on November 3, 1989 (Carol 2012: 25). This was clearly stated by the Sudanese minister of information and culture, Ali Muhammed Shammu, in Al-Dammam Al-Yawm on January 21, 1990 (TFDD 2008), as follows:

> Ethiopia's acts constitute flagrant violation of agreement [the 1902 Agreement], and the fact that Ethiopia allowing Israeli enemy to participate in construction of these dams threatens security of Sudan and Egypt, in particular, and Arab countries, in general. Ethiopia built huge dam on the Blue Nile sources 3–4 years ago [the Lake Tana regulator] without consulting Sudan. Sudan will make every effort to block suspect project.

In response to the fears of Egypt and Sudan, as well as to reports in Western and Arab media about the Israeli-Ethiopian military cooperation, Mengistu declared the following in late 1990, as cited in Bishku (1994: 51):

> Some wild allegations are already circulating. Some Arab countries allege that the Israeli army is already in Ethiopia fighting alongside the Ethiopian army, that a host of Israeli engineers is already on Ethiopian soil to construct a huge dam on the Blue Nile and stop the Nile from flowing into Egypt and the Sudan and so on. There isn't a single Israeli soldier in Ethiopia; in fact, the [Israeli] Embassy does not even have a military attaché.

Yet, by 1991, all the conditions were working against Mengistu. These can be summarized after Metaferia (2009: 69–70), indicating that the longstanding internal fighting, in addition to the misery of the famine, heavily weakened him, and on the battle ground, the liberation fronts in Tigray and Eritrea had been achieving significant successes. Externally, the U.S. was exerting great pressure on his communist regime. For example, it launched an Amharic program of the Voice of America in Ethiopia to challenge the military regime. At the same time, the Soviet Union was already under depression during the Gorbachev regime, which adopted the doctrine of superpower cooperation. Gorbachev warned Mengistu explicitly that the Soviet Union "would not continue assisting Ethiopia as it was itself undergoing radical changes in domestic policies," and advised him "to make a rapprochement with the U.S. and also seek political solutions to its domestic problems, especially to the Eritrean issue" (Metaferia 2009: 70). By May 1991, the EPRDF and the TPLF were able to topple Mengistu exiling him to Zimbabwe, while the EPLF controlled over Eritrea (Carol 2012: 25). The Tana-Beles came into a halt and was only implemented in the second half of 2000s (see Chap. 5).

4.3.7 *The Undugu Group*

After the relative success of the Hydromet survey (Sect. 4.2.6), at least from the information sharing perspective, Egypt and Sudan sought to establish an ultimate body that would gather all the riparian states. In December 1978, the representatives of the PJTC forwarded during the technical meetings of the Hydromet "a proposal for the establishment of a Nile basin commission to cooperate in the natural planning of conservation, development and allocation of water resources in the basin" (Collins 1990: 288). The other riparians were not enthusiastic for the idea, as they feared that Egypt and Sudan "would dominate by virtue of their historic and technical experience and their political and military power" (Collins 1990: 289).

In 1983, the Undugu group was formed on behest of Egypt with the initiation of Professor Boutros Boutros Ghali, Egypt's minister of state for foreign affairs (1977–1991) and the one responsible for managing the Nile case in Egypt. The founding members were Egypt, Sudan, Uganda and Central African Republic, while Rwanda and Burundi joined as full members in 1988 and 1990 respectively. Tanzania participated as an observer whereas Ethiopia "never joined," though "it supported the work" (Elemam 2010: 228). As concisely described by Waterbury (2002: 77), Ghali's philosophy in managing such a complex case is that Egypt should shift its strategy in dealing with its co-riparians, from pushing them for relying on "water alternatives" other than the Nile to collaborating with them for the creation of "multi-good" socio-economic solutions beyond consumptive water uses. Interviewed by De Chatel (2007: 143–144), Ghali summarized his experience with the water issue in general and the Undugu group in particular as follows:

> During this whole period (1977–1991), my main obsession was water. Of course, I was involved with the discussions at Camp David [with Israel], but the problem of water was always at the forefront of my concerns. … I tried to raise awareness of the importance of cooperation between riparian states over the sharing of the Nile waters; I wanted to show Egyptians that Egypt's security is related to the south, to Sudan and Ethiopia, rather than to the east and Israel. I created an organization that brought together the ministers of irrigation of the Nile riparian states … We called it Undugu, which means fraternity in Swahili, and our aim was to engender collaboration and consensus on issues related to the Nile. We held meetings in Khartoum, in Kinshasa, in Addis Ababa. But right from the beginning there was an atmosphere of mistrust and the upstream countries were suspicious of Egypt's demands. They said: 'If you want to build a new dam, it will flood our villages and displace our people. What will we receive in exchange?' They even said that they would demand one barrel of oil for each barrel of water they gave away.

In addition to his indication concerning the great difficulty of persuading the Equatorial states to build the Albert Dam in the hope of reviving the Jonglei Canal scheme and other conservancy projects in the Sudd, Ghali in his words touched upon the real problem of "mistrust" that faced multilateral cooperation in the Nile during this period. A clear example of this was the unsuccessful attempt to connect the power grids of the Nile riparians with the AHD in Egypt and the Inga Dam in Zaire, which was officially discussed in a meeting gathering the electricity ministers of all countries in Cairo on June 28, 1990 (TFDD 2008). Ethiopia did not welcome the idea, considering it an "Egyptian maneuver to preempt development of Ethiopia's huge untapped hydropower potential," and it saw the Undugu group itself as an "Egyptian Trojan horse and opposed it from the outset" (Waterbury 2002: 77–78). In addition to mistrust, a real obstacle to cooperation was the continuous internal disturbance from which the upstream riparians had suffered for long time. For example, Uganda entered into an armed conflict with its neighbor Tanzania in October 1978, which lasted for several months until the Tanzania forces, together with the Ugandan anti-Idi Amin liberation army succeeded to reach Kampala, and Amin was sent to exile shortly thereafter (Carol 2012: 23–24). In 1981, the Ugandan civil war was sparked and lasted for almost 5 years (Carol 2012: 24).

Nevertheless, although Undugu's original aim to "foster economic, social, cultural and technical ties among the Nile riparians" was not fulfilled, "it at least ensured a forum for information sharing. Undugu served as an institutional locus for sharing expertise and as a group accustomed to treating the Nile as a whole, not as less than the sum of its national parts" (Brunnee and Toope 2002: 133). Still, the serious challenges that faced Ghali in catalyzing real cooperation among riparians might justify his famous phrase, "The next war in this region will be over the waters of the Nile" (De Chatel 2007: 144). By electing Ghali to be the UN Secretary General in 1992, the Undugu group had "died" (Waterbury 2002: 78). Yet, the relative success of Hydromet survey and the efforts made by Undugu paved the way for international development partners to work on building a multilateral organization with the membership of all Nile riparians, which led to the creation of the TECCONILE and then the NBI (see Chap. 5).

4.4 Analysis and Discussion

4.4.1 Sole Utilization (T-S and S-T)

The Cold War round started with all of the riparians becoming sovereign states one after another (D). In addition, there was no longer the overarching colonial power that could manage and moderate the opposing interests of the riparians. Instead, there were two rival superpowers, the U.S. and the Soviet Union, competing over the security of international navigation in the Red Sea (G) and over expanding their presence in the region for military and economic leverage (R). Nasser, after taking power in Egypt, had an ambitious agenda for the advancement of its economy and army (C_{Eg}) in order to meet the needs of its rapidly growing population (I_{Eg}). One of the early decisions taken by the new revolutionary regime was to build a huge dam on the Nile inside Egypt, not a group of smaller dams upper the river as in the previous plans. This was motivated by the overflow of the Old Aswan Dam, the unsuccessful negotiations over the upstream projects, and more important the need to secure the stored waters away from external control (B_{Eg}). After failing to obtain funds to construct the AHD from the Western block, Nasser decided to nationalize the Suez Canal Company, which antagonized Britain, France and Israel and led to the Suez Crisis (G). The global balance between the two superpowers and the expiration of the old European colonial system (G) ended the Tripartite Aggression. However, the potential use of the Nile waters as a weapon during the crisis confirmed Egypt's fears to rely on storage reservoirs outside its borders (B_{Eg}). The Soviet Union offered providing Nasser with the required technical and financial assistance to build the AHD (C_{Eg}), but this alone was not sufficient to boost Egypt's position as the major user of the Nile waters, because Sudan at the same time was seeking to expand its water uses, and both had to coordinates over their schemes (I_{Eg} vs. I_{Su}). Yet, this was not easily amenable, and the path towards reaching a compromise passed through phases of contention (P_{Eg}-P_{Su}) and cooperation (R_{Eg}-R_{Su}), as discussed in the following two subsections. Nevertheless, the 1959 Agreement was eventually signed, regulating the "full utilization" of the Nile waters between Egypt and Sudan ($T_{Eg\&Su}$-S_{Up}).

The other riparian states did not accept the Egyptian-Sudanese agreement, but still they lacked the administrative, technical and financial capacity (C_{Up}) to challenge the status quo. Ethiopia, in particular, issued an official statement reserving its right to utilize unilaterally the waters originating on its lands for the sake of its people (I_{Et}). Through his alliance with the U.S., Haile Selassie requested American support to exploit the untapped resources of the Blue Nile (C_{Et}). Through the Bureau of Reclamation, the U.S. conducted an extensive survey (1958–1964) on the Blue Nile Basin proposing 26 reservoirs for irrigation and hydropower generation (I_{Et}). However, the proposed plan was far beyond the capacity of Ethiopia (C_{Et}) and the U.S. interests in Ethiopia (I_{US}). In addition, Ethiopia's capabilities were further weakened due to the internal conflict with Eritrean separatists and external dispute with Somalia in the 1960s (C_{Et}). With the exception of the Finchaa Dam, none of the suggested projects was completed during the Cold War round.

By the mid-1970s, the map of alliances in the whole region was completely reversed. On the one hand, the October 1973 War (R), waged by Sadat to restore Egypt's occupied lands in the Sinai peninsula since 1967, redirected Egypt's orientation to the west. In addition, Sadat emerged from the war more confident in Egypt's capability to maintain the Nile's status quo (C_{Eg}) and to implement new development schemes that would meet the growing national needs (I_{Eg}). The result was the launch of the Peace Canal, through which Egypt worked to materialize and justify its perceived rights of utilizing the Nile waters to reclaim the desert in Northern Sinai. Moreover, Sadat had the idea to extend the channel to Israel in an attempt to terminate the Arab-Israeli conflict; however Egyptian nationalists rejected this suggestion even before Israeli and Ethiopian officials did. The cluster of Egypt and Sudan also commenced the Jonglei I project to add extra waters to the White Nile, however the work was halted in 1983 due to the spark of the second Southern Sudanese civil war (D).

On the other hand, Haile Selassie was overthrown by the Derg, and Mengistu was soon able to suppress all his internal opponents and consolidate control over Ethiopia (D). He adopted communism and aligned with the Soviet Union. Mengistu did not accept the Peace Canal scheme and threatened to reduce the Blue Nile flows, but actually lacked the capacity to do this (C_{Et}). Along Mengistu's reign (1974–1991), Ethiopia witnessed several bloody events through violent campaigns against his internal rivals (D), the internal fighting against the liberation fronts in Eritrea, Tigray, Oromo and Afar (D), and the Somali war over Ogaden (R). Even during the severe drought of 1983–1985, he blocked food supplies to those territories where there were insurgent groups, which extremely aggravated the misery of the famine, and over one million people died (D).

Therefore, the capability of Ethiopia to build dams (C_{Et}) on the Nile was hindered also by Egypt but in different means during the reigns of Nasser, Sadat and Mubarak. Nasser, in the early 1960s, supported Eritrea and Somalia in their conflicts with

Ethiopia, but he downgraded his support for them by the mid-1960s to maintain his leadership position in Africa. Sadat also backed Mengistu's enemies in the late 1970s, namely Somalia. During the late 1980s, Mubarak used Egypt's right, as a downstream riparian, to block international fund to the Tana-Beles project in order to secure the Nile flows to Egypt. Furthermore, the frequent famines played a crucial role in destabilizing the political regimes in Ethiopia. This is obvious in the sheer fact that the 1973 and the 1983–1985 famines (E) represented major factors in overthrowing Haile Selassie and Mengistu respectively. The circumstances of the other riparian states in the Equatorial Lakes were no better (C_{Eq}). Uganda, for example, entered into armed dispute with Tanzania in the late 1970s (R), and thereafter suffered a 5-year civil war (D). Hence, the internal weakness of the upstream riparians in the Cold War round resulted in maintaining the sole utilization of the Nile waters by the downstream cluster.

4.4.2 Cooperation (R-R)

Egypt and Sudan began the Cold War round with the greater interest to develop their own water projects; constructing the AHD (I_{Eg}), on the one hand, and extending the Gezira scheme, heightening of Sennar Dam, and constructing the Roseires Dam (I_{Su}), on the other. However, neither of them had the financial capacity (C_{Eg} and C_{Su}) to implement the proposed plans by its own, and thus they resorted to the IBRD to fund their projects. Simply, the IBRD answered both that they should first reach an agreement over their projects, since Egypt's AHD would flood lands and displace people in Sudan, and Sudan's projects would alter the Nile flow to Egypt. The two countries conducted at first three unsuccessful phases of negotiations, in which each party was not satisfied with the other's offers and conditions over water shares and financial compensation. Although by 1957 Nasser assured the Soviet technical and financial support to build the AHD (C_{Eg}), he did not want to cause harm to Sudan without reaching an agreement in order not to disrupt his image as the leader of the Arab world (R). The internal disturbance in Sudan, which led to Abboud's military takeover in November 1958 (D), apparently facilitated the conclusion of the 1959 Agreement (R_{Eg}-R_{Su}). The eventual outcome for each country clearly indicates how rewarding payoffs of cooperation represent a whole package of benefits and costs that every side conceives as a compromise solution. On the one hand, Sudan agreed to divide the additional waters saved by the AHD after subtracting the established water rights of 1929 Agreement and the losses of evaporation and seepage. Egypt, on the other hand, agreed to give Sudan two-thirds of this remainder. In addition, the two countries pledged to share on equal basis the costs of conservancy projects and the water gains and losses. Evidently, the agreement boosted cooperation between Egypt and Sudan, and both established the PJTC to oversee water projects in each. Meanwhile, this step had widened the gap between perceptions of downstream and upstream countries concerning their water rights (B_{Do} vs. B_{Up}). From the perspective of the downstream states, the other riparians have alternative freshwater resources, and the agreement already took into account the possibility that any of the upstream riparians would "claim a share in the Nile waters," stating that in case of approving such claims, the allotted shares would be deducted equally from those of Egypt and Sudan. Yet, the informal talks between the PJTC and the representatives of Kenya, Tanzania and Uganda in 1961 reflected how difficult this could be. In contrast, the upstream states do not recognize the 1959 Agreement because they were not consulted in its making, reserving the right to utilize the Nile water that originates on their territories for the development of their countries.

Moreover, the development schemes implemented after the agreement also indicate the complex balance of benefits and costs. For instance, the AHD increased Egypt's water storage capacity, saved it from floods and droughts, expanded its irrigated lands, and generated electricity for its growing needs, all of which have been less effective by time due to the high population growth. On the other hand, the AHD led to the loss of naturally fertile soil, the erosion of coasts and delta, the disruption of ecosystems, and the displacement of local inhabitants.

Another cooperation outcome between Egypt and Sudan is the agreement upon the Jonglei I project in 1974 (R_{Eg}-R_{Su}), which was induced by the relative victorious situation of Sadat after the October War (R) the close relations between him and Nimeiri (R), and the termination of the first Southern Sudanese civil war in 1972 (D). However, the rewarding payoffs this time should be shared with the Southern Sudanese people (I_{SS}). The regional vice president perceived the project as an opportunity to develop his region, while the other parties in Southern Sudan seemed not convinced with the project. The excavation works actually started in 1978 but stopped no more than 5 years later after the spark of the second civil war in Southern Sudan (D) in response to Nimeiri's 1983 decisions to divide South Sudan and impose Islamic law across Sudan (D). The SPLA/M, which led the insurgence in Southern Sudan with the support of Mengistu, attacked the construction site and detained the international workers, which reflected the Southern Sudanese perception of cooperation over water conservancy projects in the Sudd (B_{SS}).

In light of successive environmental crises, the Cold War round also witnessed several attempts to establish multilateral and bilateral cooperation outcomes. The great rains of 1961–1964 and its destructive floods around the Equatorial Lakes and the Sudd (E) resulted in the Hydromet survey in 1967, which brought together the Equatorial riparians with Egypt and Sudan (R_{Eq}-R_{Do}), while Ethiopia and Kenya joined later in the 1970s. In the 1980s, Egypt initiated the Undugu group (R_{Eg}-R_{Ot}), which reflected a transition in its perception of cooperation with other riparians (B_{Eg}) from focusing on alternative resources other than the Nile waters to allowing non-consumptive uses of the river (I_{Ot}). For example, Egypt sought to connect the power grids of the Nile riparians with the AHD and the Inga Dam. Although they represented an important step towards institutionalizing water resources management and planning in the Nile, those multilateral programs were seriously challenged by general mistrust between upstream and downstream riparians (B_{Do} vs. B_{Up}). Another multilateral cooperation outcome is the establishment of the IGADD organization after the severe drought of 1983–1985, which covered the Horn of

Africa (E). The IGADD's agenda was not limited to coping with environmental and economic threats, but also played a political role in the Southern Sudanese problem in the post-Cold War round.

Moreover, there were two bilateral cooperation outcomes induced by environmental crises. First, the 1983–1985 drought dramatically lowered the levels of Lake Nasser and reduced the generated hydropower of the AHD (E), which made resuming the Jonglei I project of a great importance to Egypt. This pushed Mubarak to enhance relations with the Mengistu regime, since no peace could be achieved in Southern Sudan without the Ethiopian sponsorship. Second, the destructive flood of 1988 in Khartoum (E) also presented a chance to improve the rocky relations between Egypt and Sudan at that time. Nevertheless, both attempts were not effective. In the first case, this was due to the complicated relations between Ethiopia and Sudan, where each was explicitly supporting the other's insurgence groups (R), while in the second, the tension between Egypt and the Bashir regime, which took power in Sudan in 1989, was growing during the 1990s.

4.4.3 Contention (P-P)

The contention outcomes appeared repeatedly along the Cold War round, even within the downstream cluster (P_{Eg}-P_{Su}). Over 5 years before concluding the 1959 Agreement, Egypt and Sudan could not reach a compromise concerning different technical details. Each side then decided to proceed with its projects unilaterally (I_{Eg} vs. I_{Su}). In 1956, Sudan commenced the Gezira-Managil extension and 2 years later started heightening the Sennar Dam without asking the prior approval of Egypt based on the 1929 Agreement. Egypt, in turn, was able to secure the financial and technical assistance required to build the AHD from the Soviet Union (C_{Eg}). Besides, the negotiations were further complicated due to other political tensions, such as the border dispute over the Hala'ib Triangle (R) and the sensitivity of Sudanese nationalists over alignment with Egypt (D). The breakthrough in negotiations did not take place until after the bloodless military takeover in Sudan by Abboud, whose admiration for Nasser contributed significantly to the signature of the 1959 Agreement. Three decades after, relations between Egypt and Sudan again became cold after Nimeiri was ousted and the government of al-Mahdi was elected.

Contention between Egypt and Ethiopia was much more severe (P_{Eg}-P_{Et}), with apparent peaks and short reliefs. Obviously, such a prevailing state of contention was fueled by their alignment with rival superpowers and by cooperating with each other's adversaries. At the time when Egypt was constructing the AHD (I_{Eg}) with the Soviet assistance (C_{Eg}), Ethiopia, supported by the U.S. (C_{Et}), was conducting its first extensive survey on the Blue Nile (I_{Et}). Moreover, while Egypt was backing Eritrea and Somalia (R) against Ethiopia in the early 1960s, Ethiopia held close ties with Israel (R). Yet, by the mid-1960s, both had a common interest in founding the OAU (R), and the real detente came with the early 1970s, when neither was firmly aligned with any of the superpowers (G) and when Ethiopia severed relations with Israel (R). However, this was too short a period. By 1974, Egypt and Ethiopia reversed their past alignments with the poles of the Cold War after the decision of Sadat to launch peace negotiations with Israel (R) under the U.S. umbrella (G), and after Mengistu controlled over Ethiopia (D) adopting communism and allying with the Soviet Union (G). Ironically, the U.S. conducted the maintenance work required for the Soviet-constructed AHD, and the Soviet Union carried out a survey on Lake Tana to upgrade the USBR study. In 1978/1979, Sadat supported Somalia in its war against Ethiopia (R), after which he feared that Mengistu would respond using the Nile card, particularly in relation to the ongoing survey on the Blue Nile. This led both to exchange threats, which continued until 1981 because of the announcement to implement the Peace Canal project. Again, there was a short relief during the 1980s, after Mubarak succeeded Sadat (D) and while Mengistu was facing the bitter internal fighting (D) and the horrible famine of 1983–1985 (E). By the end of the 1980s, Egypt sought to block international fund for the Tana-Beles project. However, Mengistu proceeded with the project and signed a contract with an Italian construction company (C_{Et}). Nevertheless, Mengistu was overthrown in 1991 by the liberation fronts in Ethiopia and Eritrea, in that his regime was significantly weakened by the miserable consequences of the famine (D), the U.S. pressure on the regime (G), and the Gorbachev doctrine of superpower cooperation (G). Thus, along the Cold War round, the successive attempts of Ethiopia to challenge the status quo were never advanced beyond studies and plans.

For Ethiopia and Sudan, the relations deteriorated substantially with the spark of the second Southern Sudanese civil war (D) despite the fact that the first civil war was terminated through the signature of the 1972 Addis Ababa Agreement with the mediation of Haile Selassie (R). In 1983, Mengistu decided to support directly the SPLA/M that was leading the insurgence in Southern Sudan against Khartoum (D). Sudan, in retaliation, backed the liberation fronts in Eritrea, Tigray and other places in Ethiopia against the Mengistu regime (R). In 1987, Mubarak attempted to mediate between the two countries to achieve peace in Southern Sudan (R), both to prevent the breakup of Sudan and to resume the Jonglei I project after Lake Nasser had been dramatically lowered in the drought years (E), but his attempt failed. In the late 1980s, in light of the Ethiopian-Sudanese tension over local insurgence (D) and the restoration of the Ethiopian-Israeli relations (R), Sudanese officials declared repeatedly Khartoum's rejection to the Tana-Beles project and reminded Ethiopia of the 1902 Agreement, which is interpreted from completely different perspectives by the two sides. The contention between Sudan and Ethiopia (P_{Su}-P_{Et}) did not last for long time since a new phase of good relations began between the two countries after the fall of Mengistu with Khartoum's support.

Table 4.2 The outcomes of the Nile Cold War round according to the iterated PD framework

Scenarios		Does player Y utilize the river unilaterally?	
		No	Yes
Does player X utilize the river uni-laterally?	No	Cooperation (R-R)	Sole utilization (S-T)
		1959: Egypt and Sudan signed the 1959 Agreement and established the PJTC. Egypt then constructed the AHD, and Sudan implemented its Roseires and Khashm al-Girba Dams and the Managil extension of the Gezira scheme (R_{Eg}-R_{Su})	1952: Sudan was not involved in the AHD's early planning (S_{Su}-T_{Eg})
		1967: Hydromet survey was launched to study the hydrological changes at the Equatorial Lakes after the great rains of 1961–1964 (R_{Up}-R_{Do})	1959: Ethiopia and Britain (on behalf of Kenya, Tanganyika and Uganda) reserved their rights in the Nile despite the signature of 1959 Agreement by Egypt and Sudan (S_{Up}-T_{Do})
		1974: Egypt and Sudan agreed upon the Jonglei I project (R_{Eg}-R_{Su})	1961: Kenya, Tanganyika and Uganda requested informally a water share of 5 BCM and the PJTC refused (S_{Eq}-T_{Do})
		1983: The Undugu group was formed and gathered all riparians, Ethiopia joined as an observer, but the group was not particularly effective (R_{Up}-R_{Do})	1983: The SPLA/M stopped the work in Jonglei I and saw it as a project that benefited only Cairo and Khartoum but harmed Southern Sudan (S_{SS}-T_{Do})
		1990: The Nile's electricity ministers discussed the power grid connection but did not reach an agreement (R_{Do}-R_{up})	1979: Mengistu announced Ethiopia's opposition to the Peace Canal but Sadat saw it as Egypt's full right to utilize its water share (S_{Et}-T_{Eg})
	Yes	Sole utilization (T-S)	Contention (P-P)
		1952: Egypt decided to construct the AHD in 1952. It started negotiations with Sudan only in 1954 (T_{Eg}-S_{Su})	1956–1958: Sudan decided unilaterally to heighten the Sennar Dam and to implement the Managil extension, and Egypt, on the other hand, got an offer from the Soviet Union to build the AHD with no need for an agreement with Sudan (P_{Su}-P_{Eg})
		1959: Egypt and Sudan agreed upon the full utilization of the Nile waters without consulting the other upstream riparians, but the fifth article opened the door for future negotiations over their water shares (T_{Do}-S_{Up})	1958: The USBR began an extensive survey on the Blue Nile in Ethiopia (P_{Et}-P_{Do})
		1961: The informal talks of the representatives of Kenya, Tanganyika with the PJTC over the water shares of the former were not fruitful (T_{Do}-S_{Eq})	1972: Ethiopia built and operated the Finchaa Dam despite Egypt's protest (P_{Et}-P_{Eg})

(continued)

Table 4.2 (continued)

Scenarios		Does player Y utilize the river unilaterally? No	Yes
		1978: Egypt and Sudan began work on the Jonglei I project as a bilateral conservancy scheme to increase waters downstream (T_{Do}-S_{Up})	1977: Ethiopia announced its development program on the Nile at the UN Water Conference, while Sadat threatened to go to war if the Nile flows were altered (P_{Et}-P_{Do})
		1979: Sadat commenced the Peace Canal, and he had an initial idea to extend the channel to Israel, but it was opposed by Egyptian, Israeli and Ethiopian sides (T_{Eg}-S_{Ot})	1990: Mengistu proceeded with the Tana-Beles Dam despite Egypt's and Sudan's protests (P_{Et}-P_{Do})

4.5 Concluding Remarks

During the Cold War round, the game equilibrium had tended largely towards contention and slightly towards cooperation, but it ultimately remained closer to the sole utilization outcome in favor of the downstream cluster (Table 4.2). Analysis showed how the contexts had driven the equilibrium along the round. Above all, the global game of the Cold War between the U.S. and the Soviet Union had been reflected on the Nile game. The regional games of the Arab-Israeli Conflict and the Ethiopian-Somali War were also directly linked to the interactions among the riparians. In addition, the regime transformations at the domestic level had diverted the progression of the game outcomes several times along the round, such as the 1952 Revolution in Egypt, the military takeover in Sudan in 1958, the overthrow of Haile Selassie in 1974, and the spark of the first and second Southern Sudanese civil wars in 1955 and 1983 respectively.

The influences of the contexts can be seen obviously in the actions and reactions of the Nile riparians over their water development schemes, which restricted the effectiveness of using the Tit for Tat strategy along the round. In principle, the unilateral decision of the new revolutionary regime in Egypt after 1952 to build the AHD came in response to the failure of the previous negotiations over the Equatorial Nile Project and the Tana project in the late 1940s. Sudan, in turn, decided to proceed unilaterally with its plans to expand the Gezira scheme, heighten the Sennar Dam and construct the Roseires Dam. Yet, because both lacked the financial capacity to implement their projects on their own, Egypt and Sudan launched bilateral negotiations with the aim of obtaining funds from the IBRD. At the time when Egypt and Sudan were negotiating how to share the entire Nile flow among each other, Ethiopia was conducting an extensive survey to exploit the untapped irrigation and hydropower potentials on the Blue Nile, and while the Soviets were assisting in the AHD construction, the U.S. Bureau of Reclamation was administrating the Blue Nile survey. Moreover, when Egypt was supporting the Eritreans and Somalis against Ethiopia in the early 1960s, the latter was developing its cooperative relations with Israel on various aspects. After Egypt had downgraded its support to Eritrea and Somalia by the mid-1960s, Ethiopia later severed its relations with Israel in solidarity with Egypt during the 1973 War. Even in the mid-1970s, when Egypt and Ethiopia reversed their positions on the Cold War camps, each side kept employing the reciprocal strategy. On the one hand, Egypt announced the implementation of the Peace Canal, and, on the other, Ethiopia declared in the 1977 UN-Water Conference its ambitious program of irrigation and hydropower projects. In the 1980s, when Ethiopia decided unilaterally to build the Tana-Beles project, Egypt responded by using its right as a downstream riparian to block international funds to the scheme. Nevertheless, the continuous internal disturbances in Ethiopia along the round hindered the effectiveness of its reciprocal strategy. In short, it can be said that while the global context enabled Ethiopia to reciprocate Egypt's unilateral projects with conducting surveys and preparing plans, the domestic context hindered its capability to implement them on the ground.

Furthermore, reciprocity was obvious in other two scenes. The first was when the second Southern Sudanese civil war was sparked and rebels subsequently attacked the excavation work of the Jonglei I project in response to the Nimeiri's decisions to divide Southern Sudan on tribal basis and to impose the Islamic law on the whole Sudan. The second was when Sudan explicitly backed the liberation fronts in Eritrea and Tigray against Mengistu in retaliation for Ethiopia's support to the SPLA/M in Southern Sudan.

Overall, the contexts of the Cold War round made the interests of the downstream and upstream riparians more incompatible, their capabilities more independent, and beliefs more dominated by mistrust. In such an atmosphere, the attempts to establish multilateral cooperation in the basin, induced by the environmental crises of the 1961–1964 great rains and the 1983–1985 severe drought, did not exceed information sharing and technical assistance. With the fall of the Soviet Union and the overthrow of Mengistu, the game context had profoundly changed, as will be traced and analyzed in the following chapter.

Chapter 5
The Post-Cold War Round: The Failure to Build Institutionalized Basin-Wide Cooperation in the Nile Basin

5.1 The Contexts of the Post-Cold War Round

Following the collapse of the Soviet Union and the end of the Cold War in the early 1990s, the world entered into a period where there was no longer a bipolar rivalry. Instead, international development partners across the globe played a larger role, especially in underdeveloped regions, amongst which was the Nile Basin. Accordingly, the Nile riparians emerged from the Cold War with the ambition of rebuilding their economies to meet their growing needs. They were also encouraged by international development partners to open a new chapter of bilateral and multilateral cooperation. Yet, there was a competition between the downstream and upstream states over setting the foundation of the newly established cooperation, whether to base it on "causing no significant harm," in reference to the existing agreements and current uses, or on "equitable utilization" through a new legal framework involving the entire Nile Basin riparian states.

Such a competition was greatly influenced by the contextual events taking place at different levels. For instance, the opposite positions of Egypt and Sudan on the Gulf War (1990–1991) provoked tension between both countries. Furthermore, the

---

Egypt will not sign the Framework Agreement unless there is clear wording that preserves the current rights and water uses. Signing the draft agreement, with its current formula, without resolving the pending points of disagreement, does not serve the Egyptian interests and harms the Egyptian rights stipulated by the effective international agreements between Egypt and the upstream countries…

Egypt's Speaker of the People's Assembly, Fathi Sorour, on April 13, 2010, in the Sharm el-Sheikh meeting (EMFA 2010)

Some people in Egypt have old-fashioned ideas based on the assumption that the Nile water belongs to Egypt ... The circumstances have changed and changed forever ... the way forward is not for Egypt to try to stop the unstoppable. The way forward is to seek a win-win solution through diplomatic efforts.

Ethiopia's Prime Minister Meles Zenawi, on May 20, 2010 (Al-Arabiya 2010)

M. M. Samaan, *The Nile Development Game*,
https://doi.org/10.1007/978-3-030-02665-3_5

tension was exacerbated following the assassination attempt against Mubarak in Addis Ababa in 1995. However, Mubarak was more concerned with the unity of Sudan to maintain Egypt's interests in the Nile, which pushed him to restore relations with the Bashir regime shortly thereafter. In addition, Egypt inaugurated a large-scale reclamation project in the Western Desert, benefiting from its economic growth achieved after the Gulf States decided to cancel a large portion of Egypt's debts in recognition of its participation in the Gulf War. On the other hand, the emergence of China as an alternative source of financial and technical assistance in the region enabled Ethiopia and Sudan to construct national water schemes outside the NBI umbrella. Hence, the different attempts to build basin-wide cooperation ultimately reached a deadlock, given that the Nile riparians were neither able to achieve a unified vision on the new legal framework nor to adopt a common perspective on water security.

5.2 The Attempt to Institutionalize Basin-Wide Cooperation on the Nile

5.2.1 Bilateral Agreements of 1991 and 1993

Despite the explicit commitment of the 1959 Agreement whereby Egypt and Sudan would always deal with the other riparians regarding the Nile waters as one unit, each signed a bilateral agreement separately with Ethiopia reflecting differences in their perceptions of water cooperation. On the one hand, relations between Egypt and Sudan were seriously strained in light of the Gulf War (see Sect. 5.3.1). On the other hand, the nature of their bilateral relations with Ethiopia were different, since there had been already a rapprochement between the Bashir regime and the new transitional government in Ethiopia.

On December 23, 1991, Ethiopia and Sudan signed the Declaration of Peace and Friendship, while on July 1, 1993, Egypt and Ethiopia signed the Framework for General Cooperation. Both accords covered the broader relations between the signatories, but also specifically addressed water issues under multiple articles. The 1991 Declaration was premised by the Ethiopian transitional government's implicit recognition of the Sudanese support for the Tigrean, Oromo and Eritrean liberation fronts in overthrowing Mengistu (Waterbury 2002: 82). Concerning the water issue, both countries agreed upon the principle of "equitable entitlements to the uses of the Nile waters without causing appreciable harm to one another," and they pledged to "establish a joint technical committee" for data sharing and promoting cooperation, and to work for the establishment of an inclusive Nile Basin Organization (Shapland 1997: 81). In the 1993 Agreement,[1] both the new regime in Ethiopia and

[1]Available at: http://gis.nacse.org/tfdd/treaties.php?page=full&origin=river&tn=521 (accessed on October 18, 2016).

the Mubarak regime in Egypt expressed their joint desire to "consolidate the ties of friendship" and to "establish a broad base of common interests." The agreement stipulated that "the issue of the Nile waters shall be worked out in detail, through discussions by experts from both sides, on the basis of the rules and principles of international law." In addition, both countries pledged to "refrain from engaging in any activity related to the Nile waters that may cause appreciable harm to the interests of any party." The agreement also referred to cooperation over "comprehensive and integrated development scheme" and to "periodic consultation on matters of mutual concern."

Regardless of the long-term effectiveness of both accords, they clearly stated how each country formulated its interests on the Nile and perceived its water rights. This view is evident based on the following two points. First, both sides, Ethiopia, as an upstream riparian, and Egypt and Sudan, as downstream riparians, reframed the "extreme" legal terms which they had used for decades to cope with the language of the international water law (Wolf 1999). For Ethiopia, it emphasized the principle of equitable use without referring explicitly to absolute right of sovereignty to unilaterally use the Nile waters originating on its territory, as was stated in Ethiopia's previous statements of 1957 and 1978 (Sects. 4.2.3 and 4.3.2). As for Egypt and Sudan, they focused on the principle of causing no significant harm without explicitly mentioning the 1902 and 1959 Agreements (Sects. 3.2.3 and 4.2.2). Yet, their reference to "significant harm" is still based on those agreements, which has been always the critical issue in negotiations with the upstream riparians, as will be discussed in Sects. 5.4.3 and 6.3.5. Second, although Egypt and Sudan share a common understanding on the no-harm principle, each adopted a different point of view with respect to the principle of equitable water use (Dellapenna 1997: 132). Egypt, for decades, has focused on the water conservancy projects related to the White Nile, but has been always anxious about the development schemes on the Blue Nile. Thus, when Egypt signed the 1993 Agreement with Ethiopia, there was no explicit mention of the principle of equitable use, but only to the no-harm principle. Degefu (2003: 137) comments on this point, indicating that the agreement from the Ethiopian perspective favored the Egyptian side, and that avoiding to acknowledge Ethiopia's right of equitable use may challenge the agreement's legacy. In contrast, the 1991 Declaration included both principles of no significant harm and equitable water use, which is understandable due to the wide range of benefits brought to Sudan by constructing dams in Ethiopia (e.g. water regulation and flood and sediment control). This was first highlighted by Garstin's 1904 report and more extensively by the USBR 1964 study (Sects. 3.3.1 and 4.2.4).

Nevertheless, one may ask why Ethiopia signed the 1993 Agreement with Egypt although it did not state explicitly the right of equitable use. Simply, the new regime in Ethiopia was facing great challenges associated with reconstructing the country following decades of wars and disturbances, as it was necessary to prove internationally its willingness for cooperation in order "to get back into an international arena in which Egypt had become a skillful and respected player" (Waterbury 2002: 78). Dellapenna (1997: 133) indicates that the "immediate effect" of the 1993 Agreement was that "Egypt did not object to a loan application by the Ethiopians for a small-irrigation project on the Blue Nile." However, Ethiopian officials were dissatisfied with Egypt's continuous requests for additional information and further studies to ensure that such projects would cause no significant harm. They even perceived the years elapsed before the World Bank decided that the project would not harm Egypt as a proof of their "worst fears about Egyptian intentions" (Waterbury 2002: 83).

Aside from bilateral agreements in the Eastern Nile sub-basin, it is important to illustrate the position of Uganda as a major actor in the Equatorial Lakes. Uganda's primary interest in the Nile has been hydropower generation (non-consumptive water use), which facilitated its cooperation with Egypt over the construction of the Owen Falls Dam (1949–1954) and the operation of Hydromet (1967–1992) and the Undugu group (1984–1992) (Sects. 3.3.4, 4.2.6 and 4.3.7). This was extended in the early 1990s through the Exchange of Memoranda signed on May 12, 1991 between Egypt and Uganda regarding extension of the Owen Falls Dam (now the Nalubale Dam), whereby both countries explicitly recognized the previous agreements of the Owen Falls Dam, which were based on the 1929 Agreement (Salem 2011: 6). The extension scheme included the construction of a new hydropower plant, the Kiira Dam, 1 km downstream the Nalubale Dam (Fig. 5.1), aiming to add 200 MW to the existing 180 MW of the old plant in order to meet Uganda's growing energy demand (UN-Water 2006: 131). Construction work started in 1993 and was completed by 1999, while the hydropower plant only operated in its full capacity in 2007 (Wakabi 2007).

Although the Ugandan main strategy to use the Nile waters seemed compatible with the principle of no significant harm as perceived by Egypt, Mulira (2010: 155) records how, since independence, the national diplomatic and academic circles have been emphasizing Uganda's right in the equitable water use while criticizing the restrictions imposed by the Nile colonial agreements. This view can be implicitly found in the signature of the Friendship and Cooperation Agreement between Ethiopia and Uganda on November 4, 1993 (EUED 2016), in which both countries called for the "fair" utilization of the Nile's common water resources (Shapland 1997: 82). Then, why did Uganda sign the 1991 Agreement with Egypt recognizing the previous colonial treaties? Similar to the Ethiopian position discussed above, Uganda was requesting the financial support of the World Bank to implement the Kiira-Nalubale extension scheme, which required prior consultation with Egypt to reconfirm the "Agreed Curve" (Sect. 3.3.4) of 1949 and 1953 Agreements (World Bank 2001: 1, 37). The Ugandan government repeated this procedure for the Bujagali Dam (Fig. 5.1), which was launched in 2007 and completed in 2012, whereby it received a letter from the Egyptian government on May 15, 2006, "giving its no objection to the new Bujagali proposed project" (World Bank 2007: 11). Furthermore, Uganda has benefitted from its technical cooperation with Egypt, particularly in the domain of environmental protection. For instance, both countries signed an agreement in 1998, through which Egypt supported Uganda "on a grand basis" to assist it in "combating and controlling the aquatic weeds, in particular the water hyacinth in the outlets and inlets of Lakes Victoria, Kyoga and Albert, and in the Nile" (Metawie 2004: 52–53).

Fig. 5.1 Map of dams in Uganda. (Source: FAO Aquastat (2015), Maps, available at: http://www.fao.org/nr/water/aquastat/countries_regions/UGA/UGA-map_detailed.pdf)

5.2.2 The TECCONILE and the Emergence of the NBI and the CFA

Since the launch of Hydromet in 1967, there have been continuous attempts to establish basin-wide institutionalized cooperation; however these were challenged with the general mistrust of upstream riparians of Egypt's hegemonic influence on such initiatives according to their views. Nevertheless, the different riparian states had participated with varying degrees in the area of information sharing and technical assistance. By the early 1990s, whilst most of the upstream riparians were pursuing their poverty reduction and economic development plans, the liberal functionalist approach adopted by different international development partners promoted the benefits of cooperation as the primary way to access funding through multilateral mechanisms.

In December 1992, the Water Resources Ministers of Egypt, Rwanda, Sudan, Tanzania, Uganda and Zaire met in Kampala and agreed to "fold" Hydromet into the Intergovernmental Technical Cooperation Committee for Promotion of Development and Environmental Protection of the Nile Basin (TECCONILE), while Burundi, Eritrea, Ethiopia and Kenya joined as observers (Elemam 2010: 228). Supported by the Canadian International Development Agency (CIDA), TECCONILE started its activities in 1993 (Delli Priscoli and Wolf 2009: 235), whereby all the member and observer states actively participated in its meetings. However, such activities were characterized by a "continuing competitive behavior," as described by Brunnee and Toope (2002: 135). On the one hand, all the riparians were keen to participate in the manufacturing of the development plan that would be adopted by international development partners. On the other hand, the competing interests of the different riparians were apparently reflected in their behavior, as crystallized during the third ministerial meeting, which took place in Arusha in February 1995, whereby the Nile River Basin Action Plan (NBRAP) was launched (Elemam 2010: 228). Egypt's strategy was to focus on cooperative non-consumptive water projects that contribute to the upstream riparians' development without causing significant harm downstream, while Ethiopia's strategy was to achieve a multilateral legal framework that ensures the upstream riparians' right in the equitable use of the Nile waters (Waterbury 2002: 78). Eventually, the NBRAP comprised 22 projects grouped into five components: integrated water resources planning, capacity building, training, regional cooperation, and environmental protection. The fourth component of Regional Cooperation ("D") called for "the establishment of a basin-wide, multidisciplinary framework for legal and institutional arrangements," which was then named "Project D3" (Brunnee and Toope 2002: 134). At this point, the multilateral work in the Nile progressed, both formally and informally, through two simultaneous paths: studying and implementing cooperative projects, and developing a basin-wide legal framework.

As reviewed in Brunnee and Toope (2002: 138), Metawie (2004: 52, 62), Delli Priscoli and Wolf (2009: 235), and NBI (2013, 2016), progress took place along both tracks was as follows. Concerning the first track, the Council of the Ministers of Water Resources (Nile-COM) was formed in 1997, whereby the riparian states developed the NRBAP, following the recommendations of the World Bank through an International Advisory Group (IAG), leading to the creation of the Strategic Action Program and the foundation of the Nile Basin Initiative (NBI) in 1999, which superseded TECCONILE. Under the leadership of the World Bank, and the support of UNDP, CIDA and GEF, the NBI brought together all the riparians as full members, except Eritrea which participated as an observer. The NBI adopted the developed Strategic Action Program, comprising a Shared Vision Program (SVP) at the basin level and a Subsidiary Action Programs (SAPs) at the sub-basin level. The SVP's objective is the "creation of an enabling environment for investments and action on the ground, within an agreed basin-wide framework," through seven thematic projects: "water resources, the environment, power trade, agriculture, applied training, communication and stakeholder involvement, and benefit sharing" (NBI 2013). The central office that "manages" and "oversees" the NBI's activities is the NBI Secretariat (Nile-SEC), based in Entebbe, while a Technical Advisory Committee (Nile-TAC) was formed to operate as a "steering committee for the project (NBI 2013). At the lower level, there have been the Eastern Nile and Nile Equatorial Subsidiary Action Programs (ENSAP and NELSAP), which are managed by the EN Technical Regional Office (ENTRO) at Addis Ababa and the NELSAP Coordination Unit (NELSAP-CU) at Kigali respectively (NBI 2016). In 2001, ten international donors established the Nile Basin Trust Fund (NBTF), administrated by the World Bank, with the purpose of financing the NBI's activities (World Bank 2016).

Simultaneously, the riparians and donors followed the second track. In 1997, the UNDP supported the formation of a panel of experts (three from each country) to work on the D3 Project, which submitted the first draft of the Cooperative Framework Agreement (CFA) in 2000 to the Nile-COM. The NBI was regarded as a transitional body to be superseded by a permanent Nile River Basin Commission (NRBC) after accomplishing and ratifying the CFA.

The international development partners, mainly CIDA, UNDP and the World Bank, organized another informal mechanism that boosted progress along the two multilateral paths. This included the "Nile 2002 Conferences," which were held on an annual basis in different cities, starting in Aswan in 1993 (Brunnee and Toope 2002: 135). The idea was to bring together experts representing the different riparian states to informally discuss their perspectives on basin-wide cooperation. The country papers presented during the Nile 2002 Conferences, as summarized in Collins (2006) and Allam (2014: 82–86), reflected the gap between the downstream and upstream riparians in identifying the objectives of cooperation. The downstream riparians believe that cooperative programs should focus on water harvesting, groundwater exploitation, building capacity, and such non-consumptive surface water projects. In contrast, the upstream riparians view cooperation as a means of assuring their right in utilizing the Nile waters for hydropower generation and irrigated agriculture. Thus, the latter seek a new legal framework that will reshape the map of water uses across the basin. Overall, these informal meetings indicated that neither the upstream riparians were convinced with limiting their plans to non-consumptive water uses (the projects-based path favored by Egypt), nor the down-stream riparians were ready to discuss reallocating the Nile waters (the principles-based path favored by Ethiopia). Thus, the attempt to build basin-wide cooperation was faced with serious challenges, since all sides believed that it would not be possible to achieve their national ambitions under the umbrella of the NBI. Accordingly, a third path was launched, along which the riparians sought to build their national projects unilaterally; facilitated by the changing circumstances at the domestic and global levels. National water plans were prepared even lower the basin, as the long-standing cooperation between Egypt and Sudan was severely affected in the 1990s, as will be illustrated in the following section.

5.3 Challenges to the Emerging Basin-Wide Cooperation in the Nile

5.3.1 The Gulf War and the Assassination Attempt Against Mubarak

The relations between Egypt and Sudan were notably deteriorated during the early and mid-1990s. Egypt was the first to recognize the new regime in Sudan after June 1989, and it initially hoped that this change would improve the bilateral relations, which were not as flourishing as during Nimeiri's reign (Sect. 4.3.4). However, the Mubarak regime, which was already fighting terrorism internally, soon became dissatisfied with the explicit Islamist and anti-Western orientation adopted by the Bashir regime (Rubin 2014: 66–67). However, the first severe tension between both countries was during the Gulf War from August 1990 to February 1991, during which Egypt was fighting with Kuwait against the Iraqi aggression, while Sudan was supporting Saddam Hussein. Before the end of the war, press reports revealed that Khartoum might allow an Iraqi attack against the AHD using the Sudanese lands (Woodward 2003: 161). As quoted in TFDD (2008), Mubarak responded to the question of reporters in this regard on January 22, 1991, as follows:

> I will not allow under any circumstances that an inch of Egyptian territory be attacked. ... He [The Iraqi President Saddam Hussein] would not be able to hit the High Dam. Even if he brought something to Sudan, ... Sudan would pay a high price and the officials, who stood by, would pay a very high price. In this I am not joking.

In the following years, tensions between Egypt and Sudan continued to take place, particularly over the disputed territory of Hala'ib when Khartoum, in 1991, planned to grant oil concessions in the triangle (Dzurek et al. 2001) and over the Sudanese decision in 1992 to seize Egyptian assets in Sudan (Ofcansky 2003: 1018). Nevertheless, none of these were sufficient to alter water cooperation between both countries. However, the severe tension that took place due to the assassination attempt against Mubarak in 1995 and the subsequent escalation led to the disruption of the PJTC regular meetings for the first time since its establishment in 1960, and its work was not resumed until 1997 (Waterbury 2002).

As reviewed in Lunn (2004: 308), a group of assailants opened fire on Mubarak's bullet-proof car in Addis Ababa, on June 26, 1995, while he was on his way from the airport to the OAU summit meeting. Mubarak was saved and immediately returned back to Egypt. Days after, the Egyptian Islamic Group (IG) claimed responsibility, and Cairo accused the Bashir regime of facilitating the attack due to its continuous support for the Egyptian Islamist militants, the allegation of which Khartoum denied.

The crisis between both countries overshadowed two vital issues: the Hala'ib. Triangle and the Nile waters. In July, the Egyptian forces expelled the Sudanese police and officials from the Halai'b Triangle (Dzurek et al. 2001: 5–6). Sudan's interior minister stated in the press that his country would regard the Egyptian action as a violation to the bilateral border agreements and that Sudan might reconsider all its agreements with Egypt, including those related to the Nile River (TFDD 2008). The press also circulated the threats raised by the secretary general of Sudan's National Islamic Front (NIF), Hassan al-Turabi[2] indicating that "Sudan would use water as a weapon in its dispute with Egypt" (TFDD 2008). In addition, press reports covered a visit for Iraqi experts to Sudan to study the available options of altering the Nile flows to Egypt (TFDD 2008). Subsequently, Egyptian-Sudanese relations worsened, to the degree that Egypt scaled down its diplomatic representation in Khartoum and imposed visa and permit requirements on Sudanese citizens who sought visiting or residing in Egypt (Lunn 2004: 313), while Sudan also carried out similar procedures with Egyptian nationals in Sudan (Ofcansky 2003: 1018).

The incident's effects extended to Ethiopia's relations with Egypt and Sudan. Egypt blamed Ethiopia for not taking adequate measures to prevent the attempt, and Ethiopia accused Sudan of "complicity in the attack," claiming "the assailants had safe houses in Sudan and that they had received training in special camps near Khartoum" (Lunn 2004: 308). Moreover, the assassination attempt undermined Mubarak's role in Africa, whereby he did not attend the OAU (later the AU) summits in African capitals after 1995, with the exception of Abuja in 2005 when Egypt was running for a seat in the restructured UN Security Council (Jhazbhay 2012: 245). Moreover, Mubarak limited his visits even to the Nile riparians, which negatively affected Egypt's position in the basin. This is obvious in Uganda's President Youri Museveni's words to Egypt's Foreign Minister Ahmed Aboul-Gheit in August 2004, "Do you know that no Egyptian pharaoh had visited us before, although your origins are from here in Uganda and the Upper Nile?" And when Aboul-Gheit met

[2]Turabi recorded his testimony about the failed attempt of assassination and requested not to broadcast it until after his death, which took place only recently few months after he passed away in March 2016. As reported in Sudan Tribune (2016a), Turabi said that important figures in the Sudanese regime at that time were involved in the assassination attempt, but neither him nor Bashir were aware of such arrangements. Turabi added that Egyptian Islamist leaders asked him, before the attempt, to facilitate the assassination, but he refused and sought to convince them to abandon the idea.

Museveni later in another occasion, he informed him with Mubarak's response that the latter had visited him when he was a vice president, and Musevini replied, "I want the president to visit me not the vice president who became a president," as translated from Arabic by the author after Aboul-Gheit (2014: 228). The minister also indicated that the Egyptian relevant authorities had frequently resisted Mubarak's African visits for security reasons, and that when he visited Uganda in the following years, it was only a brief one (Aboul-Gheit 2014: 228).

Yet, this issue had larger dimensions, as it was linked to the problems of terrorism at the regional and global levels and to the Southern Sudanese civil war at the domestic level. Years earlier, Sudan's support for Islamists in different places around the world had worried the U.S., which preferred not to use the Cold War's conventional tools of direct intervention, especially after its experience in Somalia. Rather, it exerted great pressure on Khartoum by imposing diplomatic and economic sanctions and by coordinating with Sudan's southern neighbors to "destabilize" it (Abouyoub 2016: 328). The U.S. ambassador to Eritrea, Robert Houdek, stated in February 1995 that Afwerki, Zenawi and Museveni, the presidents of Eritrea, Ethiopia and Uganda respectively, the countries of which he called the "frontline states," were "ideal leaders in the region, adhering to the democratic principle, espousing privatization and cooperating towards achieving peaceful settlement to the regional disputes such as the one lingering in Sudan" (Abouyoub 2016: 327–328). The incident of Addis Ababa in June 1995 increased international pressure on Sudan, whereby in the following year, the UN Security Council (pushed by the U.S.) adopted three resolutions calling for the extradition of three suspects involved in the assassination attempt and imposing various sanctions against Sudan (UN 1996).

Interestingly, Egypt opposed these UN sanctions. In response, Madeleine Albright, the then U.S. permanent representative at the UN, "chastised Egypt for the weakness of its resolution" (Nilblock 2002: 206). However, Egypt had a different point of view, which can be understood from two different aspects. First, Cairo declared that such actions "would harm the Sudanese people more than the government" (Ofcansky 2003: 1018). Second, and no less important, Egypt realized that the pressing strategy adopted by the U.S. would definitely lead to weakening the Sudanese regime and thus to the separation of Southern Sudan (Woodward 2016: 93–95); the implications of which Egypt firmly rejected for decades, as it would further complicate the Nile issue. These views are clear in Mubarak's declaration on April 10, 1996, indicating that those sanctions would "lead to the division of Sudan, which we reject and refuse" (Africa Confidential 1996). Justifying Egypt's stance, the Egyptian assistant foreign minister for international and multilateral relations stated, "The Egyptian position is to increase the political pressures on the Sudanese regime so that it renounces terrorism and carries out the resolution on extraditing the suspects, without damaging the territorial integrity of Sudan or adding to the suffering of the Sudanese people" (Nilblock 2002: 206).

As recorded in Ofcansky (2003: 1018–1019), Egypt sought to improve relations with the Bashir regime and at the same time to continue talks with the Northern and

Southern Sudanese opposition combined under the National Democratic Alliance (NDA). In January 1998, Cairo and Khartoum agreed on resuming river traffic, which had been interrupted for almost 4 years, and they formed joint committee for studying opportunities of investments in both countries. In May, Khartoum agreed to return the Egyptian assets that it had seized years earlier. In June 1999, both countries declared that serious steps had been taken for restoring diplomatic relations. Simultaneously, Cairo held talks in 1997 with John Garang, the SPLA/M's leader, and with Sadiq al-Mahdi, the Sudanese late prime minister.

By late 1999, Bashir was able to firm his control over the country and began improving Sudan's relations with the external world. On December 12, he dissolved the parliament, of which Turabi was its speaker, and declared the state of emergency (Sidahmed 2011: 98). In late December, Bashir visited Cairo, seeking Egypt's support in his dispute with Turabi (Murad 1999), whereby both countries pledged to "normalize relations" and to settle the Hala'ib. issue "in an integrational brotherly context" (Dzurek et al. 2001: 6). In April 2000, Egypt appointed a new ambassador to Khartoum for the first time since 1995. In January 2000 and in August 2001, Sudan restored diplomatic relations with Eritrea and Uganda respectively, and signed several border, trade and infrastructure agreements with Ethiopia in 2000–2002 (Ofcansky 2003: 1020). On September 28, 2001, the UN Security Council adopted resolution 1372 (with the abstention of the U.S.), lifting the sanctions imposed by previous resolutions, "noting the steps taken by the Government of the Sudan to comply with the provisions of Council resolutions, and welcoming the accession of the Sudan to the relevant international conventions for the elimination of terrorism" (UN 2001).

Concerning the Southern Sudanese problem, Egypt, together with Libya launched an initiative to achieve a solution that would preserve Sudan's unity and integrity. In April 2000, Bashir attended a European-African meeting in Cairo, through which Egypt and Libya attempted to invest their relations with the Sudanese opposition to bring all sides together, working as a mediator for the "national reconciliation in Sudan" (Dawoud 2000). Against the Egyptian-Libyan initiative, there had been the effort made by IGAD, backed by the U.S. (Sect. 4.3.5), supporting the Southern Sudanese right of self-determination. As noted in Rolandsen and Daly (2016: 132–150), Khartoum preferred the Egyptian-Libyan initiative, while the SPLA/M adhered to IGAD's framework. Eventually, both sides took a significant step towards terminating the war by signing the Marchakos Protocol in 2002, which led to the conclusion of the Comprehensive Peace Agreement (CPA) on January 9, 2005, giving the Southern Sudanese a 6-year transitional period to run a referendum through which they voted in favor of independence (Sect. 5.3.1).

The regional events illustrated above had shaped the atmosphere under which the large development schemes on the Nile were implemented since the mid-1990s and throughout the 2000s. The following subsections discuss in detail the development of the Toshka project in Egypt, as well as the Tekeze Dam and Beles hydropower project in Ethiopia, in addition to the Merowe Dam and the heightening of the Roseires Dam in Sudan.

5.3.2 The Toshka Project

By the mid-1990s, Egypt's economic situation had relatively improved as a result of the cancelation of its debts to the Arab creditors and the military debt to the U.S. in recognition of the Egyptian role in the Gulf War (Azzam 2002: 114). This was also followed by writing off one-half of the net present value of the long-term concessionary debt to Western governments in the Paris Club and rescheduling the other half (Azzam 2002: 114). Statistically, according to the World Development Indicators (WDI 2016), Egypt's total external debt stocks (in current US$) had decreased from $45.64 billion in 1989 (116.61% of GNI) to $29.93 billion in 1997 (37.69% of GNI), while the total reserves (including gold, in current US$) increased from $2.5 billion in 1989 (5.47% of external debt) to $19.37 billion in 1997 (64.72% of external debt). Such a position encouraged Mubarak to adopt the construction of "national mega projects" to give a sign that Egypt would cross over its problems of terrorism and development to a new era.

Meanwhile, it was clear that Egypt had been suffering from limited water and land resources vis-a-vis rapid population growth. In this respect, the available cultivated area and the renewable surface water volume per capita have been generally decreasing (Table 5.1). In addition, Fig. 5.2 illustrates that the gap between wheat production and consumption has been widening, and thus the government has been heavily relying on imports to meet the country's urgent needs. Hence, the average decennial self-sufficiency in wheat according to Earth Policy Institute (EPI 2011) after the USDA database, decreased from 42.6% in the 1960s to 33.9% in the 1970s to 25.2% in the 1980s. To cope with these challenges, the government adopted two strategies: increasing crop productivity within the existing lands and expanding total cultivated area out of the Nile Valley and Delta. The expansion strategy depended on the reclamation of new lands in the Egyptian deserts by continuing work in the Peace Canal project in Northern Sinai, irrigating lands west of the Delta in the Northern Coast, and constructing the New Valley scheme in the Western Desert. The primary objectives were to increase the inhabitable area of Egypt (around 5% of total country area) and to increase the agriculture production both for domestic consumption and for obtaining hard currency through food exports.

To address these inherent land-water resources challenges vis-a-vis the larger question of demographic growth in the Nile valley and its overcrowded Delta (Fig. A.4), the New Valley scheme in specific aims at creating new communities of some six million inhabitants and reclaiming one million feddan in East Oweinat, the New Valley oases as well as Toshka depression (Fig. 5.3). The first and second sites rely on groundwater, while the third is irrigated by the Nile water stored in Lake Nasser with limited groundwater extraction (Wahish 1998). This mega project was "designed to create a second Nile Valley in the south of the Western Desert" (ESIS 2013a). Toshka is a depression located some 225 km south of Aswan in the Western Desert, linked to Lake Nasser by an emergency spillway through Toshka Bay (Khor) to discharge water whenever the lake level rises above 178 m, which took place in the high floods of 1996 and the following years, forming the Toshka Lakes

Table 5.1 Water and land resources in Egypt in relevance to population growth

Indicator	1958–1962	1963–1967	1968–1972	1973–1977	1978–1982	1983–1987	1988–1992	1993–1997	1998–1902	2003–2007	2008–2012	2013–2017
Cult. area (mil. fd.)	5.96	6.67	6.80	6.27	5.88	6.06	7.13	7.73	8.15	8.42	8.80	8.95
% cult. of total area	2.50	2.80	2.85	2.63	2.47	2.54	2.99	3.24	3.42	3.53	3.69	3.76
Total pop. (mil. inhab)	28.6	32.5	36.3	40.4	45.6	52.2	58.9	64.8	70.9	77.6	85.7	91.5
Pop. density (inhab/km^2)	29	32	36	40	46	52	59	65	71	77	86	91
Pop./cult (inhab/km^2)	1140	1159	1272	1532	1847	2048	1967	1996	2071	2193	2318	2433
% rural of total pop.	60.0	57.9	55.9	54.3	54.8	55.6	57.2	58.4	58.7	58.8	59.5	60.1
GDP per capita (current US$/inhab)	140	172	241	363	561	776	710	1211	1239	1675	3222	3617
Agriculture, value added to GDP (%)	N/A	28.6	31.1	27.1	19.6	20.5	16.5	17.0	16.5	14.1	11.1	11.2
Nile surface water (m3/inhab)	1681	1479	1529	1375	1217	1064	942	857	783	715	648	607
Total withdrawals (BCM/year)	N/A	N/A	N/A	48.2	N/A	N/A	N/A	58.9	68.3	N/A	78.0	N/A
% agriculture of total withdrawals	N/A	N/A	N/A	N/A	N/A	N/A	N/A	86.1	86.4	N/A	85.9	N/A
Harvested irr. temp. wheat area (mil. fd.)				1.39	1.33	1.19		2.17	2.45		3.00	

Source: Compiled after FAO Aquastat Database (2016), available at: http://www.fao.org/nr/water/aquastat/data/query/index.html?lang=en (accessed on October 28, 2016)

Total land area of Egypt is 1,001,450 km^2; and the Nile surface water per capita is calculated by dividing 48 BCM (in 1958–1962 and 1963–1968) and 55.5 BCM (since the operation of the AHD onwards) over the total population

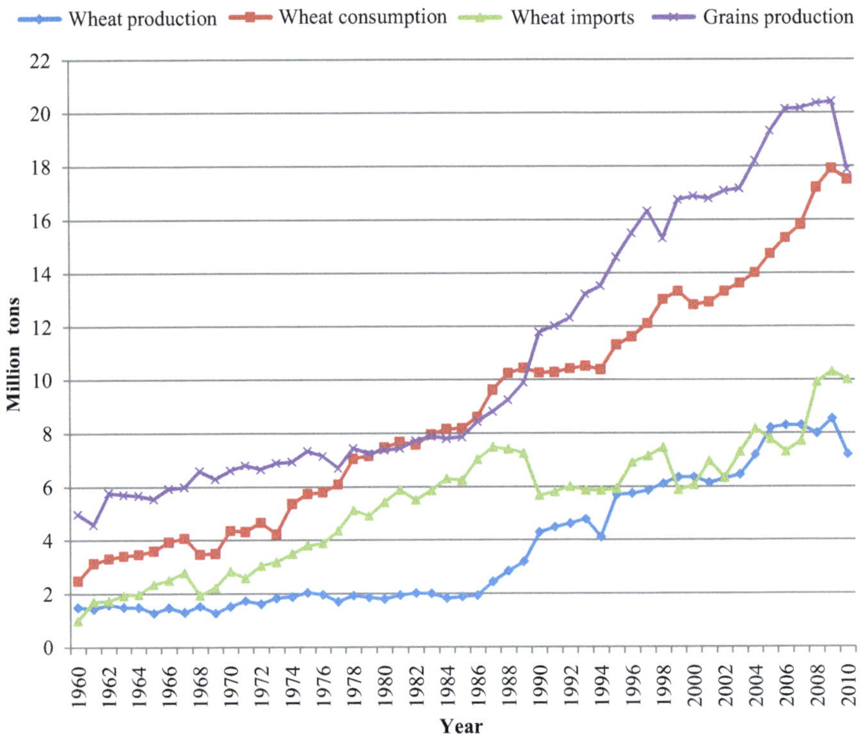

Fig. 5.2 Total grains production and wheat production, consumption and imports in Egypt, 1960–2010. (Source: Compiled after the U.S. Department of Agriculture, Production, Supply and Distribution (2011), available by the Earth Policy Institute at Rutgers University at: http://www.earth-policy.org/data_center/C24. (accessed on October 26, 2016))

(Sampsell 2014: 52). The "unprecedented" water levels of Lake Nasser in 1996 motivated the Egyptian government to implement the Toshka project, which was first suggested by the geologic and soil surveys of the Desert Reclamation Authority (1963–1971) emphasizing that a half-million acres were "arable for cultivation" in the Toshka depression (Collins 2006: 125–126). In January 1997, Mubarak inaugurated the Toshka project, whereby he declared, "A new era is opening in Egypt—the era when we go out of the confines of the narrow Nile Valley" (Baker 1997). The project comprised the cultivation of 540,000 feddans to be irrigated by surface water and 30,000 feddans by groundwater in the Toshka region (EMWRI 2013). The Mubarak pumping station, the centerpiece of the project, and the largest in the world at the time, was constructed on the western bank of Lake Nasser (about 8 km north of Toshka Bay) with 24 vertical pumps capable of pumping water up to 54 m to cope with the lowest level of Lake Nasser (EMWRI 2013). The Sheikh Zayed Canal (50 km long) was also excavated to carry the water with a capacity of 5.5 BCM/year, channeled into four areas of irrigable lands through four distributing branches (about 170 km long in total) (EMWRI 2013; ESIS 2013a).

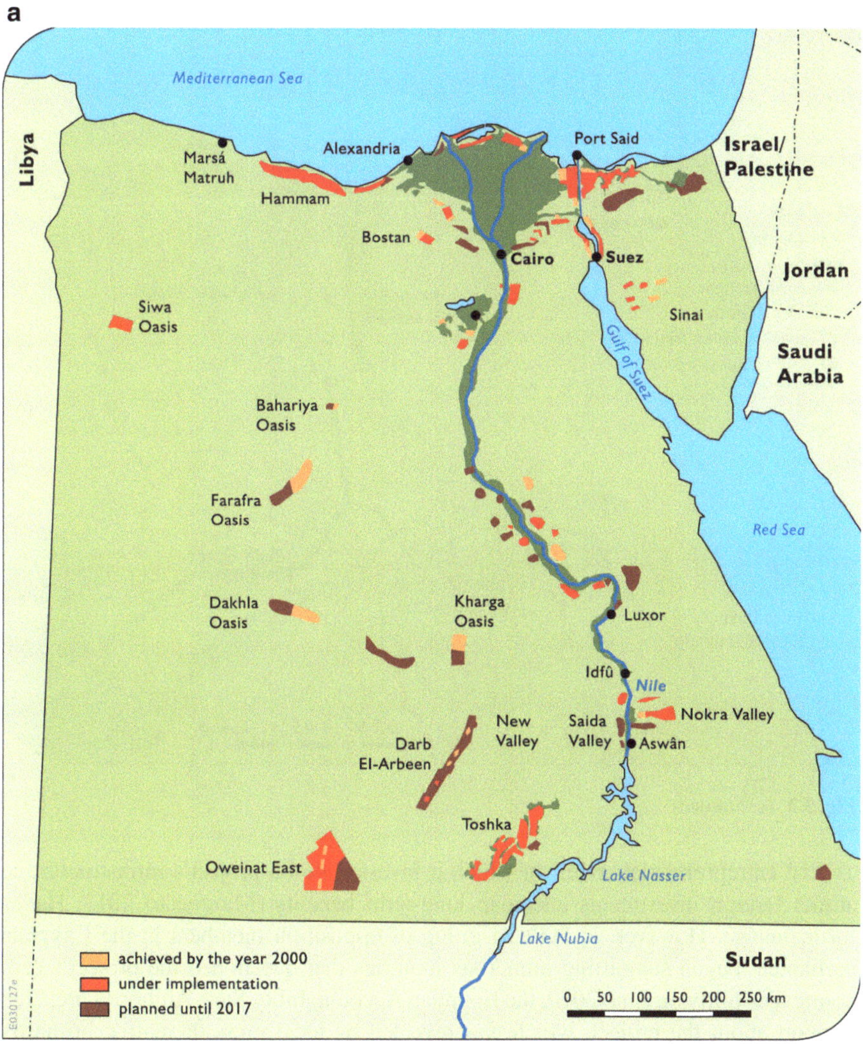

Fig. 5.3 Maps of Egypt's national water projects in general and the Toshka reclamation project in particular. (Source: The Egyptian Ministry of Water Resources and Irrigation, Planning Sector (2005: 4–6, 4–7). Water for the future, National water resources plan 2017, available at: http://extwprlegs1.fao.org/docs/pdf/egy147082.pdf. (accessed on July 11, 2018))

With a budget of more than £E6 billion, the government covered the expenses of the pumping station, the main canal and the distributing branches, and by 2005 almost 90% of the project infrastructure was described to be completed (Water Technology n.d.). The government also signed contracts with Arab and Egyptian investors to reclaim the desert lands with a total area of 343,000 feddans, whereby the investors were responsible for constructing the complementary infrastructure on their lands. The role played by the Egyptian government in the project is referred to in literature

b

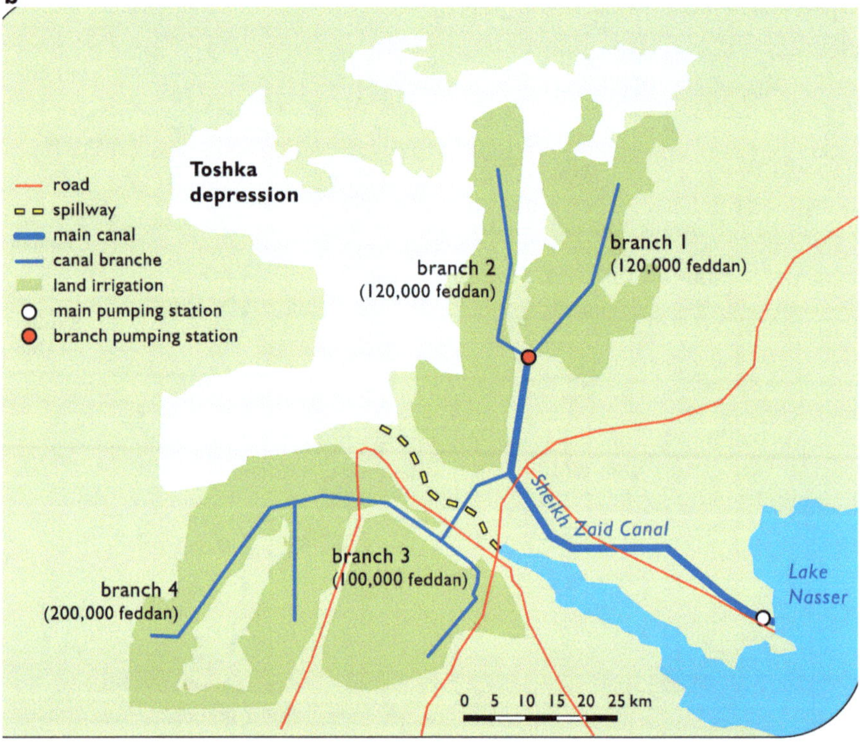

Fig. 5.3 (continued)

as "the entrepreneurial state," in which it invested in the project's infrastructure to attract foreign investments and reap long-term benefits (Mazzucato 2013; Hanna forthcoming). However, in 2006, a group of opposition members in the Egyptian parliament began submitting numerous inquiries that questioned the project's economic feasibility. In June 2010, the Egyptian Accountability State Authority prepared a report about the project, which was revealed by local press thereafter, primarily investigating the reasons behind which Toshka did not meet any of its primary objectives, given that by 2010 only 22,804 feddans were reclaimed, despite the huge public funds invested in the project (Masri 2014; Kholaif 2015). The report's findings shed light on two main factors behind the hindrance of the project. The first was the non-binding contracts that should have obliged investors to reclaim their lands according to a specific time plan, especially for the low price of lands (£E50 per feddan), while the second was the lack of consultation and monitoring by the government.

The Toshka project was not a mere national matter, but was directly linked to the Nile issue, and was subject to criticism by external sides as well as by internal ones, namely Ethiopia. The project renewed the heated debate about downstream projects and their consequences on upstream riparians, as previously triggered during the late 1950s and 1970s when Egypt announced the construction of the AHD and the Peace Canal respectively (Sects. 4.2.3 and 4.3.4). In March 1997, soon after Egypt had inaugurated the project, the then Ethiopian Foreign Minister Seyoum Mesfin sent an official letter of protest, quoted in Waterbury (2002: 84–85), addressed to the Egyptian government with copies to the UN, the World Bank and the OAU, which reads as follows:

> Ethiopia wishes to be on record as having made it unambiguously clear that it will not allow its share to the Nile waters to be affected by a fait accompli such as the Toshka project, regarding which it was neither consulted nor alerted.

In February 2005, Zenawi criticized in strong words Egypt's plan of desert reclamation, declaring to the BBC (Thomson 2005):

> While Egypt is taking the Nile water to transform the Sahara Desert into something green, we in Ethiopia—who are the source of 85% of that water—are denied the possibility of using it to feed ourselves. And we are being forced to beg for food every year.

As interpreted by Waterbury and Whittington (1999), Collins (2006), Cascao (2009), and Warner (2013), Ethiopia opposed the Toshka project because it regarded such mega projects, upon which new communities of millions of inhabitants would be created, as an attempt to foreclose upstream projects. For Egypt, the Toshka project was, and still is, a means of re-managing its annual water share to maximize the benefits generated by it. According to FAO (2016a), during the period between 1993 and 1997, Egypt's total withdrawal from the Nile water reached 58.9 BCM, a figure which increased to 68.3 BCM and 78 BCM respectively during the periods of 1998–2002 and 2008–2012 (Table 5.1). The extra volume of water above Egypt's annual share of 55.5 BCM has been provided through reusing drainage waters and pumping groundwater below the Nile Delta (Nour El-Din 2013: 16). In this respect, as reported in Ahram Weekly in August 1998, what the Egyptian government hoped to achieve through the Toshka project and other desert reclamation schemes was to redirect a portion of its share from domestic use to investments "of the export-oriented and processing varieties" (Wahish 1998). The government had developed its plan according to the conclusion of previous studies indicating that the region is a "virgin land" where fertilizers have never been used, unlike the Nile Valley and Delta, and thus its agricultural production will meet the international environmental standards and can be marketed around the world, providing Egypt with urgently needed hard currency (EMWRI 2013). In addition, in an interview with the German weekly Der Spiegel in December 1997, Mubarak had declared that "land reclamation of the desert must be pursued with the greatest urgency," in order to "provide new living space for millions of Egyptians" (Baker 1997). However, on the ground, this vision was not materialized, which was explained by Professor Rushdi Said (2006), chief geologist and the former Head of the Egyptian Geological Survey Authority, as follows:

> Desert reclamation projects will always fail unless they can cater to the needs of average Egyptians. ... As recently as the late 1990's, Egypt was still clinging to the idea that it could wrest agricultural lands from the desert by supplying them with water from the Nile. The recently-dug canals of El-Hamam along the Mediterranean coast, El-Salam [Peace Canal] in Sinai and Toshka in Nubia are but some examples of this effort. All are facing difficulties and none have attracted a substantial number of people to live in the lands they were

supposed to reclaim. ... Projects to reclaim agricultural lands from the desert were also attempted by the use of ground water that is known to exist under many stretches of the deserts of Egypt. ... The nature of the ground water reservoir ... was proved to be limited and non- replenishing. ... In addition to dwindling water resources, there were the problems of drainage and desalination, as well as the encroachment of sand dunes. ... The desert of Egypt needs an overall plan to be executed over the years to make use of its space for industrial development. The little groundwater that it holds should be preserved for use in this endeavour and not in agriculture. Agriculture should be restricted to the fertile lands of the Nile Valley and the Delta, which should be administered as an environmentally managed domain. All the industrial centers that exist at present in this domain should be moved, over the years, to the newly planned desert industrial centers.

5.3.3 *Ethiopia's Micro-Dams, the Tekeze Dam, and the Tana-Beles Project*

As illustrated in the previous chapter, the droughts of 1972 and 1983–1985 were amongst the major factors, which led to the toppling of Haile Selassie and Mengistu respectively. The post-1991 government, led by Meles Zenawi, was aware of this historic fact, and as Collins (2006: 120) describes, it "was determined to avoid these calamities even if Ethiopia had to carry out its own water plans independent of international assistance or a Nile waters agreement requiring concessions that neither Egypt nor the Sudan were prepared to make."

The new regime was a liberation movement (Sect. 4.3.5) that originated in a "deprived part of the country (Tigre)," the background by which its policy was significantly affected (Mosley 2012: 122). Joseph Stiglitz (2003: 28), the Nobel-holder World Bank senior vice-president and Chief Economist from 1997 to 2000, described Zenawi as "a man who had led a seventeen-year guerrilla war against the bloody Marxist regime of Mengistu Haile Mariam. ... A doctor by training, Meles had formally studied economics because he knew that to bring his country out of centuries of poverty would require nothing less than economic transformation." Stiglitz (2003: 28) further illustrated that Ethiopia sought to attain this goal through formulating "a rural development strategy, focusing its attention on the poor, and especially the 85 percent of the population living in the rural sector." He added, "It had dramatically cut back on military expenditures—remarkable for a government which had come to power through military means—because it knew that funds spent on weapons were funds that could not be spent on fighting poverty" (2003: 28).

Since 1992, Ethiopia has adopted economic reform and liberalization programs to meet its urgent energy and food needs with the support of various international organizations (FDRE-GoE n.d.; IMF 2004: 20). Water resources have occupied a central position in attaining this goal, not only in developing agriculture, but also in generating electricity, given that the major source of energy production in Ethiopia is hydropower (Fig. 5.4). The Water Sector Policy (WSP) adopted by the government aimed to "enhance and promote all national efforts towards the efficient,

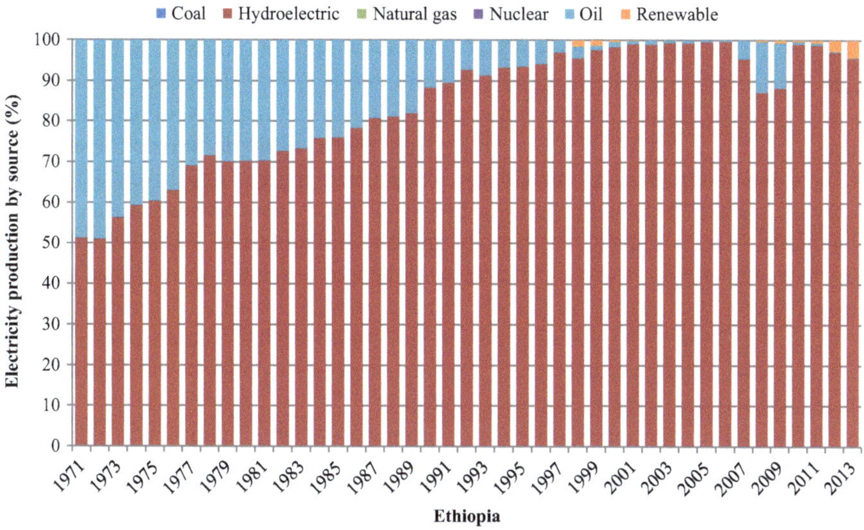

Fig. 5.4 Electricity production by source in Ethiopia. (Source: International Energy Agency Statistics (2014), available by World Development Indicators at http://data.worldbank.org/data-catalog/world-development-indicators. (accessed on November 2, 2016))

equitable and optimum utilization of the available Water Resources of Ethiopia for significant socioeconomic development on sustainable basis" (FDRE-MOWR 2001: 5). At the time, Ethiopia had two main options in utilizing its water resources; building micro-dams or large dams, each of which has its benefits and costs, as covered by Waterbury and Whittington (1999: 152–156) and Waterbury (2002: 100, 111, 126, 127). Micro-dams are typically constructed using local labor and materials, and do not need foreign technical assistance or financial aid. In addition, since they are often built with local initiatives and given that their downstream consequences are not as obvious as large dams, their political costs can be affordable to a certain extent. However, micro-dams have several problems, such as the rapid loss of water storage capacity due to the high rates of siltation, their low resilience to high floods and rainfalls, the spread of Malaria, and energy consumption for water pumping. As for large dams, the balance of benefits and costs is more complex. Although their benefits include water regulation, agricultural expansion, hydropower generation, and flood control, they entail massive negative impacts on local people and ecosystems, such as in the case of the AHD illustrated in Sect. 4.2.2. Moreover, the high rates of sedimentation and high variability of flows significantly shorten the economic viability of such large-scale projects. Furthermore, the alteration of the water flow lower the river, either during filling the reservoirs or due to the expansion of irrigation schemes, would inevitably provoke conflicts with downstream riparians. Despite all this, large dams remain in general more attractive as they directly benefit several types of actors, including host governments, international donors and private contractors, and as Waterbury (2002: 111) states, "none of these interests wants a project to be blocked on cost-benefit grounds."

Actually, Ethiopia proceeded with the two options of constructing micro-dams and large dams. The government had extensively built micro-dams in the Amhara and Tigray regional states around 1999 and 2000 (FAO 2016b), but according to official records, those were limitedly implemented in the Abay River Basin (Johnston and McCartney 2010: 14). Concerning large dams, Ethiopia initiated a series of hydropower dams, such as the Gilgel Gibe projects on the Omo River, which discharges in Lake Turkana located within Kenya. These dams have negative environmental and social impacts on the lake and the Turkana tribe, leading to potential disputes with Kenya over the project. Nevertheless, Since Kenya itself had built its own projects that already negatively affected the lake, besides its acute need for energy, the Gilgel Gibe dams have been less controversial (Salman 2011: 21–22). On the Nile, Ethiopia constructed two important large projects, the Tekeze Dam (Fig. 5.5) and the Beles hydropower plant (Fig. 4.5).

In the fall of 1997, the Ethiopian government declared that it tends to build its first mega dam on the Tekeze-Atbara tributary in the Tigray state (Waterbury and Whittington 1999: 156). Beginning with the choice of the Tekeze Dam can be interpreted from different points of view. As mentioned above, the Tigray state is the origin of the new regime in Ethiopia, and at the same time, it is a drought-prone region. In addition, the dam would not cause significant harm to Egypt in the same degree as if constructed on the Blue Nile, but it would even benefit Sudan as shown below. Interestingly, the Tekeze River, unlike the other Nile tributaries that originate from Ethiopia, was not included in the 1902 Agreement between Britain and Ethiopia (Waterbury 2002: 62–63). However, construction did not begin immediately, as Ethiopia was involved shortly thereafter in an armed conflict with Eritrea, sparked in May 1998 over a border issue. Yet, in May 2000, Ethiopia launched a successful offensive against Eritrea and was able to terminate the war (Gilkes 2003: 388).

In June 2002, Ethiopia's Electric and Power Corporation (EEPCo) agreed with a joint venture of Chinese contractors, led by the state-owned Chinese National Water Resources and Hydropower Engineering Corporation (Sinohydro), to begin the construction of the Tekeze Dam with an estimated cost of $224 million; out of which the Export-Import Bank of China committed $50 million (Foster et al. 2009: 27, 86). In general, Ethiopia as well as other governments in Africa and around the world favor Chinese dam builders amongst others for many reasons, mainly because they do not require prior approval of downstream riparians, providing "more affordable contracts" and "speedier execution of projects," without imposing "environmental, human rights or good governance conditionalities" (Cascao 2009: 262). The director of the international department of Sinohydro said, "Tekeze Dam is for Ethiopia what Three Gorges is for China," because it was the tallest arch dam in Africa at the time of construction (185 m) (International Rivers 2016a). Due to "trouble with land slides" that "destabilized the dam and delayed construction," the Tekeze Dam was an "expensive headache for the Chinese construction company that engineered and built it" (Awulachew et al. 2012: 20). In November 2009, Zenawi officially inaugurated the dam, which ultimately cost $360 million, declaring, "The new power plant will contribute a major role to the ongoing efforts to speed up development in the country" (Sudan Tribune 2009). By its operation, the

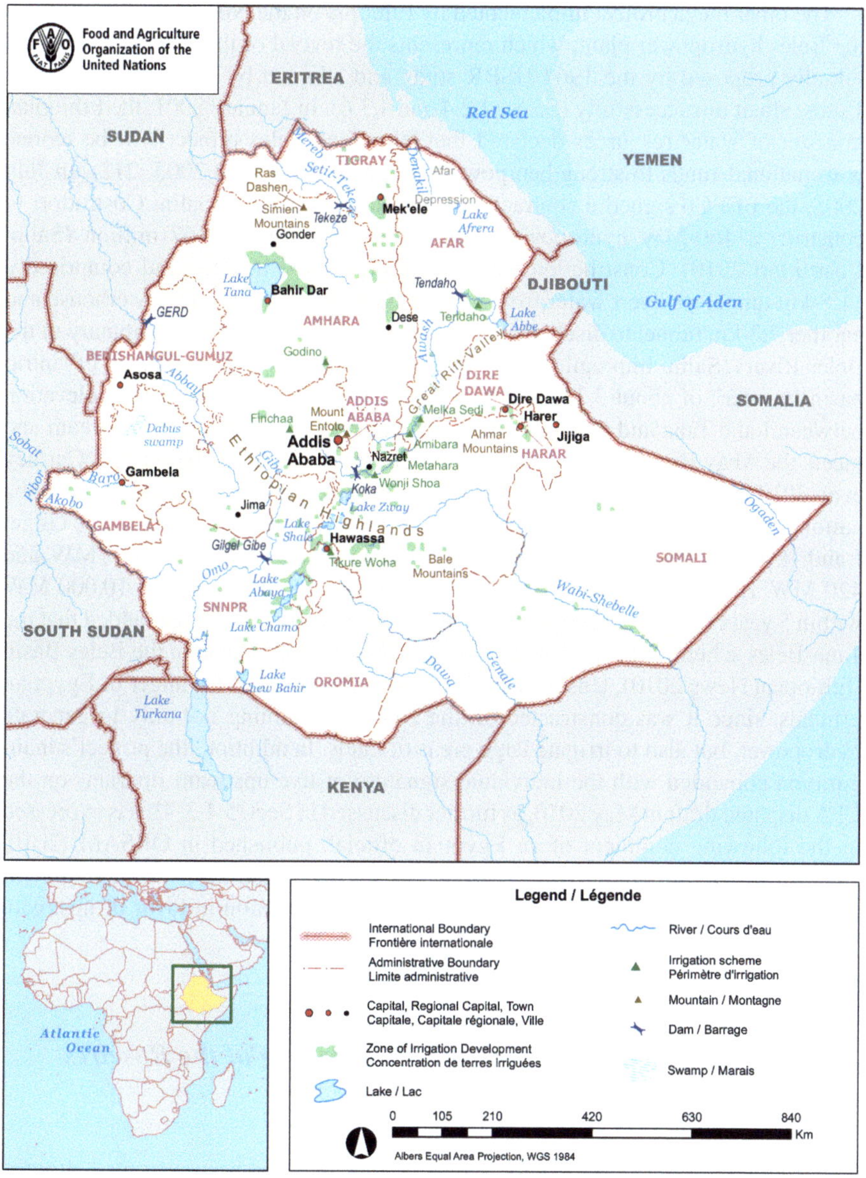

Fig. 5.5 Map of dams in Ethiopia. (Source: FAO Aquastat (2015), Maps, available at: http://www.fao.org/nr/water/aquastat/countries_regions/ETH/ETH-map_detailed.pdf)

Tekeze Dam added 300 MW to Ethiopia's national grid, and it regulated the waters flowing to lands downstream, but at the same time, disrupting ecosystems and high rates of siltation remained the project's serious problems (Awulachew et al. 2012: 20). The benefits of the dam extended to Sudan, since it controlled the large mass of silt running annually lower the river, which in the meantime caused the loss of 60% of the Khashm al-Girba storage capacity (Sect. 4.2.2).

The other mega project implemented by Ethiopia on the Nile during the 2000s is the Beles hydropower plant, which represents the revival of the Tana-Beles project initially proposed by the 1964 USBR study and adopted by Mengistu in the late 1980s, albeit unsuccessfully (Sects. 4.2.4 and 4.3.6). In January 2001, the Ethiopian minister of water resources declared that "the Tana-Beles project will be redone with national funds to strengthen power component" (Degefu 2003: 212). In July 2005, the EEPCo signed a contract with the Italian company Salini Costruttori to construct a 460-MW hydropower station with the cost of €467 million (Salini Costruttori 2010). Construction works were completed in 2010 and comprised a 11.8-km tunnel to divert water from Lake Tana to an underground powerhouse and another 7.2-km tunnel to discharge the water into the Jehana River (a tributary of the Beles River) (Salini Impregilo 2016). Hydropower is generated as a result of "intra-basin transfer" of about 3 BCM, making use of the 311-m difference in elevation between Lake Tana and the Beles River, the latter of which flows downstream and meets the Abay River before it crosses the Ethiopian-Sudanese border (McCartney et al. 2010: 5). On May 14, 2010, Zenawi inaugurated the project, indicating that the national grid capacity had reached 2000 MW, after the operation of the Gibe Gilgel I and II projects in 2004 and 2010 respectively with a capacity of 184 MW and 420 MW. He also announced that the total capacity would expand to 10,000 MW within 5 years (Ethiopian News 2010). The EEPCo's project manager added that the Tana-Beles scheme would allow for the irrigation of 140,000 ha in the Beles Basin (Ethiopian News 2010). Unsurprisingly, the project provoked the anger of Egyptian officials, since it was constructed on the Blue Nile, aiming not only to generate hydropower, but also to irrigate large areas of lands. In addition, the project's inauguration coincided with the individual signature of five upstream riparians on the CFA disputed draft in May 2010, as further discussed in Sect. 5.4.3. This is expressed in the following comment of an Egyptian official, published in ORSAM (2010: 104), "Tana Beles dam aims to provoke Egypt's anger and lead it to taking swift diplomatic behavior which would turn global public opinion in favor of upstream Nile countries."

5.3.4 Sudan's Merowe Dam, the Heightening of the Roseires Dam, and the Atbara-Setit Dam

As mentioned earlier in Sect. 5.3.1, the PJTC meetings had been disrupted after the assassination attempt against Mubarak and were resumed 2 years later when the relations between Egypt and Sudan began to improve. In late October 1998, the PJTC discussed the potential dam projects that each country planned to construct (Ofcansky 2003: 1019). In the author's interview with M.M.A., a Sudanese top water official, he cited a memorandum signed by Egypt and Sudan in 1998, in which both countries stated that they had exchanged relevant studies and plans of their potential projects, including the Toshka and Peace Canal projects in Egypt and the

Merowe, Shereik and Kajbar Dams in Sudan (Fig. 5.6). He added that the 1998 Memorandum stipulated clearly that the PJTC endorses the implementation of these schemes for the interest of the peoples of the two countries based on their agreed water shares (M.M.A. 2016c).

Decades earlier, the Century Storage Scheme of Hurst et al. (1946) had included a dam at the fourth cataract in Sudan (Sect. 3.3.4), 40 km upper Merowe city and 325 km north of Khartoum (Fig. 3.8), with the purpose of controlling floods and regulating flows to Egypt. Given Egypt's decision to build the AHD, the Merowe Dam was no longer needed by Egypt, while Sudan prioritized the construction of the Roseires and Khashm al-Girba Dams (Sect. 4.2.2). From the late 1970s to the early 1990s, three prefeasibility studies were conducted by international consultancy firms for the Merowe dam; however this time for generating hydropower (DIU 2016a). Due to the lack of both financial resources and political stability during the 1980s, Sudan was not able to implement the project. In 1992, the Bashir regime launched a 10-year Comprehensive National Strategy (CNS) for boosting food and energy security, which included the Merowe Dam (Taha 2010: 207–208).

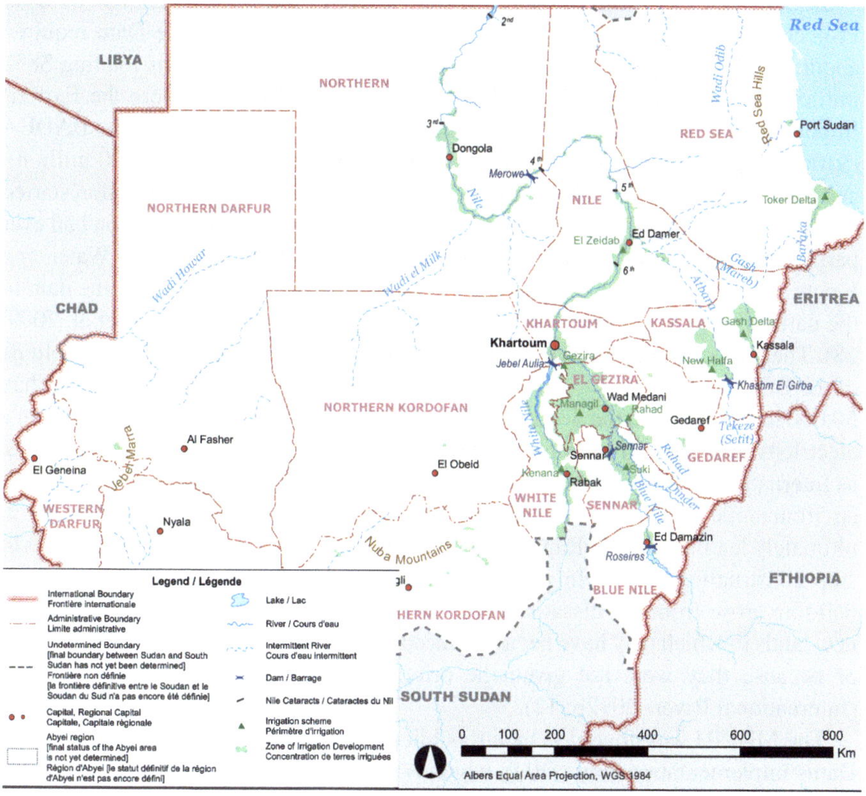

Fig. 5.6 Map of dams in Sudan. (Source: FAO Aquastat (2015), Maps, available at: http://www.fao.org/nr/water/aquastat/countries_regions/SDN/SDN-map_detailed.pdf)

Yet, carrying out the project was not possible until Sudan's economic boom driven by oil exports that only began in late 1999. Since oil reserves were located in Southern Sudan, explorations had been halted for a long time due to civil war. In addition to securing the oil fields, a major challenge was the construction of a pipeline connecting the site to Port Sudan on the Red Sea. Supported by China, the pipeline was completed in 1999, and Sudan began exporting crude oil in September of that year (Petterson 2003: 230). In the past, Sudan's exports of fuel (as percentage of merchandise exports) had ranged respectively from 0.73% to 3.47% during the peace period from 1972 to 1982, and then were completely stopped since the second civil war began in 1983 until 1999 (WDI 2016). This figure jumped to 69.33% in 2000 and reached 94.41% in 2010 (WDI 2016). Foreign direct investment (net inflows in current US$) also increased from $0.4 million in 1996 to $370.8 million in 1999 and reached $2064 million in 2010 (WDI 2016). The share of China in total lending to Sudan raised from 17% in 1999 to 73% in 2007 (Sidahmed 2013: 110). To pay back its accumulated loans estimated at about $5 billion, Sudan reached a deal with China to ship 1.5 million barrels monthly for 15 years after 5 years of grace period (Sidahmed 2013: 110).

In October 2001, Bashir issued a republican decree to establish the Merowe Dam Project Implementation Unit (MDPIU) (DIU 2016b). The Merowe Dam required about $1.2 billion to be constructed. Sudan was able to obtain loans totaling $850 million from China and the Gulf States, whereby the financers were the Export-Import Bank of China, ($400 million), the Saudi Fund ($150 million), the BADEA ($100 million), the Kuwait Fund for Arab Economic Development ($100 million), and the Abu Dhabi Fund ($100 million) (Foster et al. 2009: 28). Construction started in 2004, whereas the dam was "the largest international project that China had ever participated in," as the plant was built by Sinohydro and the China Water and Electric Corp (CWE), while the 1776-km transmission lines connecting the dam to the national grid were constructed by other Chinese companies (Foster et al. 2009: 28). The project was inaugurated in March 2009, creating a reservoir with a volume of 12 BCM and length of 174 km, which caused the displacement of more than 50,000 inhabitants and the disruption of local ecosystems. The dam doubled Sudan's electricity generation by adding 1250 MW to the grid. Nevertheless, several NGOs, as International Rivers (2012b: 12), emphasize that the project has severe social and environmental problems that previously led traditional donors to refuse financing it, ultimately leading the UN Rapporteur on Housing Rights to ask the dam builders to stop construction works. In particular, social problems associated with the dam building arose from the dissatisfaction of the displaced people, either because the new lands to which they have been displaced are not as fertile as their original lands, or because they were not given the priority to resettle around the dam's lake (International Rivers 2012b: 12).

The MDPIU was upgraded by a republican decree in September 2005 to be the Dams Implementation Unit (DIU), which is an independent body headed by Bashir himself with the membership of relevant high-level officials. The DIU's mission was to "develop and utilize water resources across Sudan according to international measures and standards" and "to develop irrigation systems to increase agricultural lands" (DIU 2016c). While the Merowe Dam was at its final stage of construction, the DIU signed a new contract in April 2008 with the same Chinese companies, Sinohydro and CWE, to enlarge the Roseires Dam by 10 m in height and 25 km in length (Sudan Tribune 2008). The project cost $396 million, which was covered by funds from the Gulf States in addition to the Sudanese government (DIU 2008). Heightening the Roseires Dam was completed in January 2013 to increase the capacity of its reservoir by around 4 BCM and its hydropower generation from 1200 to 1800 MW, in addition to doubling the generation capacity of Sennar Dam by 100% (DIU 2008). However, it also required the relocation of 22,000 inhabitants (Laessing 2013). In May 2010, the DIU signed a third contract with Chinese companies, including CWE, to build the Upper Atbara and Setit Dam to generate 135 MW, irrigate 500,000 feddans, and reduce the siltation of the Khashm al-Girba reservoir (DIU 2010). By launching the construction of the third project after the Merowe Dam and the Roseires Dam heightening, the DIU's Executive Director, Minister Osama Abdalla, declared that "three of the dreams of the people of Sudan became reality," and he thanked China for its great role as the "partner of the development in Sudan" (DIU 2010). The next projects in line are the Dal (634 MW), Kajbar (320 MW) and Shereik (420 MW) Dams at the second, third and fourth cataracts respectively (DIU 2011). Yet, these projects have been highly controversial at the domestic level in light of the demonstrations of the Nubian people who reject to be displaced and call for securing their cultural and historic sites that will be submerged under the dams' reservoirs (International Rivers 2016b). The development of water infrastructural projects in Sudan was associated with restructuring the administration of large-scale irrigation schemes, the first of which was the Gezira-Managil scheme (FAO 2016c).

Overall, the ambitious steps that Sudan has taken during the 2000s to build new schemes and to run institutional reforms have substantially boosted its capability towards utilizing its 18.5-BCM share under the 1959 Agreement. Until the 2000s, the unused portion of Sudan's share ranged from 4 to 6 BCM according to different accounts (Waterbury 2002: 132; Cascao 2009: 246, 259; Blackmore and Whittington 2008).

5.4 The Collapse of Basin-Wide Cooperation on the Nile

5.4.1 The Eastern Nile Joint Multipurpose Program (JMP)

The unilateral projects, discussed above, ran in parallel with serious attempts to implement multilateral schemes and create a basin-wide framework through the NBI and the CFA respectively, both of which, as will be shown in this section, were ultimately deadlocked.

In January 2001, the Eastern Nile Council of Ministers of Water Affairs (ENCOM) adopted the "2020 Operational Vision of the Eastern Nile," expressing the vision of the three riparians towards "strong economies which will be working together and will be globally competitive," through "economic integration" based on "joint physical, social and economic infrastructure" (ENTRO 2012: 6). To translate this vision on the ground, the ENCOM launched in February 2005 the first phase of the Joint Multipurpose Program (JMP), which includes the selection of "a coordinated set of investments" to provide "a range of transformational development benefits across sectors and countries" (NBI 2012b). The ENCOM initiated a series of assessments and consultations, amongst which was the independent study commissioned by the World Bank and conducted by the international experts Don Blackmore and Dale Whittington in the purpose of "scoping out the 'pros and cons' of different investment directions" and identifying "a first set of JMP investments" (Blackmore and Whittington 2008: 1, 2).

In their investigation, the team was searching for the projects that meet three specific criteria: "provide benefits to all, ... yield various types of benefits that can be shared by all, ... and cause no regrets" (2008: vii). The only sub-basin that met these criteria was the Blue Nile, because investments there would benefit Ethiopia by generating hydropower and expanding agriculture, Sudan by controlling floods and sediments, providing cheap electricity, introducing year-round cultivation, and improving navigation, Egypt by providing cheap electricity, and the whole basin by increasing resilience against droughts (2008: ix, 70). Conversely, investments in the Main Nile, Baro-Akobo-Sobat and Tekeze-Atbara sub-basins would be more beneficial for local settlements than for all three riparians. In particular, conservancy projects in the Baro-Akobo-Sobat sub-basin, which aim to increase flows downstream by reducing water losses at the Machar Marshes, will not be economically, socially and environmentally feasible, since "the vast majority of the water in these wetlands comes from rainfall, not over-bank spills" (2008: 41). In addition, the hydropower potential in the Baro-Akobo-Sobat and Atbara-Tekeze sub-basins is evidently limited compared to the Blue Nile, due to the latter's "high" runoff, "good" storage sites, and 1300-m difference in elevation levels (2008: 13, 48). To explore the "risk-reward tradeoffs" of the Blue Nile investments, the study used computer simulation to compare the scenarios of the "meanwhile conditions," the "evolving conditions" under ongoing unilateral projects, the development of joint water infrastructure on the Blue Nile, and reduction of flows due to climate change (2008: 54).

The study illustrated that by the operation of the ongoing unilateral projects (Toshka, Merowe, Tana-Beles, and Tekeze schemes), the meanwhile conditions would no longer persist, because Sudan's withdrawals would increase from 13.8 to 18 BCM and Ethiopia from 0.3 to 2.3 BCM, leading the AHD to operate at a lower level of water storage (2008: 30, 54). In addition, it was found that "serious disruption to Egypt's agricultural economy" would take place if Ethiopia's withdrawals reached 8.3 BCM or if annual inflows were reduced by 8%, due to climate change, as measured in Aswan, which was described as "unacceptable risk" because there would be large water and energy deficits in Egypt for several successive years (2008: 65, 67). The study also found that the limit to Ethiopia's withdrawals for manageable water deficits in Egypt would be 5.3 BCM. In this respect, the study argued that the target withdrawals can be covered for three main reasons: saving evaporation losses through building hydropower dams on the Blue Nile in Ethiopia (Fig. A.3); considering that the Nile average runoffs per year historically exceeded the 84 BCM stated in the 1959 Agreement; and improving water use efficiency in irrigation and crop mix decisions (2008: 57).

Based on the above results, the study recommended, "the anchor investment of this JMP initiative could be a large multipurpose and hydropower project on the Blue Nile in Ethiopia" (2008: 70). However, the study emphasized that "if such a Blue Nile reservoir were completed during a period of low flows, filling would need to be delayed so as not to affect adversely downstream riparians," and that "water stored in the reservoir could be used to supplement water supplies downstream during periods of drought," all of which "need to be negotiated before construction began" (2008: 57). In addition, the study indicated that there are several long-term investment potentials that should be further investigated. These include benefiting optimally from the vast water losses from surface water bodies, land surface and vegetation in the Ethiopia (around 500 BCM), reducing soil erosion on the Ethiopian Highlands through forestation, taking into account consequent evapotranspiration savings, and using irrigation water more efficiently in Egypt and Sudan (2008: 13, 69, 71).

Although it is a purely technical study, the political messages sent by Blackmore and Whittington (2008) are crystal clear, which aim to answer the following two questions. First, how does the JMP initiative respect both the principle of equitable use (favored by Ethiopia) and the principle of no significant harm (favored by Egypt)? Second, why should each of the three riparians choose to cooperate in implementing this JMP initiative?

Concerning the first question, the study recognized the rights of all three riparians to utilize the Nile waters equitably for both consumptive and non-consumptive uses (e.g. Ethiopia has the right to expand its irrigated agriculture as well as generating hydropower). At the same time, the study emphasized that such equitable uses should be reasonable and should cause no significant harm (e.g. Ethiopia's expansion in irrigated agriculture should not exceed the 5.3 BCM/year, beyond which the risks on Egypt's existing uses would be "unacceptable"). In addition, the study prioritized socio-economic and environmental sustainability in the basin (e.g. safeguarding Egypt's agricultural economy, creating clean energy, and protecting the Blue Nile watersheds and Baro-Akobo-Sobat wetlands).

With respect to the second question, the study addressed each of the riparians' needs for cooperation. As for Ethiopia, the message was that it is in great need for cooperation to be capable of exploiting its "natural advantages" since it has "little capital, a limited national market for its potential hydropower, and limited land for economically attractive irrigation schemes" (2008: 13). For Sudan, there are numerous benefits that would motivate it to endorse large multipurpose storage dam on the Blue Nile in Ethiopia, mainly for power trade, flood control, sediment management, enhanced performance of its hydropower dams and water delivery systems in its large-scale irrigation schemes (e.g. the Gezira), and improved navigation. The unmistakable message is the one addressed to Egypt: that, from the first glance, tangible benefits of cooperation may be relatively limited (mainly power trade), while direct costs are obvious (water and energy deficits, even if "small" and "manageable"), but actually cooperation remains more fruitful for two main reasons. First, the potential "large unmanageable" deficits in Egypt's withdrawals in case of expanding irrigation upstream beyond acceptable limits illustrate how "continued unilateral development is dangerous and regional cooperation is needed" (2008: 63). Second, intangible benefits of cooperation are considerable, not only to avoid high risks of upstream unilateral projects, but also to launch, together with Ethiopia and Sudan, a "cycle of increasing economic integration, specialization and trade, leading to accelerated economic growth, poverty reduction and reduced military expenditures" (2008: 18). Furthermore, cooperation is needed to mitigate the negative impacts that climate change could cause, which would be no less harmful than those caused by unilateral upstream projects.

The study was communicated to the three riparian states, hoping that it would lead to the first-ever implementation of joint project with such a scale. However, observers indicate that the relevant Egyptian authorities did not welcome it, as they "contested the findings and challenged the technical validity of the study conducted by the independent experts, following which their involvement in the JMP started to wane" (Cascao and Nicol 2016: 555). Obviously, Egypt's fears of constructing large upstream dams cannot be separated from the historical context of the Nile, which is explicitly expressed in the following lines written by the then Foreign Minister Ahmed Aboul-Gheit (2014: 254), as translated from Arabic by the author:

> I had always refused these ideas. Egypt must never be at the mercy of anyone, but waters should be stored within its borders. I never forgot that those American ideas were produced when Egypt was working with the Soviet Union to build the High Dam.

5.4.2 *The UN Watercourses Convention and the Ground of the CFA Negotiations*

Before discussing how the negotiations over the CFA came to a deadlock, it is important to illustrate first how the Nile riparians view the basic principles of international water law, which has sparked off a global debate on water rights. As noted in Sect. 2.1.2, downstream and upstream riparians in general used to advocate their "extreme" water rights of absolute territorial integrity and absolute sovereignty respectively. Through the 1966 Helsinki Rules, each side has shifted to an intermediate position; downstream riparians to the notion of no significant harm and upstream riparians to equitable and reasonable utilization (Wolf 1999). Yet, it was not easy at all to develop a convention that balances between the two notions in order to be acceptable to both sides, given that it took 23 years (1971–1994) for the UN International Law Committee (ILC) through five rapporteurs and 15 reports to submit the final draft of the UN Watercourses Convention[3] (UNWC). During this

[3]Available at: http://legal.un.org/ilc/texts/instruments/english/conventions/8_3_1997.pdf (accessed on November 19, 2016).

period, each rapporteur had a different position regarding those two notions, whether to equate them or to subordinate one to the other (Salman 2007: 3, 5). On May 21, 1997, the UN General Assembly adopted the UNWC, whereby it took the convention another 17 years to be ratified by the minimum required number of signatories (36) and subsequently entered into force in 2014 (UN 2014).

The initial voting result on the 1997 UNWC was 103 votes in favor and three against, with the abstention of 27 countries and absence of 52 (IWLP 1997). It is noteworthy that the three countries that voted against the convention were upstream riparians (Burundi, China and Turkey), and that both Egypt and Ethiopia abstained, while Sudan voted in favor of it (IWLP 1997). The fourth rapporteur, Steve McCaffrey, recorded that Ethiopia's delegation had two main reasons to abstain. First, the delegation saw that articles 3 and 4 in part I should have "required" the "harmonization" of "existing agreements" with the "basic principles" of the convention, instead of merely "calling upon to consider" doing so (McCaffrey 1998: 18). Second, The Ethiopian, Turkish and Rwandan delegations rejected the principle of "prior notification" adopted in articles 11–19 in part III, since it represents a "veto" power in the hands of the notified state (McCaffrey 1998: 23). With respect to the first point, McCaffrey (1998: 18) comments that "given the vast number and variety of existing agreements, such a requirement would have been impractical," however "this does not mean that the principles reflected in the convention will be without significance in the interpretation of existing agreements." As for the second point, McCaffrey (1998: 23) indicates that the "obligation to provide prior notification ... was accepted as a part of the convention by most delegations," which "provides further evidence that the international community as a whole emphatically rejects the notion that a state has unfettered discretion to do as it alone wishes with the portion of an international watercourse within its territory."

On the other hand, as a downstream riparian, Egypt has been anxious about the "right of equitable and reasonable utilization." McCaffrey (1998: 19–20) responds to such fears, clarifying that the right of equitable uses is constrained with a group of obligations stated in parts II and III in order to assure that the shared watercourse is used in a "reasonable" as well as "equitable" manner. Based on article 5, equitable use is restricted by considering the "interests of the watercourse states concerned" and "adequate protection of the watercourse," in addition to the "duty to cooperate in the protection and development thereof." Article 6 lists the factors that should be taken into account to attain equitable and reasonable utilization. These comprise both the natural and demographic characteristics of the basin, the "existing" as well as "potential uses," and the "availability of alternatives, of comparable value, to a particular planned or existing use." Article 7 states the "obligation not to cause significant harm" and defines four gradual steps for taking "all appropriate measures" in order to "prevent," "eliminate," or "mitigate" such harm, and eventually, where appropriate, to "discuss the question of compensation." Article 8 urges riparians to establish "joint mechanisms or commissions" to "facilitate" cooperation, while article 9 stresses that riparians "shall on a regular basis exchange readily available data and information on the condition of the watercourse." Finally, article 10 discusses the "event of a conflict between uses," emphasizing that this "shall be resolved with reference to articles 5 to 7, with special regard being given to the requirements of vital human needs." Moreover, article 14 in part III clearly states that during a period of 6 months following one state's notification to another about its planned measures, the "notifying state shall cooperate with the notified states by providing them, on request, with any additional data and information that is available and necessary for an accurate evaluation (a), and shall not implement or permit the implementation of the planned measures without the consent of the notified states (b)."

Meanwhile, the Nile riparians were working together on the D3 Project, one of the major outputs of the third ministerial TECCONILE meeting in February 1995, to develop the Cooperative Framework Agreement (CFA) (Sect. 5.2.2). With the financial support of the UNDP, the Nile-COM formed in Autumn 1996 a panel of experts (PoE), comprising three technical and legal experts from each country, to formulate a first CFA draft (Metawie 2004: 62). Initiated in 1997, the PoE took 3 years to submit its draft, and in August 2000, it was presented during the Nile-COM meeting in Khartoum (Metawie 2004: 62). Osman el-Tom, the vice-chairman of Sudan's water resources authority, reported at the conclusion of the meeting "remarkable convergence toward future cooperation." At the same time, he noted that while the PoE had reached consensus on many of the CFA parts, some disagreements remained "mainly on the principle of prior notification of planned measures and the state of the existing agreements under the new cooperative framework" (Brunnee and Toope 2002: 139).

In fact, both prior notification and the existing agreements reflect contentious issues that have persisted in the Nile Basin for a long time. For Egypt and Sudan, the 1902 Agreement (with Ethiopia) and the 1929 Agreement (with Kenya, Tanzania and Uganda) assure their right of prior notification, whereby no projects can be undertaken upper the river without their consent as downstream riparians (Sects. 3.2.3 and 3.3.2). From Egypt's and Sudan's point of view, the 1959 Agreement (Sect. 4.2.2) assures their "equitable" shares of the Nile waters, based on factors (a), (c), (e) and (g) of article 6 in the UNWC; given that they are almost hot arid countries (a), their populations are heavily dependent on the Nile (c), the agreement secures their existing uses (e), and they have no alternative freshwater resources with a comparable value to the Nile, while other riparians enjoy high availability of rainfalls and groundwater (g).

The point is seen from a completely different angle by Ethiopia and the Equatorial states. Ethiopia has repeatedly declared in earlier occasions that the 1902 Agreement is a colonial treaty and that it was not part of either the 1929 or 1959 Agreements, and thus it recognizes neither of them (Sects. 4.2.3 and 4.3.3). Tanzania has adhered to the Nyerere Doctrine since its independence, through which it considers that colonial treaties, including the 1929 Agreement, should be reviewed and renegotiated by independent states, while Kenya has officially adopted the same doctrine and Uganda has informally endorsed it (Sect. 4.2.6). In addition, Burundi, DR Congo and Rwanda were not parties to the 1929 Agreement, and they regard the colonial treaties between Britain and Belgium or Germany as no longer binding. Concerning the factors of equitable and reasonable utilization in the UNWC, upstream riparians, particularly Ethiopia, believe that other factors such as geographic and hydrological characteristics (a), the social and economic needs of riparians (b), and the potential uses of the watercourse (e) assure their rights to share the Nile waters. Even with respect to factor (g), the Ethiopian officials interviewed by the author clarified that since rainfalls are unevenly distributed in space and time across the country, it is necessary to "supplement" rainfed with irrigated agriculture to attain food security (M.M. 2016b; T.B. 2016).

Furthermore, other differences also existed, specifically over the CFA amendment (whether the articles could be modified by consensus or by majority) and the water resources to be shared equitably by riparians (whether to be limited to the 84-BCM/year surface waters or to consider the whole basin resources including the rainfalls) (Salman 2013: 23–24; Ahram 2015c). These two points are no less thorny than the previous ones. Concerning the procedures of amendment and dispute settlement, the downstream riparians emphasize that a simple two-thirds majority will undermine their voice, while consensus is viewed by the upstream riparians as a veto power in the hands of Egypt and Sudan. With regard to the water resources to be considered, literature reflects an intense debate over this issue. Phillips et al. (2008: 19, 23) showed how green water is not clearly addressed in transboundary water agreements or international water law, despite the fact that green water represents the bulk of water resources in a basin and supplies most of the world's food production through rainfed agriculture. Explaining the main reason of this issue, they quoted the statement of Dent and Kaufmann (2006), which states that "green water is ignored by engineers because they cannot pipe it or pump it, by economists because they cannot price it, and by governments because they cannot tax it." Taylor (2009) called for counting soil water in any basin water balance, while Zeitoun et al. (2010) included both green and blue water in the analysis of "virtual water flows," studying the water trade within the basin and internationally. Nevertheless, van der Zaag et al. (2002) indicated that including both the water stored in the soil (green water) and the water running in streams (blue water) may help to find the most equitable water allocations. However, they emphasized that the economic value of blue water is higher than that of green water in terms of uses and profitability, and thus there should be a relative weight assigned to each (van der Zaag et al. 2002).

The author discussed this point specifically with Professor Owen McIntyre, who is an expert on international water law. As he clarified, the UNWC refers to "accessible surface water and groundwater resources" in internationally shared watercourses. Yet, as he demonstrated, "in terms of equitable and reasonable utilization only three things matter: dependence, dependence and dependence," whereby a riparian state that is rich in green water may have less needs from the contested watercourse, and "less needs" would mean "less right." He stressed the fact that "needs" are identified based on "today's" and "tomorrow's" conditions not those of "yesterday," and this is why article 6 in the UNWC takes into consideration the population dependent on the watercourse, socio-economic needs, existing and potential uses including how efficient they are, in addition to the availability of alternative resources. Given the acute asymmetries between the Nile riparians concerning the usage of green and blue waters (Figs. 5.7 and A.5), McIntyre explained that there is no way to implement the principle of equitable and reasonable utilization but through "intense interstate engagement," since the factors defining "needs" are "subjective" not "objective" ones and thus may have different interpretations by each riparian. However, he emphasized that "from any stance, nobody would suggest for a moment that Egypt is not always going to need a huge proportion of the available Nile waters," based on "social, economic and cultural" dependence of its population, but at the same time, this should not disregard the needs of other states for development, particularly for hydropower generation.

Interestingly, this opinion almost matches what Zenawi said in his interview with the Egyptian state TV[4] on July 7, 2010, as follows:

> Ethiopians have to understand that Egypt is the creation of the Nile, ... Egypt must benefit from the Nile. This is a necessary condition. At the same time, Egyptians have to understand that Ethiopia is a country of 80 million, which has had hunger for almost 40 years now. So, yes, there is rain in some parts of the country, but there is no rain in the other parts ... Egyptians also have to understand that Ethiopians will have to benefit from the Nile. When we recognize that, it means that we will find a technical solution. What is the main content of this solution? Egypt does not have the potential to generate much electricity from the Nile. ... Ethiopia has huge potential on the Nile to generate electricity. Egypt has big potential for irrigation. Ethiopia has small potential for irrigation. So, what we do: Ethiopia focuses on generating electricity. Egypt focuses on irrigation. Yet, Egypt also raises some electricity. Yet, Ethiopia also does some irrigation. When you add that up, you will have an optimum solution for both countries. The same thing for Sudan, Uganda, and others. ... Ethiopia is not saying, 'We should divide the Nile water equally.' We did not say that. We said, 'Let us divide it equitably ... according to the needs of every nation.'

5.4.3 *The Deadlock of the CFA Negotiations*

In June 2001, the CFA Negotiating Committee was established to discuss various proposals to resolve all of the contentious issues (Woldetsadik 2016: 5). In February 2002, the committee introduced the concept of "water security" (article 14) as an "ingenious solution" to the "thorny issue of existing treaties," arguing that this would have the "advantage of relegating existing treaties to the background in favor of the more dynamic and progressive principles of international water law," as quoted in Mekonnen (2010: 430) after Amare (2009: 5). However, it seemed that while the negotiation process attempted to bypass the problem of "existing agreements," it moved in effect towards a complete impasse over "water security." Mekonnen notes that the introduced concept is "non-legal" and "potentially disruptive" (Mekonnen 2011: 362), as it would "mean anything a riparian country wanted it to mean," and while the proposal was relying on "constructive ambiguity," it ended in a stalemate because each side in the negotiations insisted on putting its own definition that would secure its interests (Mekonnen 2010: 430). Through the Nile-COM meeting in March 2006, it was clear that agreeing upon the stipulation

---

[4] Available at: https://www.youtube.com/playlist?list=PL11A98688058BBD6C (accessed on November 22, 2016).

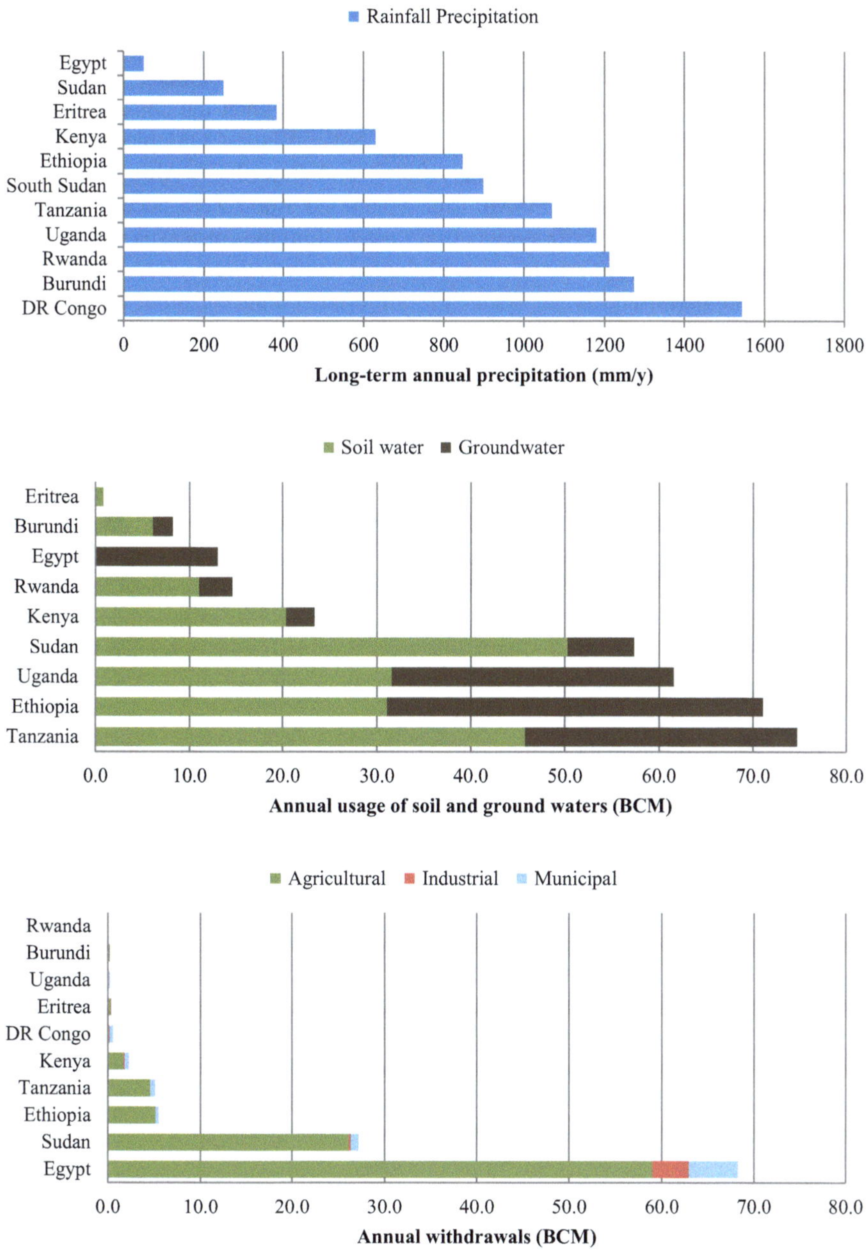

Fig. 5.7 Comparing green and blue water usage in the Nile Basin. (Source: Created by the author after Zeitoun et al. (2010: 232) (for average usage of soil and ground waters in 1998–2004) and FAO, Aquastat Database (for annual rainfall precipitation and total withdrawals in 2000–2004), available at: http://www.fao.org/nr/water/aquastat/data/query/index.html?lang=en. (accessed on November 20, 2016))

of the "water security" article became a new contentious issue to be added to the previous ones (Woldetsadik 2016: 5). Different wordings were proposed from different parties and stakeholders including the committee members, the riparian states, and even from the World Bank; yet none of them met the consensus of all parties (Ahram 2015c; Woldetsadik 2016: 6). Several Nile-COM meetings were held thereafter, in May 2006, February 2007, and June 2007. Along these, the Egyptian negotiators headed by Minister Mahmoud Abou Zeid accepted not to include the obligation of "prior notification" and to limit the consensus requirement to the amendment of core articles (e.g. of the equitable and reasonable utilization, no significant harm, and water security), while amending the other articles and resolving disputes would only require the majority approval (Allam 2014: 91). Yet, article 14 of "water security" remained highly controversial, having two specific proposals, the first defended by Ethiopia and the other upstream riparians, while the second by Egypt and Sudan. Both proposals agreed upon the first part of the article: "Having due regard to the provisions of Articles 4 and 5 (equitable and reasonable utilization and causing no significant harm respectively), Nile Basin States recognize the vital importance of water security to each of them. ... Nile Basin States therefore agree, in a spirit of cooperation: (a) to work together to ensure that all states achieve and sustain water security." The whole problem was about line (b), of which the proposal defended by upstream riparians reads, "not to significantly affect the water security of any other Nile Basin State," while the one defended by downstream riparians reads, "not to adversely affect the water security and current uses and rights of any other Nile Basin State."

At this point, it was evident that negotiations had been extremely complicated. If Egypt and Sudan agreed upon the upstream states' proposal, then nothing in the CFA would oblige the others not to disrupt their current uses, since "prior notification" was no longer an obligation and dispute resolution was based upon simple majority, considering that Egypt and Sudan are only 2 states out of 10 (now 11 after South Sudan's separation). While, if Ethiopia and the others approved the proposal of Egypt and Sudan, then they would be implicitly recognizing the existing agreements and thus foreclosing their ambitious plans of utilizing the waters upper the river. Reaching a deadlock at the ministerial level in the meeting of June 2007, the Nile-COM decided to refer the entire issue to the heads of state/government, hoping to find a satisfactory solution for all parties (Mekonnen 2011: 356). The case was no less complex at the higher level. This was explicitly reflected in the press statements by the heads of upstream states, as follows:

> The current regime cannot be sustained. It's being sustained because of the diplomatic clout of Egypt. Now, there will come a time when the people of East Africa and Ethiopia will become too desperate to care about these diplomatic niceties. Then, they are going to act.
>
> Prime Minister Meles Zenawi on February 3, 2005, to the BBC (Thomson 2005)

> Egypt should stop monopolizing the Nile waters. Egypt should sit with us and agree on new arrangements under which we should all share the use of the Nile equitably. ... The 1929 agreement was between the Egyptians and the British not between us. That was a colonial agreement. We should sit down and work together to work out a new arrangement.

President Yoweri Museveni on April 1, 2004, to a news conference (Sudan Tribune 2004)

With no particular action from the heads of state/government, several attempts were made in 2008 by the Nile water ministers and by representatives of the NBI and the World Bank, but none of these proved successful. In March 2009, the Egyptian Minister Abou Zeid was replaced by Mohammed Nasr el-Din Allam, which was unexpected decision, even to the other Egyptian officials who were working on the issue. This is clearly indicated by the then Egyptian foreign minister, who commented in his notes that changing the minister representing Egypt in negotiations at that critical moment would certainly affect the country's position (Aboul-Gheit 2014: 240). Immediately, Allam conducted several meetings with Egyptian legal experts and academics in addition to consulting an international legal expert from Britain.[5] All of them advised him that signing the CFA in its current form without stipulating the obligation of "prior notification," without including "current uses" based on "existing agreements" in the "water security" article, and without requiring consensus in amending "any" of its articles, would renounce Egypt's water rights in the Nile (Allam 2014: 93–97). In May 2009, an extraordinary Nile-COM meeting was held in Kinshasa through which the crisis was worsened (Allam 2014: 95; Woldetsadik 2016: 6–7). On the one hand, the upstream riparians suggested to annex the two previously proposed texts of article 14 (b) and to let the final decision be taken by the NBRC within 6 months after its establishment, but this was firmly rejected by Egypt. On the other hand, the Egyptian minister announced his demand to include the "prior notification," the "existing agreements," and the "amendment by consensus" in the CFA, all of which were also rejected by the upstream riparians. As for Sudan, the delegation had already withdrawn from the meeting because the entire issue was supposed to be in the hands of the heads of states. The upstream riparians passed a resolution in the meeting of Kinshasa agreeing to sign the CFA solely, a step to which Egypt and Sudan objected in a joint memorandum (EMFA 2010). It is noteworthy here that Egypt and Sudan had maintained a unified stance along the negotiations path. Even when there were differences over details, these were discussed in closed meetings; not in public ones (Aboul-Gheit 2014: 249; Allam 2014: 98; Waterbury 2002: 80).

Another Nile-COM meeting followed in July 2009 in Alexandria, in addition to a number of meetings held by the CFA Negotiating Committee and the Nile-TAC; yet none of these was capable to bridge the wide gap between the opposing positions (Allam 2014: 98–99; Woldetsadik 2013: 7). The final attempt took place in Sharm el-Sheikh on April 13, 2010, in which Egypt and Sudan presented a presidential initiative to establish the NBRC as a permanent body that would attract investments to the Nile Basin and to let the CFA negotiations continue in parallel. In fact,

[5] In his book, "The crisis of the Ethiopian Renaissance Dam," Allam mentioned that the Egyptian Ministry of Water Resources consulted an international legal expert from Britain, who remained anonymous. Allam said that this British expert advised him not to sign the CFA in its current form and confirmed to him that the Nile historic treaties are still valid and binding, especially the ones that are "border agreements" as the 1902 and 1929 Agreements (Allam 2014: 97).

this initiative had been inspired by a World Bank proposal to conclude a "simple" agreement and move forward with the establishment of the NBRC, a proposal of which the U.S. endorsed as a result of close contacts made by Egypt (Aboul-Gheit 2014: 245, 247). Nevertheless, the upstream riparians rejected the idea of continuing the negotiations believing that such a step would be fruitless and would waste all the efforts they had made since 1997 (EMFA 2010). The meeting ended without any particular progress, and it was doubtless that the negotiations had reached a complete deadlock. On May 14, 2010, four upstream riparians, namely Ethiopia, Rwanda, Tanzania and Uganda, signed the CFA in Entebbe, joined 5 days later by Kenya (Allam 2014: 103; Woldetsadik 2013: 8). Following the individual signature of the CFA by the five upstream riparians, Egypt and Sudan suspended their participation in NBI's activities (Ahram 2015c). The dimensions of the conflict and the features of the future can be seen in the statements quoted at the beginning of this chapter after Zenawi, on May 20, and after Egypt's Speaker of the People's Assembly, Dr. Fathi Sorour, during the opening of the Sharm el-Sheikh meeting. The following words were also said in Sorour's speech, which reflect Egypt's position on the CFA and the NBI (EMFA 2010).

> If the upstream countries signed this Draft Framework Agreement individually, this would be considered as a legal violation to the agreed upon Rules of Procedure of the Nile Basin Initiative, which state that the adoption of a draft agreement is done by consensus, not majority. Consequently, the agreement is outside the framework of the Initiative. … The Egyptian Government is fully aware that the issue of the Nile waters is of national security importance, related to the present and the future of the Egyptian nation, therefore it will not be allowed under any circumstances to transcend Egypt's water rights or its current usages. … We thank the development partners, donors of the Nile Basin Initiative for their continuous support for the Initiative, but we tell them that if the upstream states unilaterally signed the agreement, this will represent a death declaration for the Initiative as well as the great efforts exerted by the donors' states in supporting it.

In his interview with the Egyptian state TV, mentioned earlier (fn. 34), Zenawi was asked about the reason of signing the CFA unilaterally without waiting for reaching consensus with Egypt and Sudan, and the answer was as follows:

> We reached a stage where Sudan and Egypt were insisting that the other seven countries accept the agreements of 1929 and 1959. If we take the agreement of 1959. It says that all the water of the Nile is distributed between Egypt and Sudan. There is nothing left for the other countries. This is not a win-win concept. … The Ethiopian government in 1959 rejected this agreement because it excluded Ethiopia. … What we are saying is: 'Let's forget about this agreement. Let's design a new agreement that protects the interests of Egypt and Sudan and, at the same time, protects the interests of the others.' There was no flexibility with regard to the position of some officials in Egypt and Sudan. … So, after ten years, the seven countries felt we need to move forward. … Unless there is flexibility in our positions, it does not make sense for me and you to talk everyday about the 1959 Agreement. In the morning, I tell you, 'I am not going to accept it.' You tell me, 'You have to accept it.' And in the next morning, we will come together and repeat the same thing. … There was no progress on this point for ten years. So, the upper riparian countries became very frustrated. … Up to now, the Egyptian government is still insisting that all the other countries accept the 1959 and 1929 Agreements, which makes the other countries beggars of Nile water, not co-owners of Nile water.

The five signatories gave the other Nile riparians 1 year to join the CFA before ratifying it by their parliaments (Malone 2010). In reaction, Egypt moved at the basin and the international levels to halt the implications of that step. At the basin level, Egypt held close communications with the two remaining upstream riparians, Burundi and the DR Congo, in order to prevent the signature of the required sixth state after which the CFA if ratified by all the signatories' parliaments would enter into force. Officially, Burundi declared that it was busy with elections, which took place in May–June 2010; however, the decision to sign the CFA was not favorable at the time so as not to harm its bilateral relations with Egypt, particularly as the assistance programs were intensified after the CFA signature (Salem 2011: 7). DR Congo's position was more complex. Historically, President Nasser had greatly supported the Congolese to obtain their independence as he backed in particular Laurent-Désiré Kabila, the father of President Joseph Kabila, and Antoine Gezinga whose party led the government after the 2006 elections (Mangu 2011: 35–36). Additionally, DR Congo's interests in the Nile are significantly limited in terms of waters, both because its territory drained by the Nile is too small and given the Congo River with its incomparable water resources. While, conversely, its interests with Egypt are considerable with respect to economic and military cooperation programs (Mangu 2011: 35–36). Eventually, Burundi signed the CFA on February 28, 2011, 1 month after the spark of the January 25 Revolution in Egypt (see next chapter), while the DR Congo has not signed it to date.

At the international level, Egypt sent letters of protest to the UN, the World Bank, and other donors, and to the U.S., the EU, China, India and other countries that have investments in the Nile Basin. These letters aimed at clarifying that the CFA was opened for signature without the consensus of the Nile riparians as was agreed at the beginning of negotiations, and that no funds should be provided for upstream water projects without notifying downstream riparians in advance according to the UNWC principles and the World Bank regulations (Aboul-Gheit 2014: 259). Egypt also demanded from the Gulf States and Libya being important investors in the upstream riparians to mediate between the two sides (Aboul-Gheit 2014: 243, 265–267). It is worth mentioning here that Mubarak, according to Minister Aboul-Gheit (2014: 239, 242, 255), was not involved directly in the CFA issue due to his "old age" and "unfamiliarity" of all of its technical and legal details. However, Mubarak's general recommendation to those officials who manage the case was to maintain cooperative ties with the upstream riparians and never to let bilateral relations with any of them be severed, and he gave the issue special priority in his visits and talks with various actors (Aboul-Gheit 2014: 258). This was evident when Zenawi, in an interview with Reuters (Malone 2010) on November 23, 2010, accused Egypt of "supporting rebel groups" in an attempt to "fish in troubled waters" and "destabilize" Ethiopia and declared that Egypt "could not win a war over the River Nile." In the following day, Mubarak was keen to respond by himself, saying, "We have very amicable Relations with Ethiopia ... I was surprised by these comments because this is something we cannot do with any Arab or African country" (Reuters 2010).

5.4.4 *Why Did Basin-Wide Cooperation Fail?*

Since its foundation in 1999, the NBI has been able to carry out various programs for monitoring, information sharing, and preparing feasibility studies. However, as illustrated above, the riparian states have implemented several developmental mega projects outside the initiative. These unilateral projects have been to a large extent incompatible with the broader basin-wide plan promoted by the NBI. This was carefully investigated in the "independent evaluation report" conducted by the Stockholm International Water Institute (SIWI 2013) presented to the World Bank; as the main funding body of the initiative's activities. The report argued that the main reasons of such unilateral actions are the influences of the following factors (SIWI 2013: 22–24): the growing water, energy and food demands; the emergence of alternative sources of fund (e.g. China and Gulf States); the high national ambitions that exceed the functional, financial and temporal capacity of the NBI's implementation power; the deadlock reached by upstream and downstream riparians concerning the CFA's negotiations; and the political and economic changes taking place in many riparians.

Concluding their notes on the CFA negotiations, both Ministers Aboul Gheit (Aboul-Gheit 2014: 268) and Allam (2014: 105) indicated that Egypt should not have accepted to engage with the multilateral path of the CFA without obtaining reliable assurance from the World Bank and other development partners that none of the basic principles of international water law would be disregarded. Both mentioned that it was better to promote bilateral not multilateral cooperation with riparians, since the upstream riparians formed one block to conclude an agreement that does not recognize the obligation of "prior notification," the requirement of "consensus," and the recognition of "current uses." Their point of view was that the upstream riparians with the support of Egypt should have focused on utilizing groundwater and rainfall resources without consuming the water flow running downstream, which represents Egypt's sole source of freshwater. This approach has been completely refused by the upstream riparians; given that they considered the primary objective of the CFA to supersede the existing agreements for the equitable use of the Nile surface waters (Mekonnen 2010; Ibrahim 2011; Mekonnen 2011; Woldetsadik 2016).

In addition, Aboul-Gheit (2014: 244, 252) referred to the fact that Egypt's assistance programs to the Nile riparians such as drilling groundwater wells and building water harvesting dams, are important mechanisms of cooperation; though insufficient to build long-term trust between Egypt and its co-riparians. He specified that "trade channels with co-riparians" should have been "further deepened" to bring direct benefits to their peoples (2014: 244). He further clarified that great effort had been made from his side (2004–2011) to encourage Egyptian businessmen to do so, but he found most of them not enthusiastic enough and that some even resisted the trend to rely on imports from Africa in general and the Nile riparians in particular specifically for livestock products (Aboul-Gheit 2014: 233). Obviously, strong economic ties could have played an important role in reaching a compromise over the CFA. Contrariwise, relying only on incentives of financial support, as the case with Burundi and the DR Congo after opening the CFA for signature, was implicitly conceived as an attempt of "bribery" from the upstream riparians' perspective (Mangu 2011: 35) and an attempt of "blackmail" from Egypt's perspective (Aboul-Gheit 2014: 248, 251).

Not only were the economic ties between riparians so weak, but also the role of civil society, which was greatly undermined along the NBI's activities in general, and during the CFA negotiations in particular (Mostert 2003: 24–32). In this regard, the author interviewed a former president of the Nile Basin Discourse (NBD), which was launched in 2001 as an umbrella of local Nile Discourse Forums (NDFs) to promote the participation of civil societies in the Nile Basin. E.A. (2016) pointed out that the NBD was allowed to join the meetings of the Nile-COM and the Nile-TAC except those discussing the CFA. He added that the negotiated draft of the agreement was not shared with them and was kept as a "secret issue." Moreover, E.A. stressed the fact that the NBD was initially funded by CIDA (2001–2003) then by the UK Department for International Development (DFID) (2006–2012) and by the World Bank afterwards, while most of international donors prioritized funding governmental actors and obviously paid less attention to the activities of civil societies. E.A. emphasized that the NBD's key message that was delivered to the public through the NDFs is that the riparian states form one Nile family and that all of its members have no choice but to cooperate and to share the river without harming each other and without harming the river nor its peoples and ecosystems. However, as he explained, neither the governments nor the donors considered civil society a "real partner" or worked seriously to strengthen their role.

5.5 Analysis and Discussion

5.5.1 Sole Utilization (T-S & S-T)

By the collapse of the Soviet Union (G) and the overthrow of Ethiopia's Mengistu (D) in the early 1990s, the Nile game entered into a new round. The round's initial configuration consisted of upstream riparians seeking to rebuild their countries and stabilize political and economic conditions (D), on the one hand, and international development partners representing the sole source of financial and technical support (G), on the other. Hence, upstream riparians sought to benefit from various international assistance programs in order to upgrade their capacity (C_{Up}). However, based on the donors' rules and regulations, any support for upstream projects was particularly conditioned by the full coordination with downstream riparians. Egypt was able to maintain its position on the Nile under this initial configuration since upstream riparians were obliged to consult with it prior to construction of their schemes on a project-by-project basis. Yet, international organizations, not only facilitated bilateral coordination, but also sought to bring together all the riparians under one umbrella, creating an inclusive platform where upstream states led by Ethiopia were able to form one cluster and challenge Egypt's longstanding position in the basin (P_{Up}-P_{Do}).

Simultaneously, the downstream cluster comprised of Egypt and Sudan had entered into a period of strained bilateral relations during the 1990s. At the time when Egypt was suffering from terrorism (D), the post-1989 regime in Sudan adopted an Islamist ideology (D), supporting extremist groups in different places around the world (G). Tension between both countries was further exacerbated by the spark of the Gulf War (R) in which Egypt backed Kuwait, while Sudan aligned with Iraq. The Nile was involved in this tension, whereby the press circulated the Iraqi threats to destroy the AHD in the confrontation with Egypt. Even after the war, tensions continued over several other issues, including the disputed Hala'ib. Triangle (R) between both countries. During this tense period, their joint management of the Nile waters under the 1959 Agreement was not interrupted.

However, the assassination attempt against Mubarak in Addis Ababa in mid-1995 (R), committed by the Islamist Group, provoked a new tension that was severe enough to worsen the bilateral relations with Sudan in all fields; thus disrupting the PJTC meetings. Mubarak accused the Bashir regime of complicity in the incident, reacting by expelling the Sudanese forces from the Hala'ib. Triangle (R). In response, Sudanese officials declared that violating the border agreements between the two countries could lead to suspending all other accords, including the 1959 Agreement. The press also circulated that Iraqi experts came to Sudan to discuss the possibility of altering the Nile flows to Egypt (R). Despite all these events, Egypt sought to refrain from escalating those tensions. In the following year, it opposed the UN Security Council's decision to impose additional sanctions against Sudan due to the assassination attempt (G). Egypt's stance was motivated by the fact that the growing pressure on Khartoum would weaken it, leading to the separation of South Sudan (D). The unity of Sudan was a strategic priority for Egypt in relevance to maintaining its water interests (I_{Eg}). In addition, Egypt was aware that the ongoing multilateral interaction in the Nile required both countries to have one unified strengthened position vis-a-vis the upstream riparians (C_{Do} vs. C_{Up}). Shortly thereafter, the PJTC resumed meetings as normal, and in 1998 both countries fully coordinated their national projects. By late 1999, both Mubarak and Bashir agreed to normalize bilateral relations between their countries. Thus, it can be said that the downstream cluster was able to overcome the challenges it faced during this round, which was clear in the unified stance of both Egypt and Sudan taken throughout the CFA negotiations. Nevertheless, the sole utilization outcome, which prevailed in the previous rounds in favor of Egypt and Sudan, was seriously challenged in the post-Cold War round as will be shown in the following sections.

5.5.2 *Cooperation (R-R)*

As mentioned above, at the beginning of the post-Cold War round, international organizations played a major role attempting to catalyze cooperation among the Nile riparians. This was reflected on both bilateral and multilateral interactions in the basin.

In order to obtain funds from the World Bank, Uganda signed a bilateral agreement with Egypt in 1991 to construct its Kiira-Nalubale hydropower project as an extension to the old Owen Falls Dam (R_{Ug}-R_{Eg}), and so did Ethiopia in 1993 to carry out small-scale irrigation schemes on the Blue Nile (R_{Et}-R_{Eg}). Although Uganda's informal position emphasized that past colonial treaties should be revised (B_{Ug}), it pledged in the 1991 Agreement not to alter the White Nile flows to Egypt in accordance with the technical regulations stated in the previous Owen-Falls agreements. Ethiopia also stipulated in the 1993 Agreement not to cause significant harm to Egypt, and it did not insist to cite explicitly the equitable use principle, which contradicts with its traditional stance of reserving its right of utilizing the Nile waters unilaterally (B_{Et}). It is evident that each side in the two agreements made some concessions to reach the R-R outcome of cooperation. Egypt, on the one hand, did not object to the implementation of those water schemes upstream, while on the other, Ethiopia and Uganda accepted to respect the status quo that they were seeking to change with varying degrees. However, Ethiopian officials later expressed their dissatisfaction over the long time Egypt had taken to ensure whether the proposed scheme would cause any significant harm, and over Egypt's unilateral decision to carry out its desert reclamation project in Toshka without prior consultation with upstream riparians. In contrast to the agreement with Egypt, the Ethiopian-Sudanese 1991 Declaration had cited both the equitable use and no harm principles (R_{Et}-R_{Su}). This took place in a different context, as the post-1991 regime in Ethiopia comprised the liberation fronts that Sudan had backed in their fighting against Mengistu (R). However, the political relations between Sudan and Ethiopia were strained in the following years due to different reasons, amongst which was the assassination attempt against Mubarak. In the agreement signed between Ethiopia and Uganda in 1993, the two countries called for fair and equitable utilization of their transboundary waters (R_{Et}-R_{Ug}).

The role of international development partners was not limited to facilitating bilateral coordination, but also sponsoring multilateral cooperation. By folding Hydromet in late 1992 and replacing it with TECCONILE, the focus was on building basin-wide cooperation in the Nile (R_{Do}-R_{Up}). Because Egypt and Ethiopia had completely different understandings about cooperation (B_{Eg} vs. B_{Et}), each one sought to lead the emerging multilateral path towards assuring its own interests (I_{Eg} vs. I_{Et}). Egypt proposed a package of projects based on the no-harm principle, involving technical training, exploitation of alternative water resources (e.g. rainfall harvesting and drilling wells), and non-consumptive water uses (e.g. regional power trade). Ethiopia did not accept confining cooperation to such scopes, restricting the upstream riparians from utilizing the Nile waters. Rather, it pushed for developing a legal and institutional framework based particularly on the principle of equitable water use. Egypt was also suspicious about Ethiopia's intentions, worrying that the developed framework would ultimately result in reallocating the Nile waters. As a result, two multilateral paths emerged: the projects-based path of the NBI and the principles-based path of the CFA. Yet, mistrust between downstream and upstream riparians was prevailing over both paths (B_{Do} vs. B_{Up}), and neither of them brought a satisfactory R-R outcome that all sides could accept.

Concerning the NBI path, the subsidiary action programs in the Eastern Nile were more complicated that those ones in the Equatorial Lakes, and the JMP was a clear example. The independent experts that were selected by the three countries identified in their scoping study that large hydropower dams on the Blue Nile in Ethiopia represented an attractive set of investments with considerable benefits for all and set the limit to Ethiopia's irrigation withdrawals that would result in manageable water and energy deficits in Egypt. Unsurprisingly, this recommendation was not particularly welcomed by Egypt, since, for the latter, the concept of constructing large reservoirs upstream is directly linked to the historical events where the Nile waters were used as a weapon for threatening its security (B_{Eg}).

The CFA path was even more complex. Since the submission of the first draft to the Nile-COM in 2000, the sharp differences between the upstream and downstream riparians were crystal clear. On the one hand, Egypt and Sudan required explicit recognition of their historic water rights stated in the existing agreements, the requirement of which the other riparians firmly rejected. On the other hand, the upstream riparians insisted not to include the obligation of prior notification as stipulated in the UNWC and to emphasize only on the duty of exchanging data and information. They also insisted that the amendment and dispute settlement procedures should be based on simple majority not consensus. These requirements were also rejected by the downstream riparians, and a compromise seemed remote. Even when the riparians accepted the solution of adding an article for water security, they disagreed on the wording. Egypt insisted that the article should comprise the pledge not to adversely affect current rights and uses in the basin, which was viewed by the upstream riparians as an implicit recognition of the existing agreements that they had rejected from the beginning. The negotiations reached a deadlock and neither the ministers nor the heads of state/government were able to come to an acceptable solution to all parties. The weak economic ties between the opposing sides and the low participation of civil societies contributed to the failure of generating an R-R outcome over the CFA.

5.5.3 *Contention (P-P)*

When comparing the post-Cold War round with the Cold War round, it can be noted that, in both, the upstream riparians were not satisfied with the status quo that prevailed in the Nile in favor of the downstream riparians. For example, Ethiopia protested officially against the 1959 Agreement and the Peace Canal in 1978 and did the same against the Toshka project in 1997. In the early 1960s, Tanzania and Kenya adopted the Nyerere Doctrine contesting the validity of the Nile colonial treaties, and along the CFA negotiations, they refused to recognize these agreements. However, the important difference in the comparison is that the opposition of the upstream riparians in the post-Cold War round is obviously more effective than in the preceding one. The crucial factor behind this is the profound change in the context that led to raising the capability of upstream riparians (C_{Up}) as follows.

The status quo in the basin has been seriously challenged since the formation of the upstream cluster, induced by the termination of the global bipolar rivalry and the subsequent role of the international organizations (G), which created a playground where all the riparians could compete for their interests (I_{Up} vs. I_{Do}). In addition, when Egypt commenced the Toshka reclamation project in early 1997, a race of implementing national large-scale schemes emerged as a third path. This aimed to create facts on the ground in parallel with the two multilateral paths mentioned earlier. At the beginning, it was expected that Egypt would lead the race, since it was incomparably more stable from both the economic and political perspectives (C_{Eg} vs. C_{Ot}). It also had different sources of national and international funds, especially after a large portion of its external debts were cancelled in recognition of Mubarak's role in the Gulf War (R). In contrast, the only financial source for the upstream riparians in the 1990s was the international donors that require the downstream consent for every project upper the river as indicated above. Even Sudan did not have the capacity to utilize its full water share stipulated in the 1959 Agreement (C_{Su}), providing Egypt with additional water supply that helped it to meet the needs its rapidly growing population (I_{Eg}).

By the 2000s, all of these conditions were significantly changed, particularly after the regional armed conflicts in the basin, (including the Ethiopian-Eritrean war) came to an end (R), which obviously contributed to stabilizing their economies (D). Internationally, the emergence of China as an ideal alternative source of financial and technical assistance made it possible to implement large water projects upstream without the need for assuring prior consent from Egypt (G). These big changes in the domestic, regional and global contexts led Ethiopia, for example, to build the Tekeze Dam with the support of China and then the Beles project on the Blue Nile through an Italian construction company with national funds (C_{Et}). These projects added 300 and 460 MW respectively to Ethiopia's power grid to reach 2000 MW in total by 2010, which was a large step towards meeting the country's urgent energy needs (I_{Et}).

The conditions in Sudan also were changed. Khartoum witnessed an economic boom after the completion of the pipeline from the Southern oil fields to Port Sudan on the Red Sea in late 1999 with the support of China and by the progress in the peace process with the Southern Sudanese by 2002 (D). Besides direct cash revenues, crude oil exports allowed Sudan to receive loans from China (G). Thus, supported by China and the Gulf States, Sudan was able to implement the three largest projects it had planned since the 1960s: the Merowe Dam, the heightening of the Roseires Dam, and the Setit-Atbara Dam (C_{Su}), all of which were constructed by Chinese companies. Building these dams and restructuring the administration of its large-scale irrigation schemes, particularly the Gezira-Managil scheme, represented a dramatic transformation for Sudan towards utilizing its full water share under the 1959 Agreement (I_{Su}).

After all, it can be said that the results of the unilateral projects race did not come in favor of Egypt, and while large-scale schemes were successfully implemented upstream, the Toshka project did not achieve its main objectives. Simultaneously, the negotiations over the CFA reached a complete deadlock, and it was clear that achieving an R-R outcome over the agreement was nearly impossible. The upstream riparians led by Ethiopia decided to challenge Egypt and Sudan through opening the CFA for signature, although it was initially agreed that this step would be only taken with the consensus of all riparians. At this point, all the paths emerged in this round converged together. On May 14, 2010, on the same day that the Beles project was inaugurated, the CFA was signed by Ethiopia, Rwanda, Tanzania and Uganda, and in few days later, Kenya joined them. In response, Egypt and Sudan suspended their participation in the NBI's activities. As a result, the R_{Do}-R_{Up} outcome failed to persist in the CFA and the NBI paths. Rather, all the unilateral and multilateral paths generated P_{Up}-P_{Do} outcomes.

One last note here is that the outcomes of this round show how the rewarding payoffs of cooperation remain higher than the punishing payoffs of contention. By the suspension of the downstream riparians' participation in the NBI, the initiative lost its virtue as an inclusive river basin organization and thus lost a considerable portion of its international funds. At the same time, Egypt and Sudan lost their place in the common platform where they could oversee the development taking place in the upstream riparians. Obviously, it was better if those unilateral projects, including the Toshka project and the upstream dams, were carried out under the umbrella of the NBI and if all riparians accepted to commit to all the UNWC provisions as a whole in drafting the CFA.

5.6 Concluding Remarks

Equilibrium in the post-Cold War round had left the sole utilization outcome that favored the downstream riparians for decades to turn into contention, despite the massive effort made by international organizations to establish multilateral cooperation in the basin (Table 5.2). At first, the game contexts seemed pushing the equilibrium towards basin-wide cooperation, particularly after the end of the Cold War era and the engagement of international organizations as the major source to fund development projects in upstream riparians. Then, the relative economic growth of the latter and the emergence of alternative sources of financial and technical assistance, mainly China and the Gulf States, diverted the game progression towards the full generation of the contention outcome. What significantly distinguishes this round from the previous ones is that the contextual influences enabled the upstream riparians led by Ethiopia to employ the reciprocal strategy effectively in their favor. This took place through two stages in both the bilateral and multilateral relations.

During the early 1990s, Uganda and Ethiopia took serious steps towards fostering bilateral cooperation with Egypt through the signature of the 1991 Exchange of Memoranda and the 1993 Framework respectively. Egypt reciprocated with endorsing the construction of the Kiira-Nalubale extension in Uganda and passing the implementation of a small-scale irrigation scheme in Ethiopia on the Blue Nile. The case was different between Ethiopia and Sudan as the new transitional government in the former signed the 1991 Declaration reciprocating the Sudanese support to the

Table 5.2 The outcomes of the Nile post-Cold War round according to the iterated PD framework

Scenarios		Does player Y utilize the river unilaterally?	
		No	Yes
Does player X ulitize the river unilaterally	No	Cooperation (R-R)	Sole utilization (S-T)
		1991: Ethiopia and Sudan signed the declaration of peace and friendship, recognizing the equitable use and no harm principle (R_{Et}-R_{Su})	1997: Ethiopia sent protest letters to the UN security council and to the World Bank, protesting against the Toshka project and describing it as an attempt to create facts on the ground (S_{Et}-T_{Eg})
		1991: Egypt and Uganda signed the exchange of memoranda on the Kiira-Nalubale scheme (R_{Eg}-R_{Ug})	
		1993: Egypt and Ethiopia signed the framework for general cooperation, citing the no harm principle in particular (R_{Eg}-R_{Et})	
		1993: Ethiopia and Uganda signed the friendship and cooperation agreement that called for fair use of common waters in the Nile (R_{Et}-R_{Ug})	
		1998: Egypt and Sudan signed a memorandum stating the water projects to be implemented in each country (R_{Eg}-R_{Su})	
		1993–2010: The Nile riparians launched TECCONILE and folded it into the NBI in 1999 (R_{Do}-R_{Up})	
		1997–2010:The D-3 project began the CFA negotiations and proceeded under the NBI (R_{Do}-R_{Up})	
	Yes	Sole utilization (T-S)	Contention (P-P)
		1997: Egypt commenced the Toshka project for reclaiming 500,000 feddans in the Western Desert, constructing a pumping station and a canal with a capacity of 5 BCM. Egypt declared that this volume of waters would be provided from its current water share (T_{Eg}-S_{Et})	1997–2009: After the commencement of the Toshka project, Ethiopia announced that it would build a large hydropower dam on the Tekeze River without prior consultation with Egypt. The construction began in 2002 and was completed in 2009 (P_{Et}-P_{Eg})
			2001–2010: Ethiopia announced it would redo the Beles project. The construction began in 2005 and was completed in 2010 (P_{Et}-P_{Eg})

liberation fronts that were fighting against Mengistu. Simultaneously, tensions between Egypt and Sudan were growing in light of the Gulf War, the assassination attempt against Mubarak, and the Hala'ib Triangle dispute. Yet, in order to maintain the unity of Sudan and the cohesion of the downstream cluster, Egypt sought to relieve the tension by opposing the UN sanctions on Khartoum, and the latter responded by returning the Egyptian assets it had seized earlier in Sudan and choosing not to escalate the dispute over the Hala'ib Triangle.

Along the process of establishing basin-wide cooperation, there had been strong competition between the downstream and upstream clusters led by Egypt and Ethiopia respectively. When Egypt attempted to confine the emerging basin-wide cooperation to non-consumptive water projects, Ethiopia insisted on developing a legal framework that would allow for equitable utilization of the Nile waters by all riparians. Then, when Egypt inaugurated the Toshka project unilaterally, Ethiopia implemented the Tekeze Dam and the Beles hydropower plant outside the NBI umbrella. Although the consequences of upstream projects on downstream communities are incomparable with that of downstream projects on upstream communities, international water law urges all riparian states to notify each other concerning their plans prior to construction. Given its ability to secure national funds and international loans, Ethiopia succeeded to complete the construction of its unilateral projects. Eventually, when the negotiations over the CFA reached a complete deadlock, the upstream cluster opened the signature individually without the consent of Egypt and Sudan. The downstream cluster reciprocated by suspending their participation in the NBI activities to arrive at the contention outcome where all are subject to punishing payoffs. Nevertheless, the upstream cluster considered the move a major step for changing the status quo of the basin. As shown in the next chapter, the dramatic changes in the domestic contexts inside Egypt and Sudan enabled Ethiopia to follow that step with another larger one, which is the unilateral construction of the GERD.

Overall, the significant influence of the context in the post-Cold War round was on the balance of capabilities between the downstream and upstream riparians, in that Egypt and Sudan became no longer the sole users of the Nile waters. Such an influence further widened the gap between the riparians' interests and deepened mistrust between them.

Chapter 6
The Post-2011 Round: The New Chapter of the Grand Ethiopian Renaissance Dam and the Shadow of the Future

Before we mobilized our efforts to eradicate poverty, centuries of impoverishment curtailed our development and restricted us from exercising our right to use the resources of our own rivers. Now, thanks to the dedication of our peoples, we have safely put those times behind us. We are close to opening a new chapter through the realization of our Millennium project. Henceforward, nothing can stop us from exercising our rights; the other dams we plan to build are less challenging than this, the Millennium Dam.

The speech of the late Prime Minister Meles Zenawi on launching the GERD on April 2, 2011, as published in Meles Zenawi Memorial (2013)

6.1 The Contexts of the Post-2011 Round

January 2011 represents a dramatic turning point that has had profound effects on the Nile hydropolitics. This period witnessed the occurrence of two historic events, which reshaped the domestic challenges in Sudan and Egypt, influencing the balance of interests and capabilities in the Eastern Nile Basin. These are the voting of the Southern Sudanese in favor of separation in January 9, and the spark of the Egyptian Revolution in January 25. In April 2011, Ethiopia initiated unilaterally the construction of the GERD with a huge size, several times larger than the one

---

Egypt wants to turn the page in the history of relations between the two countries and establish a basis for mutual interest. ... [N]o one should ever feel secure about his future without the other, or build his welfare at the expense of his brother. ... I invite you today to jointly lay the foundations of a better future for our children and grandchildren, a future where all the classrooms in Ethiopia are lit and all the children of Egypt can drink from the River Nile as their fathers and grandfathers did. ... [W]e have a chance to change the course of history today by protecting the rights of our people and enhancing development and progress across the continent. ... We need to utilize our natural resources in a way that ensures equitable use by all.
President Abdel-Fattah el-Sisi's speech in the Ethiopian parliament on March 25, 2015, as compiled after the translations of Ethiopia's Embassy in Brazilia (2015), Ahram (2015b), and Daily Mail (2015)

M. M. Samaan, *The Nile Development Game*,
https://doi.org/10.1007/978-3-030-02665-3_6

indicated in previous plans and prefeasibility studies. Shortly thereafter, negotiations between Egypt, Ethiopia and Sudan began to investigate and mitigate the negative impacts of the dam on downstream communities, though in parallel, not in prior, to construction. Due to its massive benefits on the Sudanese agricultural and energy sectors, Khartoum has been supporting the GERD construction despite Egypt's grave concerns about the project's potential risks downstream. However, the successive transformations that took place in Egypt during the following years resulted in the adoption of different, even contradictory, strategies in managing the GERD issue. Consequently, the round has comprised different scenes of cooperation and contention, from the establishment of joint committees to the exchange of threats. Taking into consideration climate change uncertainties, rapid population growth, and increasing water, energy and food needs, the shadow of the future reflects the possibility of both promoting further cooperation and provoking further contention in the basin.

6.2 The Grand Ethiopian Renaissance Dam

6.2.1 The Separation of South Sudan

Based on the CPA concluded in January 2005 (Sect. 5.3.1), a referendum was held on January 9–15, 2011, in which almost 99% of Southern Sudanese people voted in favor of independence, and on July 9, South Sudan was proclaimed an independent state (Rolandsen and Daly 2016: 150, 159). As Sidahmed (2013: 110–112) illustrates, this transforming event caused Khartoum the loss of 75% of its known oil reserves. As a result, it lost $6 billion annually, which constituted the bulk of its hard currency, leading to substantial shortages in energy supply (Fig. 6.1). In addition, the external debt burden significantly increased, since Sudan depended on crude oil to repay its debts, as was the case with China. Nevertheless, the only potential advantage was transporting South Sudan's crude oil through the pipelines to Port Sudan, since Khartoum would collect shipment fees. However, during the period from July 10 to November 30, 2011, no fees were paid (calculated at $1 billion by Khartoum), leading the latter to confiscate South Sudan's oil in transit as in-kind payment. After several failed attempts of mediation and direct negotiations, South Sudan took the "drastic measure" of shutting down its oil production by late January 2012 (Sidahmed 2013: 112).

To cope with such severe economic threats, the Sudanese government adopted a series of policies. On the long term, in light of the 2008 global food crisis, the government has sought to attract foreign investments in the irrigated agriculture sector in order to benefit from the available land and water resources, particularly in the Blue Nile (Cascao and Nicol 2016: 558). Moreover, several hydropower schemes have been planned and implemented to cover electricity shortages in various sectors (Sect. 5.3.4). However, the large fluctuation in the Blue Nile flows, the high rates of

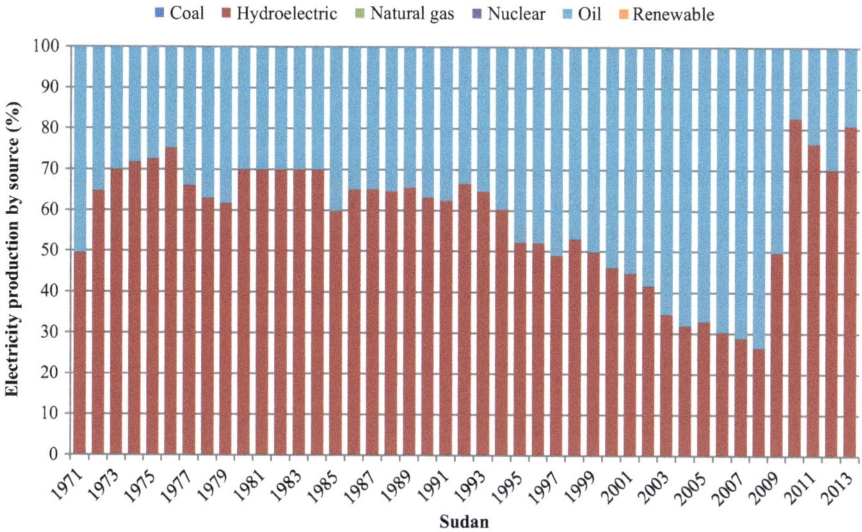

Fig. 6.1 Electricity production by source in Sudan. (Source: International Energy Agency Statistics (2014), available by World Development Indicators at: http://data.worldbank.org/data-catalog/world-development-indicators (accessed on November 2, 2016))

sedimentation, and the adverse social and environmental impacts effectively challenged the sustainability of such agricultural and hydropower schemes. On the short term, the government has decided to cut subsidies on petroleum products on a gradual basis since early January 2011. This latter policy triggered several waves of protests and clashes with security forces, the most serious of which were the ones that took place in June 2012 against the governmental austerity plan and in September 2013 against lifting fuel subsidies (Abdelaziz 2011, 2013; Abdelaziz and Dziadosz 2012). These protests came in the broader context of the so-called "Arab Spring" uprisings that sparked in Tunisia in December 2010-January 2011 and spread thereafter in other countries including Egypt.

6.2.2 *The Egyptian January 25 Revolution*

Inspired by the Tunisian Revolution, which toppled Ben Ali regime on January 14, several groups of Egyptian youth called for demonstrations using social media. During this period, Egypt's political future was unclear and mysteriously anticipated by many, as the presidential election scheduled to take place in September 2011 had provoked deep anxiety among large segments of the population who had been questioning, whether Mubarak (83 years old at the time) would remain in power for another 6 years or whether his son, Gamal Mubarak would succeed him (Sharp 2009: 6). In addition, the then ruling party was dominating the parliament elected in November 2010 with almost no representation of the opposition parties, which accused the regime of election fraud (The Guardian 2010). The activists found the date of January 25, Egypt's National Police Day, an adequate occasion to begin their demonstrations, sending a message to the regime that Egyptians need safety and security but also freedom and dignity. Young people went to Tahrir Square with no formal participation from the opposition parties and political groups, including the banned Muslim Brotherhood, which seemed "ambivalent" (BBC 2011) and adopted a "cautious approach," declaring that should any of its members participate, this would be in "personal capacity" (Awad and Zayed 2011). Clashes between protesters and police took place only after midnight and transmitted during the following days to other cities. Friday, January 28, witnessed the "Day of Rage" when demonstrations and clashes escalated and serious riots were sparked, causing the withdrawal of police forces from almost everywhere in Egypt. "Armed gangs" took advantage of the chaos and widely attacked police stations and prisons (The Guardian 2011). As a result, the military intervened and forces were deployed across the country, where people "cheered" and "welcomed" them (Aljazeera 2011a). After all his initiatives had failed, Mubarak decided to step down on February 11 and the Supreme Council of the Armed Forces (SCAF) ruled the country for a transitional period.

Since that date, Egypt has been busy with its domestic affairs. On the one hand, the successive transitional governments have been fully occupied with managing the tough economic situation. On the other, entrenched disagreements took place among different political forces over the proper roadmap to democracy in Egypt. Islamists, namely the Muslim Brotherhood who did not participate in the Egyptian Revolution on its first day and then joined later and the Salafists who did not join at all, pushed for holding the parliamentary and presidential elections first, then writing the constitution later. In fact, they wanted to take advantage of being the most organized and having the ultimate influence on people among other forces. In contrast, various groups of youth, liberals and secularists raised the slogan of "the constitution first" to assure the adequate representation of all orientations and to prohibit by law the abuse of religion in politics. Ultimately, Islamists achieved what they pushed for, after 77.5% voted in favor of their desire through the referendum over constitutional amendment held in March 2011, making use of their religious influence on people (Shukrallah and Gaber 2011). In addition, the Islamists won the majority in the first post-revolution parliamentary elections held in November 2011-January 2012 (Tarek 2012). In June 2012, the Muslim Brotherhood's candidate, Dr. Mohammed Morsi, won the presidential election with 51.73% (13.24 million votes) in the second round against Commander Ahmed Shafiq, who was Mubarak's last prime minister (Ahram 2012).

During such critical circumstances inside Egypt and Sudan, two crucial events took place upper the Nile: Burundi signed the CFA on February 28, 2011, to be the sixth state required for the agreement to enter into force (after its ratification by the parliaments of the six signatories), and Ethiopia officially announced the construction of the GERD 1 month later.

6.2.3 The GERD Official Commencement and the Question of Size and Site

On February 3, 2011, Prime Minister Zenawi announced that Ethiopia would build a "massive" dam on the Blue Nile without naming it (Stokstad 2016: 904). This was known among EEPCo's "senior managers" and the "top" government officials as "Project X" (Waldyes 2011). Only on March 30, Ethiopia's Water and Energy Minister, Alemayehu Tegenu, announced publicly, "The Great Nile dam construction is scheduled to commence presently near the Ethio-Sudan border. From this dam alone, Ethiopia expects to generate 5,250 MW" (Malone 2011a). Later, the Italian company, Salini Costruttori, which has constructed the Beles project and Gibe III hydropower plant for Ethiopia, disclosed that the company "entered into an agreement" with EEPCo on December 30, 2010, to build the new dam (Salini Impregilo 2014b). On April 2, in a groundbreaking ceremony, Zenawi officially commenced the construction of the dam, which he named "the Millennium Hydro-electric Nile Dam" (Meles Zenawi Memorial 2013). The dam is located at Guba, Benishangul-Gumuz, on the Blue Nile (Abay) river some 20 km upstream of the Ethiopian-Sudanese border (Fig. 6.2). On April 15, Ethiopia's Council of Ministers renamed the new dam to be the Grand Ethiopian Renaissance Dam (GERD) (ENA 2011).

(a) Location of the GERD on the Ethiopian-Sudanese border

Fig. 6.2 Satalite image showing the location of the GERD on the Blue Nile. (**a**) Location of the GERD on the Ethiopian-Sudanese border. (**b**) Closeup image of the GERD construction site. (Source: Google Earth (2018))

(b) Closeup image of the GERD construction site

Fig. 6.2 (continued)

The technical specifications of the GERD were upgraded thereafter. As published in Salini Impregilo (2014a), the main dam is a gravity dam in Roller Compacted Concrete (RCC), with a length of 1.8 km and maximum height of 175 m, creating a reservoir with a total volume of 74 BCM (63 BCM live storage) and surface area of 1800 km^2. The GERD is complemented with a saddle dam with a length of 5 km and maximum height of 50 m. The GERD's total capacity is 6000 MW. The power station comprises two open-air power houses located downstream of the dam on the left and right banks of the river, holding respectively six and ten turbines with installed power of 2250 MW and 3750 MW. The total 6000-MW installed capacity is to be built and operated in flexible stages, beginning with the left power house, in order to start generating electricity before completing the dam and filling the total volume of the reservoir.

The GERD site is nearly the same of the Border Dam, which was first proposed by the 1964 USBR study (Sect. 4.2.4) and was included in almost all the master plans and prefeasibility studies prepared thereafter. However, the size of the new dam is considerably greater, whereby the generation capacity and the reservoir volume are several times larger than those previously proposed. It is noteworthy that the moderate size of the Border Dam was justified in the previous studies for two main reasons: the evaporation losses and the economic feasibility. Waterbury (2002: 117) notes that the site proposed by the 1964 USBR study for the Border Dam is "in the extremely hot and arid lowlands where surface evaporation is at its maximum" in Ethiopia (Fig. A.3). He explains, "There is no doubt that surface evaporation in the Ethiopian highlands is less than it is in the border areas with the Sudan, and less than is the case at the reservoir at the Aswan High Dam. Thus, storing water at higher altitudes will yield some savings in water, but the savings may be somewhat overstated" (Waterbury 2002: 113). In their JMP study (Sect. 5.4.1), Blackmore and Whittington (2008: 48) compare the Border Dam to the Karadobi Dam, illustrating that the image of the Border Dam's site "shows less attractive elevations (inundating significant areas of arable land) and a higher surface area, which would lead to higher evaporation, as would the higher temperatures at the Border site."

The Eastern Nile Power Trade Program Study, funded by ADB and conducted by two European consultants, Electricité de France (EDF) and Scott Wilson of UK (SW) as demanded by the ENTRO, comprises prefeasibility studies for the Border and Mandaya Dams in Ethiopia and Dal Dam in Sudan (EDF and SW 2007). In September 2007, the Ethiopian Ministry of Water and Energy published on its official website[1] (FDRE-MoWR 2007) a brief summary of the study's findings, of which the paragraph concerning the Border Dam reads as follows:

> The Border site is the most downstream of the hydropower sites under consideration within Ethiopia. The topography at the site is well suited to development of a dam of moderate height of some 80 m. The USBR 1964 study recommended a dam some 80 meters in height (FSL 575 m) with a reservoir extending upstream close to the Mandaya site. This development appears near-optimal for the site since higher levels would appear to require extensive saddle dams to contain the reservoir. Observations at the site confirm the presence of good foundation rock which would be suitable for concrete aggregate or rockfill. There would be some multi-purpose benefits of river regulation and flood alleviation downstream, associated with the integrated development of the Border reservoir but not to the same extent as for Mandaya. There is likely to be some loss of flood recession agriculture along the Nile downstream in Sudan and there would be a substantial reduction in sediment delivered to Roseires reservoir. The installed capacity would be of the order of 800 MW and annual energy generation would be of the order of 7000–8000 GWh/year.

The concern indicated above by EDF and SW (2007) is what actually took place. Since the designed height of the GERD's main dam is raised to contain a reservoir with a live storage of 63 BCM (instead of 11.1 BCM as was originally proposed for the Border Dam), it was necessary to build the complementary 5-km saddle dam. The decision to increase the dam's height and to add a saddle dam meant the substantial increase of the dam's costs. Another study that commented on the economic feasibility of the dam with its original size is the comprehensive master plan of East Africa, involving the Eastern Nile countries (Fig. A.7), for the development of a unified power grid code as published by the Eastern Africa Power Pool (EAPP) and East African Community (EAC) (EAPP and EAC 2011a, b). Concerning the investigated dams in Ethiopia (Fig. 6.3), the study gave the Border Dam the least priority from an economic perspective, as follows (EAPP and EAC 2011b: 4.19):

> The large hydro plants (Mandaya, Karadobi and Border) had their entrance schedule carefully studied. There should be at least 5 years separating each plant. Also the need for all three of them has been put in question. Border was deemed unnecessary, being the most expensive of the three in terms of levelized cost [net cost to install a renewable energy system divided by its expected life-time energy output].

---

[1] Available at: http://www.mowr.gov.et/index.php?pagenum=4.3&pagehgt=1000px&ContentID=55 (accessed on December 1, 2016).

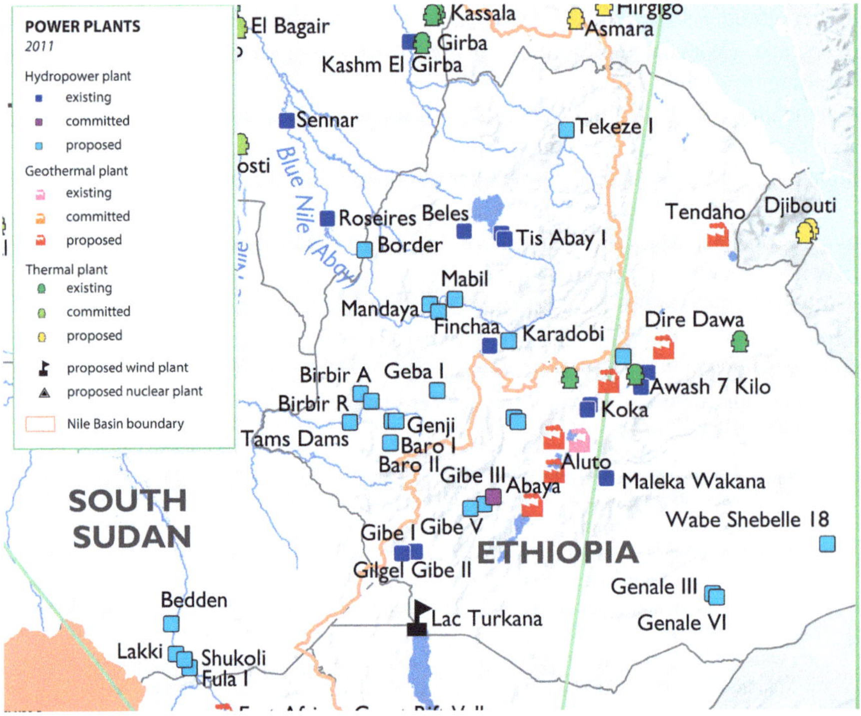

Fig. 6.3 Existing and proposed power plants based on the EAPP & EAC 2011 master plan. (Source: NBI, the State of Basin Report (2012a: 181))

The justifications presented by the previous studies and plans, shown above, raise pressing questions that deserve close investigation; For what reasons has the GERD been built at this specific site, and why was it enlarged to that extent? The answer to these questions can be deduced based on four specific points from Zenawi's speech dring the GERD official commencement on April 2, 2011 (Meles Zenawi Memorial 2013), as follows.

> Honorable peoples of Ethiopia! We have gathered here today at the largest of our rivers to witness the launch of this great project. It is rightly called the Millennium Dam. It is the largest dam we could build at any point along the Nile, or indeed any other river. More importantly the project takes the pride of place, representing an incomparable addition to our national plan for expanding power production. It will not only raise our own power-generating capacity and meet our domestic needs. It will also allow us to export to neighboring countries ...

First, Zenawi was perfectly aware that the road to construct a large dam on the Blue Nile in Ethiopia for the first time ever would be full of great challenges and high risks. Thus, it has been necessary that the dam should be as large as possible in order to present it as a national project that would bring enormous benefits to Ethiopia, and hence would deserve to bear its heavy costs. His words clearly indicate that the project's site was selected searching for the "largest dam" that could be built at "any point" along the Nile or any other river in Ethiopia. Zenawi aimed that the project would not only alleviate energy shortages and reduce blackouts all over Ethiopia, but would also generate surplus electricity for export and therefore provide cash revenues required to eradicate poverty, which as he described would be the "pride of the place." In addition, a project of such magnitude would promote national solidarity among the "peoples" of Ethiopia, which is a vital goal for the government, particularly under the shadow of the popular uprisings taking place across the region.

> Indeed, the current disposition is to make attempts to undercut Ethiopia's efforts to secure funding to cover the cost of the project. We have, in fact, been forced to rely on our own savings alone to cover the expense. The estimated cost will be 3.3 billion Euros, or 78 billion birr. As we will be financing several other projects in our plan, the expense will be an additional and heavy burden on us. All our efforts to lighten this have been unsuccessful, leaving us with only two options. Either to abandon the project or do whatever we must to raise the required funds. I have no doubt which of these difficult choices the Ethiopian people will make. No matter how poor we are, in the Ethiopian traditions of resolve, the Ethiopian people will pay any sacrifice. I have no doubt they will, with one voice, say: 'Build the Dam!' ...

Second, Zenawi was sure that the only way to build the GERD was to rely on national funds, since international donors would not finance it without the prior consent of Egypt and Sudan. Thus, presenting it in that way would persuade the Ethiopians of the need for taking the "difficult choice" and making "any sacrifice" to be capable of completing it.

> Equally, the benefits that will accrue from the Dam will by no means be restricted to Ethiopia. They will clearly extend to all neighboring states, and particularly to the downstream Nile basin countries, to Sudan and Egypt. The Dam will greatly reduce the problems of silt and sediment that consistently affect dams in Egypt and Sudan. This has been a particularly acute problem at Sudan's Roseires dam which has been experienced reduction in output. When the Millennium Dam becomes operational, communities all along the riverbanks and surrounding areas, particularly in Sudan, will be permanently relieved from centuries of flooding. These countries will have the opportunity to obtain increased power supplies at competitive prices. The Millennium Dam will increase the amount of water resources available, reducing the wastage from evaporation which has been a serious problem in these countries. It will in fact ensure a steady year-round flow of the Nile. This, in turn, should have the potential to amicably resolve the differences which currently exist among riparian states over the issue of equitable utilization of the resource of the Nile water.

Third, it is clear that Zenawi wanted the project to be large enough to bring considerable benefits to downstream riparians, particularly to Sudan. He had no doubt that Egypt would not easily change its clear position rejecting the construction of large dams on the Blue Nile in Ethiopia. Whereas, he was certain that Sudan, after the separation of South Sudan, would never lose the chance of gaining the great benefits of such a project. Sudan's consent on the project was crucial for Ethiopia to build the GERD despite the expected Egyptian opposition.

> So, the first message is that we not only have a plan, but we also have the capacity to assert our rights. The second message we want to send is that the intention to exercise our rights to use our own rivers is in order to fight poverty in our own country. It shows no malice to any of our neighbors. ...

Fourth, as shown here and in the paragraph quoting his speech at the beginning of this chapter, Zenawi saw the dam as a bridge shifting Ethiopia from the ages of incapability and poverty towards a new age of possessing the "capacity" for "exercising" its "rights." Zenawi explicitly stressed that Ethiopia conceives the Nile as one of its "own rivers" and that it has the full right to build the dam as long as any harm to other states is avoided. The confidence reflected in his words that "nothing" would "stop" Ethiopia to open a "new chapter" in the Nile story shows his complete understanding of the suitable timing to make such a move; at the time when Egypt was heavily occupied with its internal circumstances, and Sudan was in bad need for importing cheap electricity and expanding its irrigable lands. Zenawi also emphasized that because this dam would be the largest at all, the objective to build all other dams would be "less challenging."

In addition to these four points discussed above, another important element behind building such a huge dam on the Ethiopian-Sudanese border not further upstream, has been the logic argument stated by Whittington et al. (2014: 598). This contends that since "any attack on the GERD after it is filled would release a sudden and massive flood into Sudan, Egypt's lone ally in dealing with the other Nile riparians," the dam's location is in itself a considerable deterrence against the idea of attacking it, even if improbable.

6.2.4 *Reactions of Egypt and Sudan to the GERD and the Formation of International Panel of Experts (IPoE)*

On April 18, 2011, Malone (2011b) from Reuters asked Alemayehu Tegenu, Ethiopia's Water and Energy Minister, whether Ethiopia officially informed Egypt prior to the project's launch. He answered, "No, they found out from the media," adding that "Ethiopia did not inform Egypt it planned to build a huge dam on the Nile and the two countries have not discussed the issue despite fears a dispute over the river could spark war." He further indicated that, "Ethiopia would not agree to an Egyptian request to see plans for the dam unless Egypt joined the six countries that had signed the new deal [the CFA]," explaining that Ethiopia took this situation after 10 years of fruitless talks to re-negotiate the "colonial-era treaties."

The transitional government in Egypt that had been in power for only few weeks received the surprise of the unilateral announcement of the GERD and chose not to adopt a confrontational strategy, but to make use of the ongoing political change at the domestic level to relieve the tension with Ethiopia prevailing since April 2010. In the beginning of May, an Egyptian popular delegation visited Ethiopia, involving presidential candidates, representatives from political parties and youth groups, journalists, and public figures. The message sent by the 48-member delegation was that "Egypt welcomes any progress in Ethiopia provided it does not affect Egypt's water quota," said Mohammed Abul-Ghar, a prominent member of the delegation (El-Bey, Back to African roots: Popular diplomacy managed to close gaps in Egypt's relationship with Ethiopia over the Nile, 2011). In response, Zenawi announced that Ethiopia would "postpone signing the Nile Basin agreement until Egypt elects a new parliament and president" and that he would welcome the formation of "a committee of Ethiopian, Egyptian and Sudanese experts to look into the Millennium Dam project" (El-Bey, Back to African roots: Popular diplomacy managed to close gaps in Egypt's relationship with Ethiopia over the Nile, 2011). On May 13, Egypt's Prime Minister Essam Sharaf flew to Addis Ababa to meet Zenawi, and on September 17, both met again in Cairo. Through these meetings, the two prime ministers officially agreed to form the international panel of experts (IPoE) to review the GERD's design and construction documents (Aljazeera 2011b). In a joint press conference after the meeting in Cairo, Sharaf said, "We can make the issue of the GERD something useful. This dam, in conjunction with the other dams, can be a path for development and construction between Ethiopia, Sudan and Egypt." Zenawi also reconfirmed his previous declaration of postponing the CFA ratification so that "the new Egypt can study it carefully" (Aljazeera 2011b).

As for Sudan, it did not declare its position upon the GERD during the 1st months. On the one hand, Sudan had been looking forward to the construction of large dams upper the Blue Nile in Ethiopia given their multiple benefits. Taha (2010: 195) records that "contacts are continued between the Sudan and Ethiopia to execute the Karadobi Dam, with the capacity of 40 BCM and with joint financing," since the dam "would save the Sudan from the danger of floods, organize the use of water and lessen evaporation." Moreover, the problem of sedimentation in the Sudanese reservoirs has notably exacerbated. In numbers, the capacity of the Sennar reservoir has decreased from 930 to 370 MCM, and the Roseires, from 3000 to 2200 MCM (FAO 2016c). Obviously, Sudan's need for upstream dams in Ethiopia increased after January 2011 as discussed earlier. On the other hand, the GERD location on the border and its huge scale might cause significant harm to the communities lower the dam, especially in case of dam break. In order to discuss these concerns, an Ethiopian delegation led by Tegenu visited Khartoum on October 25 and met President Bashir, after which the latter announced Sudan's participation in the IPoE (Tekle 2011). Yet, Sudan's position became clear few months thereafter. On March 8, 2012, Bashir announced to the newly appointed Ethiopian ambassador in Khartoum, Abadi Zemo, that Khartoum "understands the mutual benefits the project could offer Ethiopia and Sudan," and that he would "extend the necessary support to ensure the successful completion of the massive hydropower project," as stated by the Ethiopian Ministry of Foreign Affairs (Tekle 2012). This announcement came 1 month after the completion of the 100-MW power transmission line between Ethiopia and Sudan funded by the World Bank (Tekle 2012).

As recorded in IPoE (2013), the water ministers of Ethiopia, Sudan and Egypt set the panel's structural, legal and financial regulations in the period from November 2011 to April 2012. The IPoE consisted of ten members in total: two experts from each of the three riparian states in addition to four international experts. Its mandate was to assess the potential regional benefits and risks of the GERD through reviewing the documents provided by the Ethiopian government. However, the establishment of the IPoE was based upon two conditions stipulated by Ethiopia, namely that the IPoE work would take place in parallel with the GERD construction without halting the project at any step, and that the IPoE findings are not obligatory but advisory for the three governments (FDRE-MFA 2013). The IPoE work took place from May 2012 to May 2013, conducting six regular meetings and four field visits to the GERD site. On May 31, 2013, the IPoE submitted its final report to the three governments addressing three main topics: dam safety and engineering; water resources and hydrology; and environment and socio-economics.

To date, the IPoE final report has not been officially published by any of the three governments. Yet, the International Rivers Network (IRN) published a leaked copy of the IPoE final report[2] online without its appendices on April 2014 (International Rivers 2014a). In general, International Rivers (2012a: 3–5) notes that the GERD planning and implementation phases have been managed through a "top-down approach," where various groups of stakeholders and affected people have not participated in the decision-making process of the dam, whereas the project's technical details were not announced to public. Thus, by publishing the leaked copy of the IPoE report, information about the GERD and its anticipated impacts were made available for the first time, since no one but the IPoE, being a neutral side established by the three countries, was given the chance to review and comment on the project's technical documents.

6.2.5 *The IPoE Findings and Recommendations*

In its introductory section, the IPoE final report recognizes the "good faith" shown by the Ethiopian government, inviting Egypt and Sudan to assess trilaterally the GERD's documents (IPoE 2013: 1). However, throughout its different parts, the report describes most of the design and construction documents as "basic," "preliminary," "outdated," "scattered," "temporarily submitted," or "not made available" (IPoE 2013: 14, 15, 20–43). Examples of this are as follows. Commenting on the "Main Report (Volume I: Basic Design)," submitted by Ethiopia in July 2012, the IPoE states that "the information given in the report with respect to the water resources and power generation aspects is very basic" (2013: 38). The IPoEs comment on the "Hydrological and Reservoir Simulation Study," submitted in July 2012, is that "the analysis presented is very basic, and not yet at a level of detail, sophistication and reliability that would befit a development of this magnitude, importance and with such regional impact as the GERD project" (2013: 37). Another comment on the same study is, "Whilst the GERD can stabilize downstream water supply, the report also shows that this may result in the GERD being drawn down to the minimum operating level for about 15 consecutive years," indicating the challenge of balancing water releases and hydropower generation (2013: 36). In addition, the IPoE indicates that "a study of the impact of a dam break was also not

[2] Available at: https://www.internationalrivers.org/sites/default/files/attached-files/international_panel_of_experts_for_ethiopian_renaissance_dam-_final_report_1.pdf (accessed on December 4, 2016).

presented," but still "ongoing" (2013: 42). Moreover, the IPoE report questions the profitability of the dam's size in terms of economic feasibility against the negative impacts downstream. It notes, "The economic justification is not given with respect to the installed capacity of 6,000 MW, particularly with consideration to the apparent low load factor and the cost of transmission to the main load center" (2013: 38). The report also notes that the dam size was designed to achieve optimal power generation, but did not consider the negative environmental and socio-economic impacts downstream (2013: 38, 40).

Concerning the specific IPoE findings, the report endorses the preliminary design preferences in terms of international standards and regulations of dam safety. Nevertheless, it recommends reconsidering several technical issues about "extreme flood events," "bottom outlet," "sliding stability," and "pressures" due to "groundwater" and "excavation works" (2013: 20–34). With respect to the GERD's environmental impacts, the IPoE report shows that the dam aims to manage floods, control sediments, regulate flows, reduce evaporation losses, generate clean energy, and enhance the Nile resilience against the probable extreme events under climate change through increasing the basin's water storage capacity. However, the report expresses serious concerns about the adverse environmental impacts of the dam and its reservoir, as follows (2013: 39–41). The "methane gas content" and "dissolved oxygen depletion" due to the "decay of flooded vegetation and soil" will deteriorate water quality, affecting the "fisheries" and "biodiversity" in the downstream reservoirs. Besides, the "vegetation clearance," to be "carried out prior to first impoundment," is significantly challenging in technical and financial means. Water quality in the reservoir is also subject to the risks of "eutrophication" and "organic matter content" in sediments trapped upstream the dam, which requires advanced forecast models rather than the "classic empirical" ones used in the preliminary studies. "Deteriorated water quality" will adversely affect the "downstream aquatic biodiversity and fisheries" and "fish stocks," particularly those related to the "Roseires reservoir." Although evaporation losses are to be reduced comparing the site characteristics of both the GERD and AHD (Fig. A.3), losses due to "deep percolation during first filling of the reservoir" are not taken into consideration. In spite of the dam's potential benefit of flood management, adverse impacts on "riparian forest" and "seasonal replenishment of groundwater" along the Blue Nile and Main Nile River are not addressed. The dam's documents also do not address the losses of "suspended organic matter," "riverbed and banks erosion."

As for the socio-economic aspects, the IPoE report lists the benefits that the GERD aims to bring to the riparians, which can be grouped by country as follows (2013: 35–43). Ethiopia will benefit from hydropower generation for domestic development and regional power trade, water storage, fishery industries, navigation improvement, and recreational projects. For Sudan, generated benefits include importing low-cost energy, securing population from devastating floods, expanding irrigated agriculture, saving dredging costs in its reservoirs due to sedimentation, and enhancing the hydropower generation capacity and extending the lifetime of its dams. Benefits to Egypt are mainly importing energy and reducing sediments in Lake Nasser.

Against these benefits, the report underlines the expected risks to the local and downstream communities (IPoE 2013: 7, 36–43). It is mentioned that the project will have negative impacts on "local population," "household/livelihood assets," "public infrastructure," and "social services," all of which are addressed by the project's preliminary studies, outlining a range of "mitigation and livelihood enhancement measures." The GERD first impoundment and operation strategies, aside from expected agricultural expansion in Sudan, will all cause a "fall in the water levels of Lake Nasser." As a result, possible "reduction of power generation" and "irrigation deficit" may negatively affect Egypt. In Sudan, the negatively affected stakeholders involve "fishing communities" around the Roseires Dam and those relying on traditional "recessional agriculture" and on "brick industry."

Based on the findings listed above, the IPoE recommended carrying out further sophisticated technical investigations, specified in two main studies that should be conducted jointly by the three countries. The first is a hydrological simulation study that takes into account the different scenarios of filling the reservoir in "dry years" and operating the dam under "climate change" conditions. The second is a transboundary environmental and socio-economic study that quantifies all potential downstream impacts, extending the study zone to comprise the whole Eastern Nile Basin from Lake Tana and Tekeze River in Ethiopia to the Nile Valley and Delta in Egypt, including the Roseires and Merowe Dams in Sudan (IPoE 2013: 37, 42). The IPoE also recommended further studies concerning dam safety and engineering to be conducted by the Ethiopian side.

In addition to the IPoE report, there have been several studies assessing the GERD from different aspects: Veilleux (2013), addressing it socially; Jeuland and Whittington (2013) and Beyene (2011), economically; King (2013) and Kantoush (2013), hydrologically; Batisha (2013), geologically; and Chen and Swain (2014), Hammond (2013), Samaan (2014), and Tawfik (2015), politically. The general findings of these studies are largely consistent with the IPoE report. Moreover, in November 2014, a group of international experts published an independent report[3] on the GERD through holding a workshop convened by the Massachusetts Institute of Technology (MIT). The MIT report is based on the full recognition of Ethiopia's right to construct dams on the Blue Nile for national economic development and regional integration. At the same time, it stresses on the necessity of involving Egypt and Sudan in every phase of the project to assure that no harm would be caused to downstream communities. Similar to the IPoE, the MIT report urges the three countries to agree upon the strategies of first filling, annual operation and power trade. In addition, the MIT report recommends that further studies should be conducted on the GERD's downstream impacts, especially for the "loss of recession agriculture lands" that will affect "tens of thousands of hectares" in Sudan and the "salinization" that will impact "millions of hectares" in Egypt (MIT 2014: 9). However, the point underlined by the MIT report with special emphasis is the serious threat associated with the GERD's saddle dam for its "extensive" scale and "catastrophic" risks in case of any dam safety problem (MIT 2014: 6).

[3] Available at: http://jwafs.mit.edu/sites/default/files/documents/GERD_2014_Full_Report.pdf (accessed on December 16, 2016).

6.2.6 The Tension over Diverting the Course of the Blue Nile

From May 24 to 28, 2013, few days before the IPoE submitted its final report, Egypt's President Morsi visited Addis Ababa to participate in the 21st session of the African Union summit. During the visit, Morsi met Ethiopia's former Prime Minister Hailemariam Desalegn, who succeeded Zenawi, after the latter had passed away on August 20, 2012. As reported in Ahram Weekly (Leila 2013a), Desalegn responded to Morsi's concerns over the GERD in their meeting, asserting that Ethiopia "would never harm either Egypt or Sudan." Few hours after Morsi returned to Cairo on May 28, Ethiopia announced beginning the diversion of the Blue Nile course as part of the GERD construction work (Leila 2013b). The diversion step was described by the Ethiopian government as "a mere re-routing of the water, a normal procedure in building a dam and no different from the established norms of the science of dam engineering" (FDRE-MFA 2013; Berhane 2013).

However, since it was undertaken without prior arrangement, this step caused tension with Egypt and Sudan but with varying degrees. The Turkish Anadolu Agency circulated the statement of the Sudanese ambassador in Cairo, commenting on the issue on the same day, "There are continuous calls between the Egyptian and the Sudanese authorities to look into Ethiopia's sudden and shocking decision." He also indicated that the two countries "may call for intervention by the Arab League" (Ahram 2013b). Two days later, the Sudanese Ministry of Foreign Affairs "denied" Ali's statement to Anadolu and affirmed that neither the diversion step nor the GERD construction "impose any threat to Sudan" (Sudan Tribune 2013a). In Egypt, Morsi's spokesman Omar Amer commented immediately following the Ethiopian announcement that the step had no negative impacts downstream, declaring, "Any project on the Nile requires diverting the course of the water passageways before starting construction" (Ahram 2013a). On the contrary, the headlines of the Egyptian newspapers heavily criticized Morsi's weak reaction, especially since Ethiopia took the decision of diverting the most substantial watercourse for Egypt without mentioning the topic to Morsi during his meeting with Desalegn few hours earlier. This is particularly true given the mainstream opinion among experts in Egypt pushing for the halt of the GERD construction work until the downstream countries would assure that the GERD in its massive scale would not harm them. Such headlines included the following, as translated from Arabic by the author: "Morsi visited Ethiopia, then the course of the Blue Nile was diverted" (Abou-Amra and Abnoudi 2013); "The activists blame Morsi for the crisis of the Renaissance Dam" (Abdel-Hadi 2013); and "After the solemnity of 40 years since the Sadat era, Ethiopia dare to Egypt during the reign of Morsi" (Ashra 2013).

Meanwhile, popular dissatisfaction and anger had been increasing among Egyptians against Morsi and the Islamists since the day when the latter used violence to suppress the demonstrators in front of the presidential palace in early December 2012 (Al-Monitor 2012). Those were protesting against Morsi's decree through which he seized "sweeping new powers" and put his decisions "beyond any legal challenge" (Awad 2012). In late April 2013, a group of youth, calling themselves "Tamarod" in Arabic or "Revolt" in English, launched a petition campaign to "withdraw confidence" from Morsi and hold early presidential elections (Taha 2013). By late May, millions of people had already signed the Tamarod form, and the group declared their target to collect more than 15 million signatures by the first anniversary of Morsi's presidency in late June in order to exceed the number of votes he got in the 2012 presidential election (Taha 2013; Mouterde 2013; BBC 2013d). Hence, the way and timing of diverting the Blue Nile course in relevance to Morsi's visit to Ethiopia put great pressure on him and aggravated his situation at the domestic level.

On June 3, Morsi called for a meeting with a group of Egyptian politicians from different backgrounds to discuss the issue of the GERD.[4] At the beginning of the meeting, the main findings and recommendations of the IPoE final report were presented, highlighting the negative impacts that the GERD could cause downstream, as well as the insufficiency and incompletion of the dam's design and construction documents. As the attendees were not informed that the meeting was broadcast live by state TV, they "spoke their minds unreservedly" (BBC 2013a). Some proposed "military action" as a "decisive response" to what one of them called Ethiopia's "declaration of war," while others suggested sending "special forces to destroy the dam," and supporting "rebel groups" in Ethiopia to be used as a "bargaining card" in negotiations (BBC 2013a). Not only did the attendees criticize Ethiopia but also Sudan for not aligning with the Egyptian position. Although Morsi did not comment on these suggestions and explained that these statements did not represent Egypt's official point of view, his Aide for Political Affairs, Pakinam el-Sharkawy, apologized later in the evening. She said, "It was initially planned that the national meeting would be recorded and aired the following day as is usually the case, but due to the importance of the topic it was decided at the last minute to air the meeting live ... I forgot to inform attendees of the changes" (Ahram 2013c). One may guess that, in an attempt to calm down the people, Morsi wanted the meeting to be a public declaration showing that he was taking the GERD matter seriously and that he would react strongly on the Ethiopian step. However, ultimately, the meeting had a severe adverse effect. Six days later, Morsi adopted a threatening tone in a televised speech quoted in BBC (2013b), saying:

> Egypt's water security cannot be violated at all ... As president of the state, I confirm to you that all options are open ... If Egypt is the Nile gift, then the Nile is a gift to Egypt ... If it diminishes by one drop, then our blood is the alternative.

It is noteworthy that the GERD drove both the Egyptian and Ethiopian sides to stress on their ownership of the Nile, in that Zenawi had called it on the inauguration day of the dam, one of Ethiopia's "own rivers," whereas Morsi here called it "a gift to Egypt."

The Ethiopian response was immediate. On the next day, in an interview aired on the state TV, Desalegn commented on Morsi's threats stating, "All options include a war. I do not think they will take that option unless they go mad ... I urge them to abandon such an unhelpful approach and return to dialogue and discussion"

---

[4] Available at: https://www.youtube.com/watch?v=kdyMi1hrpoA (accessed on December 5, 2016).

(Pflanz 2013). He also emphasized, "Nothing and no one will stop the dam's construction" (Milas 2013). Moreover, the Ethiopian parliament unanimously took its decision on June 13 to ratify the CFA. On that day, the spokesman of the Ethiopian government declared, "We delayed it as a gesture of good will to the people of Egypt until a former elected government was in place ... We have a principled stance on the construction of dams. We are determined to see our projects brought to completion" (Maasho 2013).

The exacerbated tension between Egypt and Ethiopia motivated non-governmental bodies to seek a relief between the two countries. Amongst these was the Coptic Orthodox Church in Egypt. On June 11, the Coptic Pope held a meeting with the church leaders to discuss the role they could play in the crisis (Ahram 2013d), benefiting from the historic relation between the Egyptian and Ethiopian churches (fn. 12). A visit by the Ethiopian Patriarch to Cairo was scheduled on June 17 to further discuss the issue. Yet, given the exchange of hostile words between the two countries, this was postponed 4 days before the planned date (Ahram 2013e). At the official level, Egypt's Foreign Minister, Kamel Amr, visited Addis Ababa on June 18 to meet his Ethiopian counterpart, Adhanom Tedros as an attempt to ease the tension. They discussed the steps to be taken after the submission of the IPoE final report and agreed to hold further meetings together with Sudan, paving the way for conducting the technical studies recommended by the IPoE (BBC 2013c). Apart from the tension over the GERD, the Nile-COM held a meeting in Juba on June 20. Except for Egypt, all the Nile riparians attended, including Sudan that decided to resume its full participation in the NBI, in addition to South Sudan that had been a full member in the initiative since July 2012 (Sudan Tribune 2013b).

6.2.7 The Ouster of Morsi and a New Transitional Period in Egypt

The political events in Egypt were rapidly developing. The number of people who signed the Tamarod form exceeded the campaign's target, and its organizers expressed their hopes to "transform the millions of signatories into a huge rally against the Morsi regime" (Mouterde 2013). On June 30, mass protests "flooded into streets" across Egypt, demanding "Morsi quit" and voicing "their anger at the Brotherhood, which they accuse of hijacking Egypt's revolution and using electoral victories to monopolize power and impose Islamic law" (Fayed and Saleh 2013). Following Sunday's massive demonstrations, Tamarod group gave Morsi an "ultimatum to resign by 17:00 on Tuesday or face a campaign of complete civil disobedience," and they urged "state institutions including the army, the police and the judiciary, to clearly side with the popular will as represented by the crowds" (BBC 2013d). At the time, the bitter hostility between these state institutions, on the one hand, and the Muslim Brotherhood, on the other, was obvious. This is mainly due to the fact that the latter aimed to dominate over these public institutions, changing their leadership (e.g. the SCAF's leaders in September 2012 and the prosecutor general in November 2012) while putting them under siege several times (e.g. the Supreme Constitutional Court in December 2012 and the Media Production City in March 2013) (Abdel-Alim 2013). Morsi's implicit threat to launch a war against Ethiopia and explicit intension to participate in Jihad against al-Assad regime were also important reasons behind the anger of the military. A security source told Reuters in this regard, "It reached a point where we began to be worried about putting important national security reports in front of someone we perceived as a threat to national security" (Saleh and Taylor 2013).

On the following day, the military announced an ultimatum giving Morsi 48 h to resolve the crisis and satisfy people's demands, but he declared his rejection to resign in a televised speech on Tuesday, saying, "If the price of protecting legitimacy is my blood, I'm willing to pay it" (Kirkpatrick and Hubbard 2013). Protesters remained in the streets and occupied large squares in Egyptian cities. On Wednesday, July 3, almost 2 years and half after the January 25 Revolution, the military decided once again to intervene. Headed by General Abdel-Fattah el-Sisi, the military held a meeting with leaders of all political and religious forces in Egypt, including Al-Azhar (the top Islamic institution), the Coptic Orthodox Church, Tamarod group, liberal parties, and the Salafist Nour Party (the second largest Islamist party after the Muslim Brotherhood). Upon this meeting, Morsi was ousted and a new transitional period began under the presidency of Adli Mansour, the then president of the Supreme Constitutional Court (Perry and Saleh 2013). Unlike January 25, when Mubarak was overthrown, after the military had supported the popular protests, the ouster of Morsi has been viewed from two controversial perspectives; whereby most of Egyptians called it the June 30 Revolution while many international sides called it the July 3 Coup. The Western nations firmly opposed Morsi's ouster, and the African Union decided on July 5 to suspend Egypt's participation in the AU's activities "until the restoration of constitutional order" (BBC 2013e). Nevertheless, Egyptians were more concerned about the position of the Gulf States, which immediately declared their support to the revolution, as well as over Sudan's position due to its potential impact on the GERD issue, since the latter shared the same Islamist background with the Morsi regime. However, Sudanese officials affirmed the "internal nature" of the events in Egypt (Sudan Tribune 2013c).

Consequently, in light of such dramatic political changes, Egypt was again fully absorbed in domestic affairs. Besides political instability and economic pressure, a new wave of terrorism was sparked in retaliation to ousting Egypt's first Islamist president. The wave began specifically on August 14, when security forces broke up the Brotherhood's 47-day armed sit-ins in Rabaa and Nahda squares, Cairo (Maged 2015). Reuters reported that more than 200 were dead, including at least 43 police, as declared by Interior Minister Mohamed Ibrahim (Saleh and Finn 2013). The instant response, within the following days, was attacking 42 churches, out of which 37 were burned or damaged, as well as dozens of other Christian religious institutions across Egypt (Human Rights Watch 2013). On the long term, terrorism has been concentrated in Northern Sinai, but also military and police officers, judges, and civilians have been subjected to frequent attacks in the entire country; though decreasing over time.

6.2.8 A Stalemate over the Implementation of the IPoE Recommendations

Talks over the GERD were continued only few months later when Khartoum hosted three sequential meetings on November 4, December 8–9, 2013, and January 4–5, 2014. The main objective was the establishment of a tripartite technical committee that would be responsible for the implementation of the IPoE recommendations. In late October 2013, after a meeting of the Supreme Committee for the Nile Waters in Egypt (SCNW), the newly appointed Prime Minister Hazem el-Beblawi commented[5] that the GERD "can be transformed into a useful project for all sides." He also declared that the Egyptian delegation, participating in the Khartoum talks, would present a "comprehensive vision for cooperation that secures the interests of all sides." However, the Khartoum talks were apparently unsuccessful and reached a deadlock (Bloomberg 2014; Hailu 2014). In an interview with the Official Ethiopian Press Agency, Fekahmed Negash, the Director of Boundary and Transboundary Rivers Affairs at the Ethiopian Water Ministry, said that the Ethiopian delegation "automatically" rejected the Egyptian proposals during the discussions series because these would "harm the national interests, sovereignty and development of Ethiopia" (Hailu 2014). The Egyptian proposals were to suspend the GERD construction until the IPoE recommendations would be implemented, to involve international experts in the new committee, to make the committee's decisions obligatory for all the three countries, and to pledge in a written commitment that the GERD would not affect the water shares of Egypt and Sudan. The Ethiopian delegation clarified that they could not approve these proposals. They insisted that halting the construction works is a non-negotiable matter, that there is no appropriate reason to internationalize the committee, and that making the decisions of a technical committee obligatory for Ethiopia violates its sovereignty. As for the issue of water shares, they emphasized that it should be taken "to the right platform" that includes all the Nile riparians, referring to the CFA (Bloomberg 2014). As a result, the meetings were terminated without achieving any progress. Commenting on the event, the then Egyptian Water Minister, Mohammed Abdel-Moteleb, declared, "We have exhausted all opportunities to negotiate with Ethiopia because of the intransigence of Addis Ababa" (Bloomberg 2014). On an earlier date, after the first meeting in November 2013, President Mansour criticized Ethiopia's decision to build a huge dam on the Blue Nile from the beginning. "Ethiopia says the purpose of establishing the dam is to produce energy which could be generated by other means or different types of dams that could use engineering systems that do not harm downstream countries," said Mansour (ESIS 2013b).

On February 11, the Egyptian and Ethiopian Water ministers held a meeting in Addis Ababa, but this also proved fruitless (ESIS 2014a). Moreover, Desalegn sent a clear message to Egypt during a local press conference in the same week

[5]Available at: https://www.youtube.com/watch?v=ftNEe0iRXBA (accessed on December 9, 2013).

(FDRE-MFA 2014; Chimp Reports 2014), stating that Ethiopia and Egypt have "no other option except dialogue and negotiation to provide a win-win solution over the GERD." About the possibility that Egypt might internationalize the case, he answered that there is "no international court responsible for investigation or judgment on water issues and such a move could have no result." In response, Abdel-Moteleb declared in early March that "contacts would be made with international institutions and organizations in charge of financing big water projects with the aim of bringing Ethiopia back to the negotiating table to reach a conciliatory solution that serves development in Ethiopia and makes no damage to Egypt's water share" (ESIS 2014b). Furthermore, late that month, the Egyptian Foreign Ministry circulated an official paper to all Egyptian embassies abroad and foreign mass media about Egypt's perspective on the GERD (EMFA 2014a, b). The concluding remark of the paper states the following:

> [R]egrettably, despite the absence of complete environmental and hydrological impact assessment studies, Ethiopia has continued the construction process at the dam site in violation of all the well-known international legal principles regarding projects and/or constructions on international rivers ... However, absent of such an agreement [that assures the interests of all countries as referred to earlier in the statement] and the continued construction by the government of Ethiopia of the said dam, creates a fait acompli that potentially constitutes a significant threat to Egypt's national and water security.

It is worth noting that the term "fait accompli" was used before in the Ethiopian statements protesting against the Peace Canal and Toshka projects (Sect. 4.3.3 and 5.3.2). Despite that the impacts of unilateral upstream projects are incomparable to those of downstream ones, this indicates how each side views the other's national schemes as creating facts on the ground that harm its own interests, and that each side reserves its complete right to assess whether its own projects harm the other or not.

Concerning the Egyptian criticism against the GERD over the violation of international water law, an Ethiopian top water official, M.M. (2016b), told the author, "We are not working to harm the peoples in the other countries. The Nile is our common resource, and we should share it for the common benefit of our peoples in the whole basin." With respect to the obligation of prior notification in particular, the author asked T.B. (2016), an adviser to the Ethiopian water ministry, about why Ethiopia did not notify its co-riparians prior to construction of the GERD. He answered, "Ethiopia, in the case of the GERD, did more than notification" through inviting national experts from the downstream countries in addition to international experts in order to review and comment on the dam's design and construction documents. He also emphasized, "So far in the Nile basin, the only structure on which consultation and information sharing were done was the GERD." He gave an example of the Aswan High Dam (AHD) in which Ethiopia raised its concerns about it and indicated that the dam will "foreclose its future development," but Egypt replied that it is free in managing the waters reaching it. He added that "exchange of information is reciprocal," and that "no one has notified Ethiopia before" in building the other dams in Egypt and Sudan. He also clarified, "Ethiopia believes in cooperation, but cooperation is a multilateral process," and principles should be respected by all sides. He ended his answer to this question saying, "There is no mechanism that binds or forces Ethiopia to notify the downstream countries, except the CFA, which is not yet ratified by all countries," but in the absence of such a mechanism, when "Egypt was ready for cooperation, Ethiopia invited it and Sudan for consultation on the GERD."

From the Egyptian perspective, the projects implemented downstream have no significant or appreciable harm on upstream riparians in contrast to projects built upper the river that may alter the flows to downstream riparians in both quantity and quality (Ahram 2015c). Concerning this point, Professor McIntyre (McIntyre and interview by 2016) explained to the author that the obligation of prior notification comprises two important conditions, "to notify properly, early before construction and with sharing full impact assessment studies, and to notify in good faith, to be ready to modify the project and bear costs of reaching a compromise." He clarified that these two conditions apply to both upstream and downstream projects, since the latter may negatively affect the health of the entire basin and may foreclose development upstream. It is also important to note that the Egyptian official paper mentioned above was written few months prior to the ratification of the UNWC by minimum required number of signatories to enter into force, which took place only on August 17, 2014. Thus, the author also asked Professor McIntyre whether a riparian state can resort individually to international arbitration when it believes that one of its co-riparians has violated the international water law. McIntyre (2016) said that international arbitration can be held only if the concerned riparians jointly accept it or if there is a binding agreement on the disputed issue. In the GERD case, it is evident that Ethiopia sees no need to internationalize the issue. Furthermore, it already challenges the validity of the 1902 Agreement, and does not recognize the 1929 and 1959 Agreements. This makes the scenario of arbitration obviously challengeable. In addition, the 1993 Egyptian-Ethiopian Agreement contains an explicit article about causing no significant harm, but Ethiopia believes that it has not violated the principle, since it argues that the dam would harm neither Egypt nor Sudan.

Tension between Egypt and Ethiopia over the GERD increased in April 2014, after International Rivers had published online a leaked copy of the IPoE final report (International Rivers 2014a). Ethiopia accused the non-governmental organization of being backed by "Egyptian financers" as part of an international campaign against the GERD, the claim of which International Rivers (2014b) quickly rejected. Simultaneously, the Egyptian government made another attempt by offering to take part in financing and managing the GERD, but Ethiopia firmly refused the offer. The spokesman of the Ethiopian Foreign Ministry, Ambassador Dina Mufti, explained that the GERD "belongs to Ethiopians and the government of Ethiopia" and "there is no room for Egyptian administration of the dam," since it is "a matter of sovereignty and not of aid and cooperation" (Berhane 2014). Ethiopia had taken this decision since the first day of the GERD official commencement, when Zenawi declared that his country would go for the "difficult choice" of building the dam with domestic funds. As illustrated by the International Monetary Fund (2013: 6), the difficulty of this choice is due to the risk of "squeezing" domestic financial sector, considering that "the ongoing large public investment projects rely heavily on domestic financing and would lead to a large accumulation of public debt." The cost of construction, which may reach $5 billion, represents about 12% of Ethiopia's annual output and it absorbs about 27% of the banks' loan books accessible for private borrowers (Maasho 2014). Nevertheless, Ethiopia relies on the expected revenues of exporting hydropower generated by the GERD to its neighbors, as Sudan, South Sudan, Kenya and Djibouti (World Bulletin 2014). Besides, there are Chinese companies that are supporting Ethiopia in carrying out the high voltage lines connecting the dam to the grid (Maasho 2014).

The position of Sudan in the issue remained completely different from that of Egypt. During the inauguration of the power linkage network between Sudan and Ethiopia on December 4, 2013, in Gedaref state, President Bashir announced publicly his country's full endorsement to the GERD and that this position is based on economic not political reasons. He clarified that Sudan will benefit directly from the electricity generated by the Ethiopian dam and indirectly through raising the power generation capacity of Sudanese hydropower projects after the construction of the dam. The Sudanese president and the Ethiopian prime minister signed, on that day, an agreement strengthening ties between the two countries through the establishment of railways projects and free trade zones (Sudan Tribune 2013d; Al-Haj 2013).

On February 18, 2014, Ali Karti, Sudan's Foreign Minister, confirmed that his country is neither taking sides with Egypt nor Ethiopia in the dispute. He said, Khartoum would continue its efforts to bridge the gap between the two countries (Sudan Tribune 2014a). At the same time, Sudan did not dismiss its concerns regarding the GERD's potential negative impacts, which was clear in Karti's statement 3 months thereafter indicating, "Sudan would suffer the most if constructing the dam is done without environmental studies to prove its safety or economic feasibility to the three main Nile Basin countries." He added that when his government realized that "there was a slackening in examining these issues," it formed "a national committee to study all aspects of the dam with the right to cooperate with any of the national committees in Ethiopia or Egypt" (Sudan Tribune 2014b).

6.3 The Search for a Solution

6.3.1 A New President in Egypt and a New Strategy over the GERD

On June 8, 2014, el-Sisi was inaugurated as Egypt's new president after he won 23.78 million votes (96.91%) in the presidential election that was held on May 26–28 (Ahram 2014b). During his electoral campaign, the Nile occupied a central position in el-Sisi's statements. Concerning his vision on national development, el-Sisi announced his intention to revive the land reclamation plans across Egypt's deserts (Fig. 6.4), including the Toshka and Peace Canal projects (Ahram 2014a), which had been witnessing serious problems since the 2011 Revolution. Regarding the GERD, el-Sisi declared as a presidential candidate his strategy in resolving the issue, focusing on the completion of the required studies that assure filling the reservoir and operating the dam without harming Egypt, besides broadening

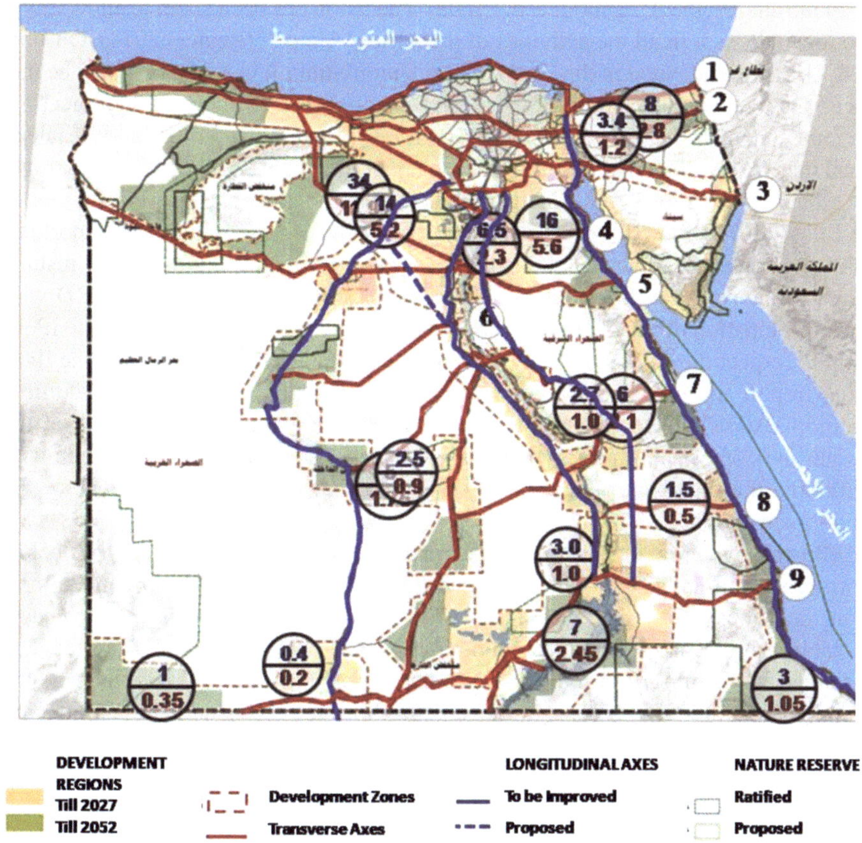

Fig. 6.4 Map of future urban development in Egypt. (Source: Egypt's General Organization of Physical Planning (2010), The Strategic Plan of Egypt)

bilateral cooperation with Ethiopia. This is stated clearly in the following phrases said in his televised interview with Sky News Arabia[6] on May 11, as translated by the author:

> It is extremely important that our relations with Ethiopia and all the Nile Basin countries be so good. Unfortunately, this had not been the case in the past. Our message to them should be clear: we cannot focus only on our interests and overlook yours. Total number of population in Ethiopia is very similar to us, and they aim to achieve development and progress for their people. We understand this and we are not opposed to it, but even we are very willing to cooperate over it. At the same time, they should understand that water for Egypt is a life or death issue. It is therefore a vital issue for both of us, and we must find a solution together. Concerning the Renaissance Dam, we view the adverse impacts on the downstream countries during the reservoir filling and dam operation as potential threats that must be addressed in further studies. Upon these, we must reach a binding agreement that assures the success of such a huge project without causing harms to us.

[6] Available at: https://www.youtube.com/watch?v=OFnZkKtut9w&feature=youtu.be. (accessed on December 11, 2016).

On June 17, the AU announced, "Today a decision has been taken to allow Egypt to regain its seat in all the activities of the African Union" (Reuters 2014). On June 26, el-Sisi participated in the 23rd African Union Summit in Malabo, where he met with Desalegn on the margin of the event (Ahram 2014c). Following their meeting, a joint statement[7] was issued (EMFA 2014c), stipulating that el-Sisi and Desalegn had met "to set the foundation for a new chapter of enhanced bilateral relations and regional cooperation." The statement indicated that both countries would establish a Bilateral Joint Commission with the aim of exploiting the "enormous opportunities and potentials of working together." It was also agreed to "immediately resume their participation in the trilateral technical committee regarding the GERD with the participation of Sudan, in order to implement the recommendations of the IPoE, and to respect the outcomes of the joint technical studies recommended in the IPoE final report throughout the implementation phases of the project." In addition, Ethiopia stipulated to "avoid any potential adverse effects of the GERD on the water uses of Egypt," and Egypt committed itself to "a constructive dialogue with Ethiopia that takes into account the developmental needs and aspirations of the Ethiopian people."

Based on the Malabo statement, the Water ministers of the three countries held a two-day meeting in Khartoum, on August 25–26, 2014, through which three important decisions were taken (Ahram 2014d; Salman 2014). First, it was agreed to establish the Tripartite National Committee (TNC), which consists of four experts from each country (12 in total). Second, the mission of the TNC is to manage the implementation of the IPoE recommendations through hiring an international consultancy firm that will conduct the two required studies of the hydrological simulation and the environmental and socio-economic impacts of the GERD. Third, the then Egyptian Minster, Hossam Moghazi, pushed for shortening the period of conducting the needed studies from 1 year to 6 months. These three decisions obviously reflected a sensible compromise that dealt with the disputed points previously raised during the Khartoum talks (Sect. 6.2.8). In return for the Egyptian implicit recognition of the project as a fact on the ground that could not be halted, the Ethiopian side accepted to hire an international consultancy firm to implement the IPoE recommended studies; though under the management of a committee of national experts. However, the shortened period of 6 months was evidently too ambitious, as the process of choosing an international consultancy firm that would satisfy the three countries was certainly difficult. Actually, the nomination of the firms took exactly 2 years, from the first TNC round, on September 20, 2014, to the signature of the technical contracts, on September 19, 2016.

Following the August 2014 meeting, a series of meetings were held in Addis Ababa, Cairo and Khartoum, which addressed both technical and legal aspects. Technically, the meetings aimed to study the presented offers by the different international consultancy firms. On the margin of one of these technical meetings,

---

[7] Available at: http://www.mfa.gov.eg/english/embassies/egyptian_embassy_albania/mediacenter/pages/NewsDetails.aspx?Source=6781921f-3993-444a-859e-ee26ce851de8&newsID=942cac21-0b94-4eeb-a0ae-a1f678c7faea (accessed on December 12, 2016).

specifically on September 22, the Egyptian and Sudanese ministers, invited by their Ethiopian counterpart, conducted a tour in the GERD's construction site, which marked the first time for an Egyptian official to visit the project (Ahram 2014e). Legally, efforts were underway to conclude an agreement to be signed by the three countries stating the general principles upon which the trilateral cooperation would be founded. On March 3–5, 2015, a preliminary draft of the proposed agreement was reviewed by the foreign and water ministers from the three countries to be submitted to the heads of state/government (Sudan Tribune 2015a).

6.3.2 *The Agreement on the Declaration of Principles (DoP)*

On March 23, 2015, Bashir, Desalegn and el-Sisi signed in Khartoum the Agreement on Declaration of Principles on the GERD[8] (DoP). The DoP comprises ten principles committed by the three countries. The first and ninth principles set the foundations of cooperation between the three countries, involving "common understanding" of both "upstream and downstream water needs in its various aspects" and recognition of "sovereign equality" as well as "territorial integrity." The second principle stipulates that "the purpose of GERD is for power generation, to contribute to economic development, promotion of transboundary cooperation and regional integration through generation of sustainable and reliable clean energy supply," while the sixth principle gives Egypt and Sudan the "priority ... to purchase power generated from GERD" as a sort of "confidence building." The third and fourth principles state respectively the obligation "not to cause significant harm" and the right of "equitable and reasonable utilization." In the seventh principle, the three countries pledge to provide the TNC with needed data and information "in good faith and in a timely manner." The eighth principle states the appreciation of the three countries to the "efforts undertaken thus far by Ethiopia in implementing the IPoE recommendations pertinent to the GERD safety" and urges Ethiopia to continue the "full implementation" of the IPoE recommendations in this regard. The tenth and last principle indicates that the three countries may "jointly request for conciliation, mediation or refer the matter for the consideration of the heads of state/heads of government" in case of unsettled disputes. Nevertheless, the most important principle is the fifth, which reads as follows:

> The Three Countries, in the spirit of cooperation, will utilize the final outcomes of the joint studies, to be conducted as per the recommendations of the IPoE Report and agreed upon by the TNC, to: (a) Agree on guidelines and rules on the first filling of GERD which shall cover all different scenarios, in parallel with the construction of GERD. (b) Agree on guidelines and rules for the annual operation of GERD, which the owner of the dam may adjust from time to time. (c) Inform the downstream countries of any unforeseen or urgent circumstances requiring adjustments in the operation of GERD.

---

[8] Available at: https://hornaffairs.com/2015/03/25/egypt-ethiopia-sudan-agreement-on-declaration-of-principles-full-text/ (accessed on December 13, 2016).

> To sustain cooperation and coordination on the annual operation of GERD with downstream reservoirs, the three countries, through the line ministries responsible for water, shall set up an appropriate coordination mechanism among them.
>
> The time line for conducting the above mentioned process shall be 15 months from the inception of the two studies recommended by the IPoE.

The difficult compromise made by Egypt and Ethiopia in the DoP can be clearly observed, not only in relevance to the 4-year tension over the GERD, but also in light of the deadlock over the CFA. In this respect, the main reason behind which Egypt has refused to sign the CFA is that the latter gives upstream riparians the right to equitable and reasonable utilization without stating the obligation not to alter the current water uses downstream (Sect. 5.4.3). Yet, in the DoP, Egypt committed itself to the principle of equitable and reasonable utilization, as copied from article 4 in the CFA, which is unprecedented since Egypt has not signed to date the CFA, nor ratified the UNWC. At the same time, there is no mention in the DoP to historic rights or existing agreements. In contrast, it recognizes the water needs of both upstream and downstream riparians and emphasizes the sovereign equality in the basin as well as the territorial integrity. Evidently, Egypt has agreed to this compromise in return for Ethiopia's commitment to respect the outcomes of the joint studies and to conclude the agreements of the first filling and annual operation. In addition, the GERD's purpose is clearly declared to target power generation, which was an Egyptian concern that the water in the reservoir may be used for irrigation, despite Ethiopia's affirmation that the ground surrounding the reservoir is not suitable for agriculture. Moreover, Ethiopia pledges in the DoP to follow the three gradual steps of not causing significant harm; to eliminate, to mitigate and to compensate.

The signature of the DoP was preceded and succeeded by gestures of improved bilateral relations between Egypt and Ethiopia. On March 13, 2015, Desalegn participated in the Egyptian Economic Conference held in Sharm el-Sheikh (ENA 2015a). In his speech[9] during the conference opening, Desalegn directed his words to el-Sisi, saying, "The Nile constitutes a common destiny: we either sink or swim together. You know, Mr. President, we chose to swim together" (Ahram 2015a). On the same day of signing the DoP in Khartoum, el-Sisi started a three-day visit in Addis Ababa, which is the first official state visit by an Egyptian president to Ethiopia in 30 years (ENA 2015b; Sudan Tribune 2015b). During the visit, el-Sisi met Ethiopia's President Dr. Mulatu Teshome and the then Prime Minister Desalegn in addition to delivering a speech in the Ethiopian parliament,[10] parts of which are stated at the beginning of this chapter.

Those words are of great significance and need to be closely scrutinized extracting two substantial ideas. The first is that Egypt, after long ages of being the sole user of the Nile waters, has ultimately realized that water security cannot be achieved

[9] Available at: https://www.youtube.com/watch?v=MznnXnN42Ek (accessed on December 13, 2016).

[10] Available at: https://www.youtube.com/watch?v=BDfXYlHM-qY (accessed on December 13, 2016).

by maintaining one's own interests and leaving the other incapable of reaping benefits from the shared resource, as this will lead the latter to retaliate 1 day and harm the interests of the former. At the same time, el-Sisi declares that it will not allow the others pursue their interests and promote their welfare "at the expense" of Egypt. Thus, the second idea is that Egypt views the "foundation of a better future" as encouraging the generation of hydropower upstream so that "all classrooms in Ethiopia are lit" and then the Nile waters are released downstream so that "children of Egypt can drink from the River Nile as their fathers and grandfathers did." Here, el-Sisi emphatically states that his willingness to cooperate over upstream dams and to respect the principle of "equitable utilization" never means that Egypt will renounce its historic water rights.

In addition to the official events, there has been an exchange of visits by popular delegations from Egypt and Ethiopia, as an attempt to cement ties between the two countries (Fahmy 2015). As for Sudan, it continued playing its role as a mediator, not only in water issues but also for broader cooperation in various fields. In a trilateral summit in Sharm el-Sheikh, on June 10, 2015, Bashir suggested the establishment of a supreme joint committee to strengthen ties between the three countries at the leadership level (Sudan Tribune 2015c). This was followed by another summit, again in Sharm el-Sheikh, on February 20, 2016, where the three leaders agreed to continue the work of institutionalizing cooperation through the supreme joint committee (Ahram 2016a).

6.3.3 Sharp Technical Differences over the GERD

Despite the growing rapprochement between Egypt and Ethiopia at the official level, as shown above, there were major differences between the two countries over fundamental technical issues. In specific, these were centered on three main points: selecting the international consultancy firm, reducing the GERD storage capacity, and increasing its bottom outlets.

As mentioned earlier, the TNC had conducted several meetings since September 2014 to select the international consultancy firm that would conduct the IPoE recommended studies. Only in April 2015, the TNC reached a decision identifying the BRL French group and the Deltares Dutch firm; whereas, while Ethiopia was pushing for the selection of the former, Egypt supported the latter (Daily News Egypt 2015a; Tekle 2015). The initial problem was therefore how to coordinate the work between the two firms and how to define the scope of each. By late July, the TNC came up with a solution to assign the primary role (70% of tasks) to BRL, while Deltares will take the secondary role (30% of tasks) (Hussein 2015). The Dutch firm was not satisfied with the arrangement and decided to withdraw its offer in September. During the ninth TNC round, on November 7–8, Egypt "rejected the implementation of the technical studies of the Ethiopian Renaissance Dam by only one company" (ESIS 2018).

Concerning the size of the reservoir, there has been a sharp difference over whether or not the IPoE recommended studies should involve a reconsideration of the dam height and the storage capacity. This issue was publicly discussed in a televised interview[11] with the three water ministers of Egypt, Ethiopia and Sudan on an Egyptian private channel in mid-October 2014. The first question for the three ministers was about the general vision adopted by each country over the GERD. The Sudanese Minister Moataz Mousa clarified his country's initial support for the project, as the construction of upstream dams has been a long-standing proposal since a century ago, but at the same time, he emphasized that Sudan is pushing for the completion of the IPoE recommended studies to assure the mitigation of any negative impacts caused by the dam. Nevertheless, the answers of the Egyptian and Ethiopian ministers were focusing on the dam capacity, reflecting the wide gap between their points of view, as follows:

> We are not opposed to the dam in itself but to its storage capacity. We believe that the dam's original capacity of 14 BCM was acceptable, in that it would not be harmful to Egypt. However, Ethiopia announced the construction of the dam with an increased capacity of 74 BCM, which will cause, from our point of view, significant harm to us, especially during the period of first filling due to the consequent large deficit in Egypt's water share. Thus, we seek to reach a compromise with Ethiopia based on conducting the IPoE recommended studies, which will investigate the reasonable size of the reservoir and the period of first filling. ... Our fixed position, since the beginning of the negotiations, is that the capacity of the dam must be reduced.

The then Egyptian Water Minister Hossam Moghazi (translated from Arabic by the author).

> The Ethiopian peoples have constructed the GERD to compact poverty in the country and in the region. This dam is designed in the way that will not cause significant harm to the downstream countries. Rather, it will substantially benefit them. We are here not to discuss the reduction of the volume, the reduction of the height. We are here to discuss how to implement the IPoE recommendations that were given to the three countries. These are the hydrological modeling and the socio-economic impacts.
>
> The then Ethiopian Water Minister Alemayehu Tegenu.

The Egyptian position is founded on the following phrases stated by the IPoE final report in its comments on the GERD's design and construction documents submitted by Ethiopia:

> The dam size is consistent with the inflow at the GERD project site, but should also be considered in relation to the impacts downstream. Similarly, account should be taken of future upstream storage. (IPoE 2013: 38)

> The report [environmental and social impact assessment report submitted by Ethiopia] noted that the dam height was determined on technical criteria, such as energy production, reservoir filling, dam site morphology and upstream dam site location, so the optimization of the dam height does not consider environmental and socio-economic impacts downstream. (IPoE 2013: 40)

---

[11] Available at: https://www.youtube.com/watch?v=FeDUEsEABuI&feature=youtu.be (accessed on December 14, 2016).

In August 2015, after almost 1 year had passed since the formation of the TNC, Egypt's spokesman for the GERD negotiations and the adviser to the water minister, Dr. Alaa Yassin, expressed to Al-Monitor (Hussein 2015) his anxiety over the lost time that would lead the dam capacity to be a non-negotiable issue even if the studies recommended reducing it, as follows:

> Our official position is that this dam is harmful to Egypt, and its storage capacity has no technical or economic justification. The differences remain unresolved, and a great deal of time has been consumed. We were supposed to finish the studies in no more than six months, but around a year has passed without signing the contract related to the consultants that will conduct the studies.

The height and capacity were not the only differences over the dam design, but also the number of bottom outlets. Egypt requested from Ethiopia redesigning the GERD to add two bottom openings in order to address the downstream concerns about water deficits resulted from filling the reservoir and operating the dam in its proposed huge capacity. Increasing the bottom outlets to four instead of two would assure that in case of sequential dry years, larger amounts of water could be released and thus reduce the deficits downstream. This request was particularly aligning with the MIT report (2014: 8) that stated the following:

> At this time, we do not know the elevation (location) or capacity of these low-level release outlets in the GERD, or how precisely they can be operated. If the capacity of these low-level outlets is too small, or the elevation at which water can be released is too high, the GERD may not have sufficient operational flexibility to meet reasonable downstream demands during both the period of filling and during periods of prolonged drought.

In addition to the differences shown above, the delay in implementing the recommended studies, while construction has been progressing steadily, brought another problem about whether Ethiopia would start filling the reservoir only after the completion of the studies or whether the first filling would start directly once the first construction phase is accomplished. The author asked T.B. (2016), the adviser to the Ethiopian water ministry, about this specific point. He answered that even before the IPoE recommended carrying out these hydrological studies, Ethiopia studied different strategies of filling the reservoir in order to "minimize the impacts on the downstream countries." He added that the Declaration of Principles stated that the countries would agree on guidelines and rules on the first filling of the GERD in parallel to its construction, after the approval of studies and reports by the Tripartite National Committee. He also clarified that hopefully the studies will be ready shortly, but even if it needed longer time, the first filling can run "in parallel" to that technical process according to his interpretation for the DoP. He explained that the GERD is a concrete gravity dam and designed to be filled in stages without the need to wait until the whole structure is completed. Obviously, this interpretation contradicts with the one adopted by the Egyptian officials, as they insist on the fact that although the DoP allows for conducting the joint studies and concluding the agreements in parallel with the construction, it stipulates that the first filling will only take place based on the outcomes of the studies.

In contrast to the Egyptian-Ethiopian technical differences, there has been a clear understanding between Ethiopia and Sudan over various technical issues. For instance, M.M.A. (2016c) told the author that Sudan acknowledges the great effort made by Ethiopia to remove the vegetation cover at the reservoir site in order to avoid the deterioration of water quality that may negatively affect fishing in Sudan's reservoirs on the Blue Nile. However, the most serious concern for Sudan is the dam safety, particularly what is related to the saddle dam, which is concisely explained in the MIT report (2014: 6) as follows:

> We are concerned that the risks posed by the GERD's extensive saddle dam may not have been fully appreciated or analyzed. The design, construction, safety features, and maintenance of the saddle dam need to be rigorously scrutinized by all parties potentially affected. ... Problems with the GERD's saddle dam would entail significant maintenance costs for Ethiopia, while a significant rupture of the dam would have potentially catastrophic consequences. The latter would be an economic disaster for Ethiopia, due to the loss of both hydropower and infrastructure, and an economic and humanitarian disaster for Sudan, where the waters released by a saddle dam failure would flow.

About this vital issue, M.M.A. affirmed to the author that Sudan, through continuous visits to the dam, is satisfied with the measures that Ethiopia has taken so far in following the IPoE recommendations concerning the dam safety.

6.3.4 *The Khartoum Document and the Signature of the Technical Contracts*

The picture remained unclear, regarding the technical differences shown above, until the tenth TNC round in late 2015, when two crucial tripartite meetings were held in Khartoum, on December 11–12 and 27–28. There, the six ministers of water and foreign affairs from the three countries attended in order to find a political as well as a technical solution for the pending points. Concerning the international consultancy firms, it was agreed that the French Artelia group would replace the Dutch firm to be responsible of 30% of tasks. The three countries signed a document containing the "summary and the outcomes" of the December 27–28 meeting, in which they named the French BRL and Artelia groups for conducting the technical studies and the British Corbett firm for arranging all legal issues, and they reaffirmed the commitment of their countries to the DoP. This was then named the "Khartoum Document" (Salman 2016). Egypt considered the document an assurance that the fifth principle in the DoP will be fully respected by Ethiopia, meaning that the reservoir will not be filled but according to the outcomes of the prospective studies (Sudan Tribune 2015e).

With respect to the dam design, it became evident that none of the basic characteristics of the project will be modified, not only in terms of height and capacity, but even the bottom outlets. Two weeks after the signature of the Khartoum Document, Ethiopia declared its rejection to the Egyptian request about the latter point. Bizuneh Tolcha, Public Relations Director at the Ethiopian Ministry of Water and Irrigation, told the state-run Ethiopian Broadcasting Company, "The decision to build two openings came following intensive studies, and Ethiopia does not need to redesign the dam project" (Alaa el-Din 2016).

The news on the GERD aroused the anxiety of public opinion in Egypt, particularly when Ethiopia announced, on December 26, 2015, diverting the Blue Nile to its natural course to run through the GERD, giving a strong message that there would be no delay in the completion of the construction work (Daily News Egypt 2015b). On December 30, el-Sisi commented in a public speech, "I totally understand the concern of Egyptians ... Do not worry about the Ethiopian dam ... I have not led you astray before and I will not lead you astray now" (Ahram 2015d).

Although it was agreed in the Khartoum Document to "expedite" the implementation of the studies, the signature of the technical contracts with the selected firms was postponed until September 19, 2016 (Ahram 2016b). This was "because of the need of the tripartite technical delegations of the three countries to return to their governments for further consultation on the wording of the contract to comply with the legal requirements of each country" (ESIS 2018). Not only did the complex review process at the national level consume the period of almost 9 months, but also did the tensions over political bilateral issues. For instance, whilst reviewing the contracts, during a meeting held in February 2016, the Egyptian delegation strenuously objected to the maps prepared by the French groups because the Hala'ib. Triangle was drawn inside the Sudanese border (Hussein 2016). As discussed in the previous chapters, the border dispute between Egypt and Sudan over the Triangle has persisted since the independence of Sudan in 1956 (Sect. 4.2.2) and reached its peak in 1995 after Mubarak's assassination attempt (Sect. 5.3.1). In recent years, the tension has been exacerbated due to the decision of each country to consider the triangle an electoral circle in its national elections and due to Sudan's frequent complaint to the UN Security Council against Egypt about the issue (Badawi 2016; Sudan Tribune 2016b).

Another example is the tension between Ethiopia and Egypt due to the Oromo demonstrations held in November 2015 against the decision of the central government to expand the boundaries of Addis Ababa, annexing parts of the Oromo lands. These events then turned into further protests against political and economic marginalization (BBC 2018b). Ethiopia has been suspicious that Egypt may be playing a role in backing such protests, and Egypt had strongly denied Ethiopia's accusation in many occasions. In October 2016, the Ethiopian Foreign Ministry summoned the Egyptian ambassador in Addis Ababa who affirmed that Egypt "does not intervene in the internal affairs of other countries" (Ezzidin 2016; Ahram 2016c). In another matter, the official visit of Israel's Prime Minister Netanyahu to Ethiopia on July 7, 2016, "provoked a storm of comments" in Egyptian media on the "successful Israeli encirclement in the Nile Basin" (Raslan 2016). During the visit, Netanyahu acknowledged, in a joint statement, Desalegn's wording that "Israel has a special place in Ethiopia and Ethiopia has a special place in Israel." Netanyahu also declared that there is "a plan for concrete cooperation in areas that are important for both our peoples," whereby the "first and most important is water," the field of which "Israel has a great expertise" (IMFA 2016).

6.3.5 The Disagreement over the Inception Report by the Consultancy Firms

One year later, in October and November 2017, several TNC and ministerial meetings were held to discuss the "inception," also referred to as the "preliminary" or "introductory," technical report submitted by the French firms. During all of these, the three countries "did not reach any agreement on adoption of the introductory report," as declared by Egypt's Water Minister Dr. Mohammed Abdel-Ati (ESIS 2018). He also expressed "Egypt's concern over the future talks with Sudan and Ethiopia after they rejected the introductory report" (ESIS 2018). Reuters reported, after Sudan's Water Minister Mr. Moataz Mousa, that "Egypt was unwilling to accept amendments to the report put forward by Khartoum and Addis Ababa" and that "Sudan and Ethiopia had expressed concern over several points, especially the proposed baseline from which the study would measure the dam's impacts" (Awadalla and Knecht 2017). According to Sudan Tribune (2018a), the Sudanese side rejected the report because "the two firms, BRL and Artelia, ... did not observe the 1959 Agreement." Egypt, on the contrary, accepted the report and rejected the Sudanese and Ethiopian proposed amendments, since the latter do not recognize its "current share of the Nile river" (Aman 2018). In fact, setting the baseline according to the current uses satisfies neither Sudan nor Ethiopia, because the former has not yet made use of its full share of the 1959 Agreement, while the latter does not acknowledge both the 1929 and 1959 Agreements that have been governing the use of the Nile waters for decades.

The disagreement was too wide to the extent that negotiations were completely stalled. Again, Egyptians felt deep anxiety, particularly due to the time wasted in the protracted talks. "It's clear they don't want to reach conclusions quickly. We believe they probably want to start filling the dam and complete construction while there are still some ongoing discussions," said Mahmoud Abou Zeid, Arab Water Council Chair and the former Water Minister who had participated in the CFA negotiations until March 2009 (Awadalla and Knecht 2017). As reported in Ahram (2017), el-Sisi, few days after the negotiations had been stalled in November, provided a "message of reassurance that Egypt's 100 million people needed to hear," saying, " Rest assured that no one can touch Egypt's share of [Nile] water, ... Egypt's share of the Nile's water is a matter of life or death for the nation, ... we are capable of protecting our national security, and water to us is a question of national security. Full stop."

As an attempt to find a solution, in late December 2017, Egypt officially proposed "the participation of the World Bank as a technical party with an impartial view to decide on the differences" (ESIS 2018). Desalegn rejected the proposal, clarifying, "Seeking professional support is one thing, transferring [arbitration] to an institution is another thing. So we told them that this is not acceptable with our side" (Maasho 2018a). This rejection came 3 days after Desalegn's visit to Cairo, where he and el-Sisi made positive comments to reporters. "The Nile Basin enjoys great resources and capabilities that makes it a source of interconnection, building and development, not a source of a conflict," declared el-Sisi. Desalegn responded, "We must make sure that this great river never becomes an object of competition, mistrust or conflict" (Reuters 2018b). Nevertheless, such statements were not enough to ease the tension.

Aside from the technical disagreement, tension was growing between Sudan and Egypt over different political issues. These included the Sudanese-Turkish agreements to rebuild an Ottoman port city on the Red Sea and construct a dock to maintain civilian and military vessels, signed during the visit of President Tayyip Erdogan to Khartoum in late December, which also coincided with a simultaneous visit by the chief of the general staff of Qatari army. Since Turkey and Qatar are the major backers of Muslim Brotherhood and have been antagonizing Egypt's post-2013 regime, Egyptian media launched a "severe campaign" against the Sudanese government to which Khartoum responded by recalling its ambassador to Cairo (Sudan Tribune 2018b). The tension was further exacerbated when Sudan deployed troops along the Eritrean border, admitting that this "came as result of military threats from Eritrea and Egypt against the country," after different reports had claimed that an Eritrean military training center on the border "received Egyptian troops with logistical gears and vehicles" (Sudan Tribune 2018c). In response, el-Sisi affirmed that Egypt does not "interfere in other countries' affairs" and will never "fight its brothers ... I am saying this as a message to our brothers in Sudan" (Reuters 2018a).

At this critical moment, a tripartite summit between the leaders of Egypt, Ethiopia and Sudan was held in Addis Ababa on January 29, on the margin of the AU's 30th Ordinary Session of the Assembly. In response to the reporters' question about the GERD crisis, el-Sisi said, "None of [us three] countries ... will be harmed. ... Egypt's interests are one with Ethiopia's and also one with Sudan's. We are speaking as one voice. ... There is no crisis," and Bashir concurred, saying, "There is no more crisis" (Ahram 2018a). Following the summit, Egypt's Foreign Ministry announced that the three leaders agreed to "resolve all disagreements" on the GERD "within one month" (Ahram 2018a). However, in a televised address on February 15, Desalegn announced, unexpectedly his surprise resignation, saying, "Unrest and a political crisis have led to the loss of lives and displacement of many. ... I see my resignation as vital in the bid to carry out reforms that would lead to sustainable peace and democracy" (BBC 2018a). On April 2, Abiy Ahmed was elected as Ethiopia's new Prime Minister to be the first country's Oromo leader, which is the largest ethnic group in Ethiopia and has been at the center of the anti-government protests that were sparked in 2015, as mentioned above. The new 42-year-old PM fought when he was a teenager against Mengistu in 1990. He was appointed as a Minister of Science and Technology in 2016 and got his PhD in Peace and Security Studies in 2017 from Addis Ababa University (BBC 2018b).

Under such dramatic changes, the technical talks were resumed. Two nine-party meetings were held on April 6 and May 15 in Khartoum and Addis Ababa respectively, in which all of the water and foreign ministers as well as the heads of intelligence agencies from the three countries participated (Ahram 2018b). The second meeting produced a significant "outcome document," in which the three sides agreed on five concrete points.[12] The first and the second aimed at fostering cooperation between the three states in general, through holding a "tripartite summit" every 6 months and founding the "Tripartite Infrastructure Fund" to be responsible for implementing various developmental projects. In the third and fourth points, it was agreed that every country would send its comments and observations on the inception report to the two consultancy firms, providing them a period of 2 weeks to respond to all the queries and to arrange for a meeting within 1 week to discuss the responses with representatives of both. The fifth, and the last point, recommended establishing the "National Independent Scientific Research Study Group" (five experts from each country) with the mandate of "discussing and developing various scenarios related to the filling and operation rules," in accordance with the equitable use and no-harm principles, and that it should submit the "outcome of its deliberations" within 3 months. Similar to the previous documents, the agreed schedules were too ambitious. Nevertheless, the document represented a serious effort to break the recent deadlock.

On the margin of the second nine-party meeting, Abiy Ahmed met the high-level Egyptian delegation, who informed him about el-Sisi's official invitation to visit Cairo. This actually took place on June 10, shortly after the beginning of el-Sisi's second term, whereby the Nile was at the top of the agenda for both leaders. As reported by Reuters (2018c), in a joint press conference, elSisi said, "We have come a long way in building confidence and strengthening bilateral cooperation." Ahmed affirmed, in Amharic, "We will take care of the Nile, and we will preserve your share and we will work to increase this quota, and President el-Sisi and I will work on this." In good spirit, Ahmed fulfilled el-Sisi's request to "swear to God," in Arabic, before the Egyptian people, "We will not harm Egypt's waters." Notably, this is an unprecedented move, whereby an Ethiopian leader not only pledges to recognize Egypt's Nile share, but also to work jointly on increasing it. This was not just an emotional statement induced by Cairo's warm welcome. Ahmed's declaration was first made during his visit to Khartoum on May 3, in which he and Bashir asserted that the GERD would not "harm Egypt's share of water supplies" (Daily Mail 2018).

Tracking Ahmed's moves since taking office reflects that his recognition of Egypt's Nile share is part of a broader strategy centered on regional cooperation and economic integration. Few days before his visit to Khartoum, he had been in Djibouti, where he reached a deal to take a stake in its port on the Red Sea in return for giving Djibouti the option to take stakes in Ethiopian state-owned firms, including EEPCo that owns the GERD (Maasho 2018b). It is no surprise that this was Ahmed's first foreign official visit, as Djibouti is the major trade gateway of his landlocked country, and it had been seeking investors for the port since it terminated Dubai's state-owned DP World's concession 2 months earlier (Maasho 2018b). Besides, the Chinese-built 756-km electrified rail project connecting the two countries had officially started operations this year (Ethiopia's Embassy

[12] Available at: https://www.infoafrica.it/wp-content/uploads/2018/05/accordo-nilo.pdf (accessed on July 13, 2018).

in Brussels 2018). In addition, the deal follows an agreement to acquire a stake in Port Berbera in the breakaway Somali region of Somaliland (Maasho 2018b). Nevertheless, as reported in Washington Post (Schemm 2018), the real "historic breakthrough" was Ahmed's pledge, on June 5, to give up the disputed border town Badme to Eritrea, accepting the 2002 decision of the international boundary commission, created under the UN 2000 peace agreement that terminated the 2-year armed confrontation between the two countries. After almost two decades of military standoff, Ahmed and Afwerki exchanged then official visits, on July 8 and 14, normalizing bilateral relations and agreeing that Ethiopia would "look to start using Eritrea's Red Sea ports." The mediation by Saudi Arabia and United Arab Emirates "probably helped create the breakthrough."

As covered by Reuters (2018d), the UAE's Crown Prince, Sheikh Mohamed bin Zayed Al-Nahyan visited Addis Ababa in mid-June and pledged a total of $3 billion in aid and investments to Ethiopia, $1 billion of which would be deposited in the Ethiopian central bank to "ease a severe foreign currency shortage" caused "partly" by "spending on big infrastructure projects." Noteworthy, the visit was "welcomed" with "cautious optimism" by Egyptian commentators, expecting that it "could put pressure on Addis Ababa to exercise flexibility in the dam negotiations" (El-Bey 2018). Overall, the ongoing changes in East Africa raise hopes that long-standing disputes in the region can be settled, including even the civil war in South Sudan (Schemm 2018). Yet, it is too early, until this moment (mid-July) to anticipate how this will be reflected on the Nile game.

6.4 The Shadow of the Future

6.4.1 Expanded National Projects

From the national perspective, each country in the Nile Basin is facing serious long-term challenges with respect to rapid population growth and uncertain climate variability. As shown in Figs. 6.5 and 6.6, it is inevitable that the total water resources per capita in each riparian will decrease dramatically to the extent that most of the riparians will be water scarce states by the last quarter of the century. NBI (2012a) explains that the threat of water shortage could be exacerbated due to the considerable uncertainties associated with different scenarios of climate change, whereas it is unknown whether rainfalls within the basin will increase or decrease and to what extent extreme events of floods and droughts will be exaggerated. However, most of these scenarios agree on the fact that air temperature will increase, causing larger evapotranspiration losses and reducing crop productivity. Hence, the basin is subject to threats of the occurrence of more famines and floods, particularly in upstream states. Moreover, the fourth assessment report[13] of the Intergovernmental Panel on

[13] Refer to the IPCC's fourth assessment report, in which Sect. 6.4.2.1 is devoted to "Deltas" under climate change risks, available at: https://www.ipcc.ch/publications_and_data/ar4/wg2/en/ch6s6-4-1-2.html (accessed on December 28, 2016).

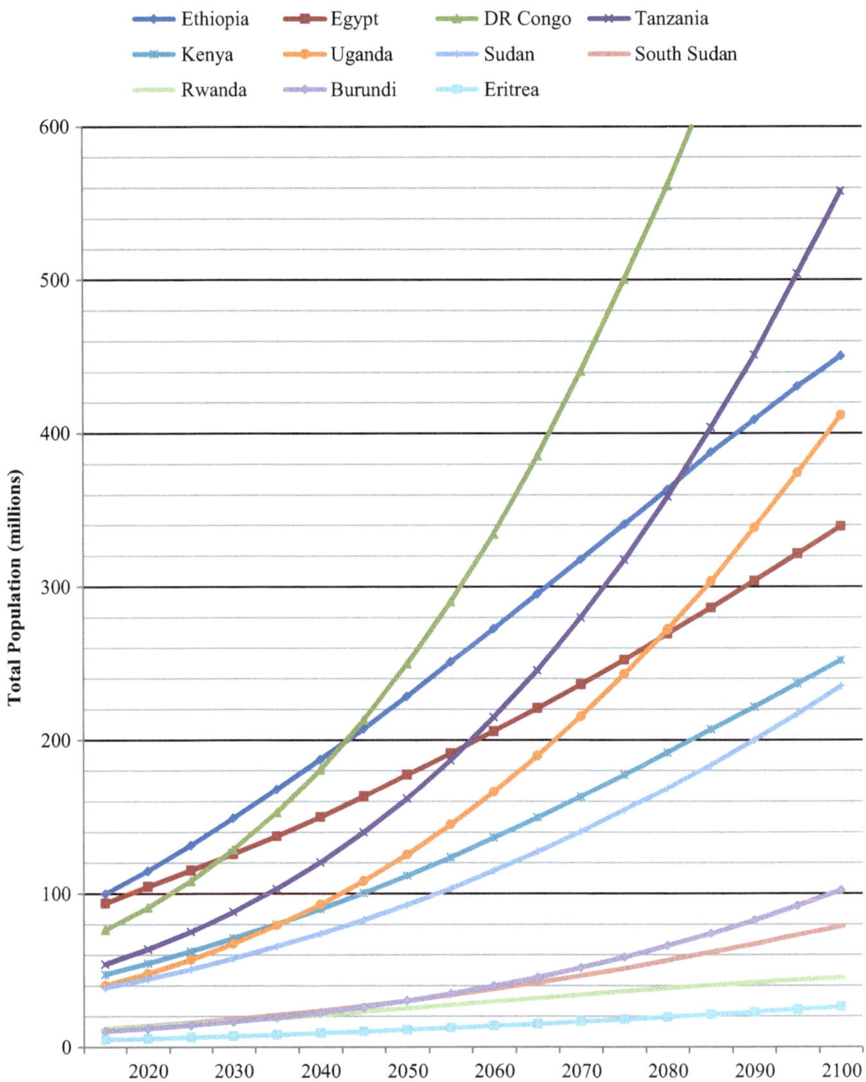

Fig. 6.5 Probabilistic population projection until 2100 based on the UN World Population Prospects, the 2017 Revision (upper 95% prediction interval, 2015–2100). Note: The projection of the DR Congo's population in 2100 is 852 millions, which is not presented in the chart. (Source: Created by the author after the UN Population Division Datasets, available at: https://esa.un.org/unpd/wpp/Download/Probabilistic/Population/ (accessed on July 18, 2018))

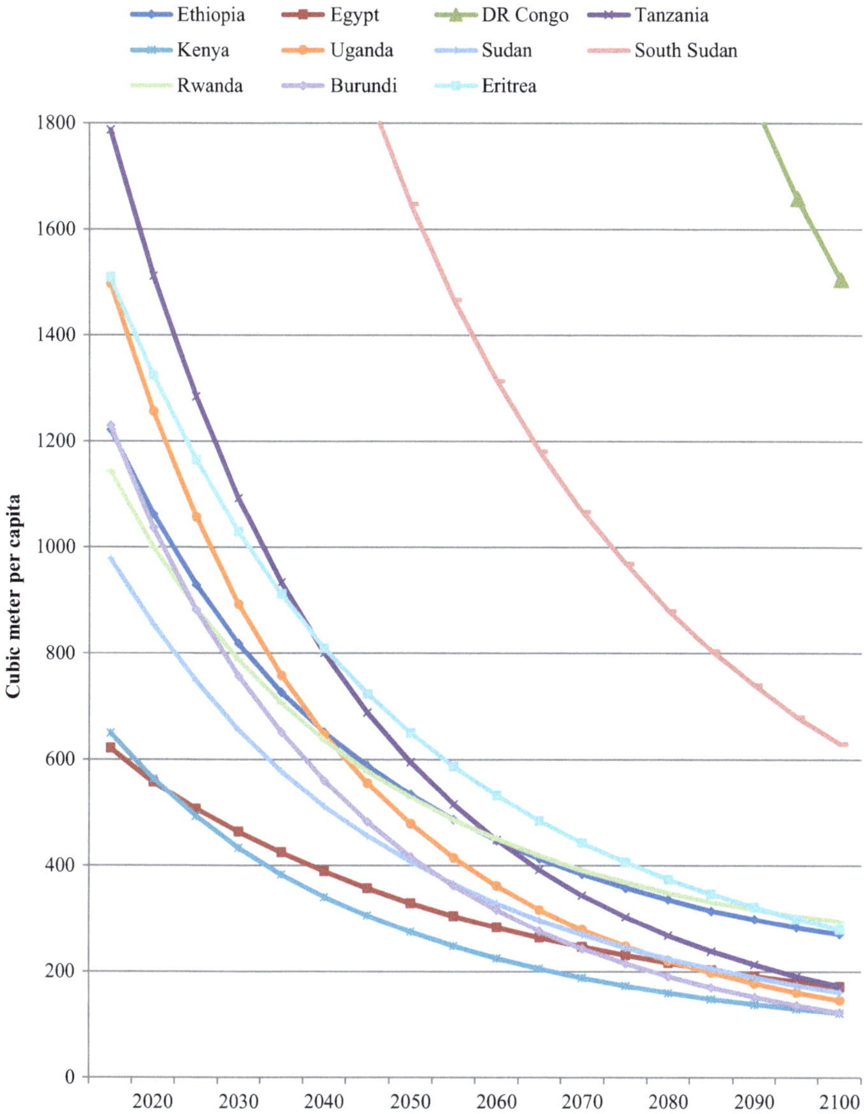

Fig. 6.6 Total freshwater resources per capita (surface and ground waters). Note: The DR Congo and South Sudan are not fully presented in the chart since their values decrease respectively from 16,838 in 2015 to 1505 in 2100 and from 4166 in 2015 to 630 in 2100 (cubic meter per capita). (Source: Created by the author after the FAO, Aquastat Dtabase (for total renewable freshwater resources) and the UN Population Division (for the probabilistic population projections until 2100 based on the 2017 Revision), available at: http://www.fao.org/nr/water/aquastat/data/query/index.html?lang=en; https://esa.un.org/unpd/wpp/Download/Probabilistic/Population/ (accessed on July 18, 2018))

Climate Change (IPCC 2007) identified the Nile Delta in Egypt as one of the world's three extreme "hotspots for vulnerability." As indicated by NBI (2012a: 209), it is anticipated that "a 1-meter sea-level rise could cause a loss of 4,500 km^2 of cropland, and displace some 6.1 million people in the Nile Delta," in addition to increasing the salinity of coastal lands.

In order to cope with such risks, each of the Nile riparians has developed its own plans on the Nile as discussed below.

As mentioned earlier, Ethiopia's plans on the Nile comprise numerous dams, the largest of which are on the Blue Nile. As late Prime Minister Zenawi said during the GERD official commencement, "Henceforward, nothing can stop us from exercising our rights; the other dams we plan to build are less challenging than this, the Millennium Dam" (Meles Zenawi Memorial 2013). The other dams, shown in Table 6.1, include the Beko Abo Dam (2100 MW), Karadobi Dam (1600 MW), and Mandaya Dam (2000 MW). These were first introduced by the 1964 USBR study (Sect. 4.2.4) and were reassessed in the 2000s. For example, the Beko Abo Dam is added with a storage capacity of 37.5 BCM. In addition, the proposed height of the Mandaya Dam has been increased from 164 to 200 m, creating a larger reservoir (49.2 BCM instead of 15.93 BCM) that would extend upstream close to Karadobi Dam; thus obviating the need for the originally proposed Mabil Dam (FDRE-MoWR 2007). In total, these dams together with the GERD would increase Ethiopia's storage capacity on the Blue Nile to over 200 BCM.

The author asked T.B. (2016) whether Ethiopia would build these dams jointly with its co-riparians or unilaterally as in the GERD case. He answered, "There is a plan, and it will be implemented. I am not sure how the downstream riparians would contribute to the implementation of the plan, and whether it will be built jointly, but I confirm that Ethiopia will follow the international standards and norms, such as prior impact assessment and so on." He added, "Within the Blue Nile Gorge, all the dams mainly aim to generate hydropower, but there will be supplementary irrigation schemes on other tributaries upstream." As recorded in Blackmore and Whittington (2008: 24), the Ethiopian plans comprise the cultivation of over one million hectares on its tributaries in the Nile Basin. Still, in light of the "currency crisis that might last for 15 or 20 years" if the Ethiopian government continued to expand its mega plans, as Abiy Ahmed said shortly after taking office (Reuters 2018d), Ethiopia would be in bad need to attract foreign investments in order to implement its Nile projects; a process which primarily requires full coordination with the downstream riparians. However, the matter is not that simple, considering that the three Eastern Nile riparians have not yet, since 2011, agreed how to fill the GERD's 74-BCM reservoir nor and how to operate the hydropower generation without causing significant harm to downstream communities and ecosystems. The major problem of dam operation would be during successive dry years, which takes place more than one time a century in the Nile, whereby reservoirs along the river would shrink and need to be filled in again, similar to what occurred during the 1980s (Sect. 4.3.5). A tradeoff is then generated between water regulation and power generation (IPoE 2013: 36).

Table 6.1 Hydropower dams on the Blue Nile in Ethiopia included in the EAPP & EAC Master Plan

Plant	Source	MW	FSL (m)	Dam height (m)	Tail-water (m)	Storage (MCM)
Karadobi 2006	MoWR	1600	1146	250	910	40,200
Beko Abo 2007	MoWR	2100	906	110	795	37,500
Beko Abo	ENTRO	1000	906	110	795	
Mabil	USBR	1200	906	171	740	13,600
Mandaya	USBR	1620	741	164	580	15,930
Mandaya 2007	ENTRO	2000	800	200	605	49,200
Border 2007	ENTRO	1200	580	90	490	14,470

Source: EAPP & EAC (2011c: A-5)

As for Sudan, the completion of the GERD will provide the country with millions of feddans suitable for irrigated agriculture on the Blue Nile. As shown in the recent trends of large-scale land acquisitions (LSLA) in different location across its territory (Hanna 2016), Sudan will attract further investments in the agricultural sector benefiting from water regulation downstream of the GERD. Countries as China and Saudi Arabia already began signing investment contracts with the Sudanese government in the agricultural sector on the Nile (Sudan Tribune 2016a). Moreover, the controversial Sudanese dams with respect to local Nubian inhabitants, namely Dal, Kajbar and al-Shereik Dams, have already secured funds from the Saudi Arabia (Sudan Tribune 2015d). In this regard, the author asked M.M.A. (2016c) whether Sudan's water share will be sufficient to implement the planned expansion in irrigated agriculture on the Nile. He said that Sudan is fully committed to the 1959 Agreement with Egypt and to the water shares stipulated in it to each country, and that Sudan has the right to define its priorities in utilizing its share. He added that this process runs in coordination with the Egyptian side through the PJTC, and that the joint stations installed along the Nile can dispel Egypt's fears about this issue. Nevertheless, this means that expanded irrigation is Sudan would rely on water reuse, which requires taking appropriate measures concerning water quality.

In Egypt, the Nile share of 55.5 BCM/year in addition to deep groundwater (2 BCM/year), rainfall (1.3 BCM/year), and sea desalination (0.2 BCM/year) represent no more than three quarters of the country's actual withdrawals that exceed 81 BCM/year (Nour El-Din 2013: 16). The remainder is supplied through the reuse of agricultural drainage and the Delta's shallow groundwater (16 BCM/year and 6.2 BCM/year respectively) (Nour El-Din 2013: 16). Yet, all this does not meet the population's needs, given that more than 32-BCM virtual water is imported annually in crops and livestock as estimated by Zeitoun et al. (2010: 237), and such a gap will increase over years. In this respect, future national plans in Egypt are also directly linked to the Nile, both in increasing crop productivity in the old Nile Valley and Delta and in expanding the irrigable lands across the country's deserts. As shown in Fig. 6.4, the desert reclamation project adopted by the current government aims to add over four million feddans to Egypt's uninhabitable land in the Western,

Eastern and Sinai Deserts, in order to create new communities, obtain hard currency, and boost food security (Ahram 2014a). The underground water resources will supply about 88% of the project's needs, while the remaining portion is supplied by the Nile waters, particularly in the Toshka project (Antar 2015). In an interview with the author, M.A.S. (2016a) explained that the cultivated lands represent no more than 40% of the total project area, while the remaining parts will be for housing, service and industrial uses. He clarified that the project will pass into two phases: utilizing water resources directly for agricultural purposes in the first phase, and utilizing reused and recycled waters in agriculture after the regions are inhabited and industrialized in the second phase. He also emphasized that Egypt has learned lessons from the past, whereby land reclamation this time is based on a holistic vision of creating comprehensive communities and relying on both underground and surface water resources, since the success of this project is inevitable to combat the challenges of poverty and unemployment that feed terrorism and instability. On the long term, M.A.S. noted that the project aims to contain the rapidly growing population, since Egypt's population is expected to increase by more than 60 millions by 2050, and to host millions of the Delta's inhabitants that might be displaced in case of sea level rise.

In the Equatorial Lakes sub-basin, Uganda has an ambitious plan for constructing new hydropower dams. In addition, Kenya, Tanzania and Uganda also seek to expand their irrigation projects on Lake Victoria. Furthermore, the need for reviving the water conservancy projects at the Sudd swamps still persists, and once peace is restored in South Sudan, different sides will push for the scheme; either Egypt and Sudan to increase the White Nile runoffs, or the South Sudanese themselves to exploit irrigation potentials in their country.

In short, it is evident that every country has its own national plans and ambitious on the Nile. However, as Whittington (2004: 19) points out, the extensive implementation of such plans would cause the loss of the Nile's "unique" and "rich" environmental assets (Fig. A.6), such as the Blue Nile canyons, the forests on the Ethiopian Highlands, the Sudd swamps, and Lake Victoria. These are viewed by environmentalists as "international public goods," which are of great importance, not only to "the people in their immediate environs or even in the Nile Basin," but also to "people throughout the world" (Whittington 2004: 19).

6.4.2 *Prospects of Basin-Wide Cooperation*

"We have already noticed that water demand exceeds supply in the Nile Basin. Then, what can we do? If each country just moves unilaterally, we will need another Nile to be built," J.N. (2016), a former executive director of the NBI Secretariat told the author in an interview. The NBI was originally established to bring together all the riparians to agree upon the optimum way of utilizing the Nile's natural resources, providing benefits for all with least possible environmental and social costs. However, as explained by J.N., the NBI is no more than a transitional mechanism for promoting cooperation and building capacity, while the major aim is to construct the NRBC that involves all the riparians, which would be the institution that owns power of planning and managing water resources from a basin-wide perspective. He indicated that the NBI is still playing its preliminary role even after Egypt had suspended its full participation in the NBI since mid-2010 and after the NBRF was closed in late 2014. He clarified that all the riparians, still regard the NBI as a forum for cooperation even with limited participation of Egypt in the past 2 years, especially that the NBI is currently receiving financial support from the EU, Germany and other parties.

The question here is about the destiny of the CFA, particularly after it has been signed by the majority of riparians, given the signature of Burundi in late February 2011, followed by the ratification of the Ethiopian parliament in June 2013, the Rwandan during the same year and the Tanzanian in 2015. The author asked M.A.S. (2016a) about Egypt's position concerning the CFA. He emphasized that Egypt still views the agreement in its current form as "unfair" to the downstream countries. Egypt objects to the CFA mainly because it does not recognize the obligation of prior notification as stated in the UNWC, its amendment and dispute settlement measures are taken by simple majority, as it gives the upstream riparians complete right to utilize the Nile waters at the expense of existing uses downstream (Sect. 5.4.3). The author asked M.M.A. (2016c) about the same issue, and he pointed out that the Sudanese objection is based upon the same reasons raised by Egypt, in addition to the fact that the upstream riparians opened the CFA signature without giving adequate time and making further attempts to reach a satisfactory compromise. He added that Sudan does not see any problem if the CFA enters into force and a commission is established accordingly, whereas in this case, its role would be confined to water management upper the Nile, such as Lake Victoria Commission or other similar organizations. He also excluded that such a commission would set new water allocations in the Nile Basin.

Interestingly, Egypt's late Foreign Minister Aboul-Gheit mentioned in his notes that Zenawi informed him, during a visit to Addis Ababa following the CFA signature in 2010, about Ethiopia's vision over the issue, which is close to what M.M.A. told the author. Zenawi declared to Aboul-Gheit (2014: 261) that the relation between Ethiopia and Egypt is "everlasting" similar to "Catholic marriage," and although Ethiopia views the 1959 Agreement as "unfair," since it entitles Egypt and Sudan to the entire volume of the Nile waters, Ethiopia would not ask for cancelling it. Zenawi explained that after the CFA would enter into force, there would be two "bodies" in the Nile; Egypt and Sudan, on one side committed to the 1959 Agreement, and the other countries on the other side gathered under the CFA. In this case, the role of the NBI, as indicated by Zenawi, would be essential in bringing together both sides to implement potential cooperative projects.

Concerning Ethiopia's perspective on water reallocation, T.B. (2016) told the author that Ethiopia is in great need to supplement its rainfed agriculture with irrigation projects in order to protect its people from frequent famines and climatic variability. He noted that, through the NBI, riparians should discuss the water demands of each and focus on projects that can save waters across the basin, such as the construction of dams on the Ethiopian Highlands. "We should try to meet everyone's need through water saving and conservation projects," T.B. said, "The reallocation of water could be the very last priority."

Thus, it cannot be said that the refusal of Egypt and Sudan to join the CFA would foreclose basin-wide cooperation in the Nile. There are two specific examples in this regard. First, Egypt, Ethiopia and Sudan have already taken steps towards linking their national power grids, especially that the DoP has given priority for Egypt and Sudan to import the GERD's generated electricity. The supreme joint committee established by the three countries at the leadership level is expected to push forward regional integration projects in energy, food and trade sectors. Second, Egypt has proposed the Victoria-Mediterranean navigational project for connecting the upstream landlocked riparians with the Mediterranean Sea. Ethiopia would be linked to the project in two ways; through the Baro-Akobo-Sobat tributary, and through land routes from Ethiopian cities to Lake Victoria. As M.A.S. (2016a) informed the author, all relevant parties have welcomed the project, whereby the prefeasibility studies have already been completed. Nevertheless, Ethiopia's enthusiasm to participate in the proposal should be viewed in light of the recent ports' deal made by Abiy Ahmed with Djibouti, Somaliland and Eritrea.

6.5 Analysis and Discussion

6.5.1 Sole Utilization (T-S & S-T)

In January 2011, two historic events occurred at the domestic levels in each of Sudan and Egypt, namely the Southern Sudan referendum and the Egyptian Revolution, both of which have had profound effects on the balance of interests (I) and capabilities (C) and opened a new round in the Nile game.

By the separation of South Sudan (D), Khartoum lost the lands where most of its oil resources existed and consequently lost its main source of hard currency. As an urgent measure for bridging the gap between energy supply and demand, the central government took gradual steps towards lifting subsidies on oil products, which provoked several waves of demonstrations against the regime (D). On the long term, Sudan has viewed the Nile as the most attractive option to meet its severe energy shortages and to increase the hard currency revenues (I_{Su}). The plan is to advance the construction of hydropower dams and attract foreign investments in the agricultural sector, particularly in light of the recent international food crisis (G). However, the hydrology of the Nile typically imposes direct challenges for the implementation of such a plan. On the one hand, the high siltation rates significantly reduce the efficiency and profitability of both the hydropower dams and irrigation schemes, while on the other, the high seasonal and annual fluctuations of the flows between floods and droughts hinder foreign investments in irrigated agriculture. In this respect, Sudan has been fully aware that the ideal solution for these problems is the construction of large upstream dams in Ethiopia so as to control sediments, manage floods, and regulate flows (I_{Su} and I_{Et}).

In Egypt, the sudden change of the Mubarak regime that persisted for 30 years (D) led the country to be extraordinarily busy with domestic affairs, and consequently reduced its capacity to deal effectively with external events (C_{Eg}). In few years, Egypt witnessed multiple waves of uprisings, several transitional governments, and repetitive constitutional, presidential and parliamentary elections. Such political instability has been reflected on Egypt's management of vital water issues (C_{Eg}). In effect, six Egyptian water ministers have succeeded one another since the January 25 Revolution to date.

Under these exceptional conditions, Ethiopia announced unilaterally the construction of the GERD on the Blue Nile at the site where the Border Dam was proposed by all the previous plans; from the 1964 USBR study to the prefeasibility studies conducted through the NBI. However, the new dam is several times larger in total capacity (74 BCM instead of 14 BCM) than originally planned, although its original size was defined based on specific economic and climatic criteria. On the one hand, the average air temperature at the dam site on the border with Sudan, albeit lower than that at the AHD site, remains relatively higher than that on the Ethiopian Highlands, which increases total evaporation losses. On the other hand, increasing the dam height requires the construction of a large saddle dam, pushing up the total cost of the generated electricity and overlooking the socio-economic and environmental risks downstream; a point of which the IPoE and the MIT reports have emphasized taking into account the local communities and ecosystems along the Nile Valley in Sudan and Egypt.

Yet, the reasons behind this decision clearly reflect political not technical considerations, all of which were implicitly alluded to in Zenawi's speech during the project's inauguration on April 2, 2011. In his first words, Zenawi described the GERD as the largest dam that could be built at any point on the Blue Nile or even on all other rivers in Ethiopia. Maximizing the dam size to be as large as possible assures several advantages of great importance to Ethiopia. As Zenawi mentioned, such a massive dam represents a national project that deserves tremendous sacrifices from the Ethiopians to fund it domestically (C_{Et}), especially that the expected objection from Egypt to the project would block international funds. In addition, Zenawi wanted the dam to be large enough to bring considerable benefits to Sudan in order to ensure the latter's political support for the project. Evidently, Zenawi was aware of Ethiopia's need to form a new cluster with Sudan in order to take such a step on the most important Nile tributary for Egypt ($T_{Et\&Su}$-S_{Eg}). Moreover, Zenawi pointed out that nothing would stop Ethiopia from building the dam and that the construction of all the other dams would be facilitated thereafter. These two points indicate his deep awareness of the excellent timing of the decision in relevance to the political instability in Egypt (C_{Et}) and the economic crisis in Sudan (I_{Su}). Furthermore, the large generation capacity of the dam (6000 MW) allows for exporting electricity to neighboring countries, which brings cash revenues to Ethiopia (D) and assures regional consent over the project (R). In fact, after almost 7 years from Zenawi's speech, Ethiopia has been able to achieve what it sought. The GERD has become a reality with the funds raised by Ethiopia's citizens (D) and its friends across the world (G), with the explicit support of Sudan (R), and by signing purchase power agreements with neighboring countries (R).

Overall, the GERD, in its maximized capacity, reflected the new balance of capabilities between Ethiopia and Egypt (C_{Et} vs. C_{Eg}) and the common interests between Ethiopia and Sudan (I_{Et} and I_{Su}), but also indicated the large gap between the beliefs of Ethiopia and Egypt over the Nile (B_{Et} vs. B_{Eg}). In his speech, Zenawi named the Nile as one of Ethiopia's "own rivers," while Egypt's Morsi, more than 2 years later, expressed his anger over the GERD and described the Nile as the "gift to Egypt" as Egypt is the "gift of the Nile."

In short, for Ethiopia, the GERD has manifested the end of the sole utilization of the Nile waters by the downstream cluster ($T_{Eg\&Su}$-S_{Et}). Conversely, Egypt has conceived the way in which the GERD has been constructed as an attempt by Ethiopia to reap all benefits with Sudan and make it bear all costs ($T_{Et\&Su}$-S_{Eg}). This is particularly because the decision was made without prior notification and construction has been progressing in high rates without waiting for the implementation of complete impact assessments. Yet, Egypt has not accepted to let the game outcome be turned into sole utilization in favor of the Ethiopian side, entering into several rounds of negotiations with the aim of transforming the outcome into cooperation (R_{Eg}-$R_{Et\&Su}$) or even into contention (P_{Eg}-P_{Et}), as will be shown in the following subsections.

6.5.2 *Cooperation (R-R)*

Only 1 month after the official commencement of the GERD, Egyptian popular and official delegations visited Ethiopia to discuss the procedures assuring that the dam would cause no significant harm to Egypt. Ethiopia, then, invited Egypt and Sudan to form the International Panel of Experts (IPoE), involving national and international experts to closely investigate the dam's design and construction documents. Egypt requested halting construction until the IPoE would issue its report and considering its recommendations binding on all three countries; however Ethiopia did not accept both requests. The IPoE took a year, from May 2012 to May 2013, to issue its final report, which clearly indicated that most of the submitted documents were preliminary and did not befit the massive magnitude of the project and the vitality of the Blue Nile. While the report had emphasized the benefits for all three countries (mainly hydropower generation, sediment control and flood management), it also shed light on the potential negative impacts on downstream communities. In this respect, the general risks of the dam include the disruption of ecosystems, the deterioration of water quality, and the reduction of water flows downstream; particularly in dry years. More specifically, Sudan will face the loss of recessional agriculture, fisheries and brick industry, while Egypt will suffer from deficits in water and energy supplies and the salinity of vast lands in the Delta; especially those vulnerable to high sea level rise and climate variations. Yet, the catastrophic risk is the occurrence of disastrous floods in case of dam failure. Accordingly, the IPoE recommended that the three countries would jointly conduct two studies; one for the hydrologic simulation, and another for the environmental and socio-economic impacts of the dam, while providing Ethiopia with specific recommendations concerning the dam safety.

The IPoE findings and recommendations confirmed Egypt's suspicions that the dam construction had started without preparing all required documents, and that its capacity was maximized without conducting adequate studies (B_{Eg}). Such suspicions were further fuelled when Ethiopia speeded up construction without waiting for the implementation of the IPoE recommendations and without prior arrangement with Egypt (B_{Eg}). As a result, any potential cooperation outcome seemed far away, especially under the exacerbated tension between Egypt's Morsi and Ethiopia's Desalegn. The June 30 Revolution and the ouster of Morsi (D) provided the transition government in Egypt with a new chance to relieve the tension with Ethiopia. However, along three meetings in Khartoum between November 2013 and January 2014, Egypt and Ethiopia were not able to come to a middle ground. Egypt repeated its request to halt the construction until implementing the IPoE recommended studies and to respect the outcomes of these studies as binding on all parties. It also requested to add international experts to the new committee that would manage the implementation of the recommended studies. All these propositions were again rejected by Ethiopia, leading the negotiations to reached a complete deadlock.

Once more, the election of el-Sisi as a new president in Egypt paved the way for restoring negotiations. He has employed a new strategy; to focus on reaching a binding commitment over the first filling and dam operation and assigning the task of implementing the recommended studies to international consultancy firms. Ethiopia welcomed Egypt's new strategy, whereby a group of serious steps have been taken towards achieving an R-R outcome; including the Malabo Statement in June 2014, the establishment of the Technical National Committee (TNC) in August 2014, and the Declaration of Principles (DoP) in March 2015. Moving from the T-S to the R-R outcome required that Ethiopia would fill the reservoir and operate the dam only in a way that causes no significant harm to Egypt, as explicitly stipulated in the DoP. Such steps were associated with serious attempts to broaden bilateral and trilateral cooperation in the Eastern Nile sub-basin (I_{Eg}, I_{Et} and I_{Su}), through establishing commissions at the leadership level to promote regional integration projects. In addition, several visits by popular delegations have been made with the purpose of strengthening ties between the three countries at the lower level.

Nevertheless, there have been sharp technical differences between Egypt and Ethiopia along the process, mainly over selecting the international consultancy firm, reducing the reservoir capacity, increasing the dam bottom outlets, and starting the first filling only after the completion of the recommended studies. Ethiopia's stance was fixed over these points, insisting that the French BRL group, not the Dutch Deltares firm, would take the primary role in implementing the IPoE recommended studies, that the reservoir capacity is a non-negotiable matter, and that there would be no increase in the bottom outlets. Finding technical solutions to such differences was not enough, which motivated the ministers of foreign affairs from all three countries to join the ministers of water resources in the talks of December 2015. The result was the Khartoum Document, in which they named the French BRL group as the primary consultant and they reaffirmed their commitment to the DoP. The technical contracts were signed in September 2016 with the consultancy firms that would conduct the recommended studies.

However, a deeper disagreement has occurred since November 2017 over the baseline used in measuring the significant harm during the first filling and the dam operation. Egypt accepted the baseline as depicted in the inception report of the consultancy firms to be the current water uses, while both Sudan and Ethiopia rejected it and requested amendments, on which Egypt disagreed. This time, the heads of intelligence agencies joined the water and foreign ministers in the recent round of negotiations in April-May 2018; whereby the effort to solve the disagreement is still ongoing. Moreover, the surprise resignation of Desalegn and the election of Abiy Ahmed (D) have opened a new horizon for reaching a compromise, given that the new Ethiopian prime minister is employing a broader strategy of regional cooperation, to the extent that he verbally recognized Egypt's Nile share and normalized relations with Eritrea.

Obviously, each time the talks over the GERD are deadlocked, the Egyptian negotiator is faced with two alternatives, whether to provide concessions or not. Given that the dam has already become a reality and that Ethiopia is speeding up construction, the cost of finding no solution is too expensive to Egypt, which justified its readiness to make concessions in many cases, as long as its current water share is secured.

Aside from the GERD, although the downstream and upstream states have not yet resolved the CFA deadlock, the NBI is still functioning as a forum for promoting cooperation in the basin, especially after Sudan has resumed its full participation since 2013 while Egypt has been partially participating in some of the NBI activities. Moreover, there are other attempts to carry out cooperative projects among the Nile riparians, such as the navigational scheme proposed by Egypt to connect Lake Victoria with the Mediterranean Sea, which will benefit all landlocked upstream riparians, including Ethiopia through the Baro-Akobo-Sobat River. However, such attempts to generate further cooperation outcomes over the entire Nile Basin will remain challenged given the completely different visions of riparians on their water rights. As indicated in el-Sisi's speech to the Ethiopian parliament in March 2015, Egypt believes that the equitable utilization of the Nile waters comprises the right of upstream riparians to generate electricity and the right of downstream riparians to receive the flows without any sort of reduction. On the other side, confining the rewarding payoffs of cooperation to non-consumptive water uses have been rejected by Ethiopia and the other riparians for decades.

6.5.3 Contention (P-P)

The contention outcome has been generated in several situations along this round. The first serious confrontation took place when Ethiopia diverted the course of the Blue Nile to advance construction few days before submitting the IPoE final report and few hours after Morsi's meeting with Desalegn in Addis Ababa. This step provoked the anger of the Egyptian public opinion, because it would foreclose any modification to the dam design in accordance with the IPoE findings, and because

Desalegn did not mention the matter in his meeting with Morsi. Most of the Egyptian media vented their wrath on Morsi, as his spokesman emphasized on the same day that this was a normal step and a mere technical matter. At that time, Morsi was already facing a large petition campaign organized by the Tamarod youth group, criticizing heavily his domestic policies and calling for an early presidential election (D). In order to show his concern over the issue, Morsi held a meeting with a group of politicians representing different political forces in Egypt. Not informed that the meeting was broadcast live, the attendees came up with several hostile suggestions on how to deal with Ethiopia concerning its strategy to speed up construction regardless of the harm may be caused to Egypt. Amongst these were the suggestions to launch an offensive against the GERD and to support opposition in Ethiopia to be used later as a bargaining card in negotiations. Even Sudan, despite its ideological alignment with Morsi's Muslim Brotherhood group, was explicitly offended by these statements. Few days after the meeting, Morsi threatened Ethiopia in a public speech, declaring that all options are available and that blood would be the alternative in case of reducing a drop of Egypt's Nile water. Ethiopia responded forcefully to these threats through Desalegn's comment indicating that nothing and no one could halt the dam construction and that Egypt would not go for war unless its leaders went mad. In addition, the Ethiopian parliament ratified the CFA; in spite of suspending this decision for more than 2 years in an attempt to find a solution with the new regime in Egypt over the pending matters. Yet, the dramatic events that took place in Egypt and led to the ouster of Morsi (D) prevented the exacerbation of the tension.

One year later, el-Sisi was elected as Egypt's president. In parallel with his new strategy to reach a compromise with Ethiopia over the GERD, Egypt's widely supported new leader declared that he would revive the previous desert reclamation schemes, including the Peace Canal and Toshka projects, both of which Ethiopia had protested against during the 1970s and the 1990s respectively, claiming that they created facts on the ground that foreclose development upstream. However, this time Ethiopia did not officially protest, as development in Ethiopia has already become a reality since the 2000s that can no longer be foreclosed. Yet, Egypt sent a clear message to Ethiopia indicating that the Nile remains a major solution to meet the massive needs of its rapidly growing population (I_{Eg}), and that no Egyptian regime would accept to let Egyptians settling around the Nile Valley and Delta or in the new communities in the desert be harmed because of upstream projects.

As for the GERD, the road to implementing the recommended studies from June 2014 to date has been full of sharp technical differences between the three countries, and the contention outcome could have been generated in more than one occasion, particularly in December 2015 and November 2017. Each time, other political tensions had overshadowed the water negotiations, such as Ethiopia's accusation to Egypt of supporting Oromo protests in 2016 (D), the Egyptian-Sudanese dispute over the Hala'ib Triangle (R), the Turkish and Qatari military cooperation with Sudan (R), and the Israeli investments in East Africa (R). However, Egypt has been aware that reaching a deadlock would mean that Ethiopia might fill the reservoir and operate the dam without mitigating the risks of the GERD downstream (I_{Eg}). At the same time, Ethiopia's decision to build the GERD and other mega projects with national funds resulted in acute shortage of hard currency (C_{Et}). Here, comparing rewarding payoffs of cooperation against punishing ones of contention is obvious for both sides. This explains why el-Sisi has been keen to avoid the escalation of tension with Ethiopia and Sudan and the willingness of Abiy Ahmed to "preserve" Egypt's Nile share, but to "increase" it. Nevertheless, the possibility of generating the contention outcome cannot be dismissed. Simply, construction is more likely to be completed before accomplishing the technical studies, leading to a difficult situation where Ethiopia will need to fill the dam and begin generating hydropower, while Egypt will insist that Ethiopia complies with the pledge stated in the DoP, to run the first filling based on the findings of the studies.

The shadow of the future in the Nile Basin reflects further potential contention as well as cooperation outcomes. The great uncertainties of climate change exert additional pressure on the basin, including the alteration of rainfall patterns, the increase of evaporation losses, the reduction of crop productivity, and the submergence and salinity of lands in Egypt's Delta. Furthermore, all of the riparians have ambitious plans on the Nile in order to cope with the high rates of population growth and increasing needs in water, energy and food sectors. Ethiopia aims to proceed with implementing its large dams on the Blue Nile. By the completion of the GERD, Sudan will be able to attract international investments in the agricultural sector, while Egypt is moving forward with its desert reclamation schemes. For the Nile Equatorial riparians, there are plans to expand irrigated agriculture and build new hydropower dams. If those different national plans are not implemented from a basin-wide perspective, the Nile riparians will yield punishing payoffs of the contention outcome, which, need be neither equivalent nor comparable. These include the disruption of ecosystems and traditional livelihoods, fragmentation of regional peace, and diminishing opportunities for further cooperation, all of which will compromise the right of next generations to live in a healthy, developed and peaceful basin.

6.6 Concluding Remarks

As shown in this chapter, the game equilibrium has been swinging in the post-2011 round over the GERD between the cooperation and the contention outcomes among the three Eastern Nile riparians (Table 6.2). The round has run under an extremely different context compared to all the previous ones. During the post-Cold War round, Ethiopia began to achieve political stability and economic growth that enabled it to implement its national water projects, either with national funds or through loans from China. However, during the post-2011 round, the political instability in Egypt following the January 25 Revolution, led it to be fully absorbed in domestic affairs and reduced its capacity to deal with external threats. In addition, the independence of South Sudan changed the Khartoum's calculations and priorities, whereby constructing large dams in Ethiopia became of great importance more

Table 6.2 The outcomes of the Nile post-2011 round according to the iterated PD framework

Scenarios		Does player Y utilize the river unilaterally?	
		No	Yes
Does player X utilize the river uni-laterally?	No	Cooperation (R-R)	Sole utilization (S-T)
		May-Sep. 2011: Ethiopia and Egypt agreed upon the establishment of a trilateral committee (the IPoE) to assess the design and construction documents of the GERD (R_{Et}-R_{Eg})	Apr. 2011 to date: Egypt viewed the GERD as an Ethiopian step to build a huge dam on the Blue Nile to reap all benefits with Sudan regardless of the negative impacts downstream, especially that the dam size is maximized with no adequate economic justification and that the construction work has been speeded up before the completion of the required studies (S_{Eg}-$T_{Et\&Su}$)
		Mar. 2012 and DecCooperation (R-R)	
		May–Sep. 2011: Ethiopia and Egypt agreed upon the establishment of a trilateral committee (the IPoE) to assess the design and construction documents of the GERD (R_{Et}-R_{Eg})	
		Mar. 2012 and Dec. 2013: Sudan announced its explicit support for the GERD (R_{Su}-R_{Et})	
		Jun. 2014: Egypt and Ethiopia agreed upon resuming negotiations over the GERD (R_{Eg}-R_{Et})	
		Aug. 2014: Egypt, Ethiopia and Sudan formed the TNC ($R_{Et\&Su}$-R_{Eg})	
		March 2015: The leaders of the three countries signed the DoP ($R_{Et\&Su}$-R_{Eg})	
		Dec. 2015: The water and foreign ministers from the three countries signed the Khartoum Document ($R_{Et\&Su}$-R_{Eg})	
		Sep. 2016: The three countries signed the contracts of the joint studies with the BRL and Artelia French groups ($R_{Et\&Su}$-R_{Eg})	

(continued)

Table 6.2 (continued)

Scenarios		Does player Y utilize the river unilaterally?	
		No	Yes
	Yes	Sole utilization (T-S)	Contention (P-P)
		Apr. 2011: Ethiopia commenced the construction of the GERD without prior notification (T_{Et}-S_{Eg})	Jun. 2013: Egyptian politicians mentioned hostile suggestions against Ethiopia (P_{Eg}-P_{Et})
		May 2013: Ethiopia diverted unilaterally the course of the Blue Nile (T_{Et}-S_{Eg})	Jun. 2013: Morsi threatened that all options are available. Ethiopia's Desalegn responded forcefully and the Ethiopian parliament ratified the CFA (P_{Eg}-P_{Et})
		Dec. 2013: Sudan announced its support for the GERD ($T_{Et\&Su}$-S_{Eg})	Apr.–Dec. 2015: Egypt and Ethiopia disagreed upon the international consultancy firms that would conduct the IPoE required studies (P_{Eg}-P_{Et})
		Apr. 2014 and Jan. 2016: Ethiopia rejected Egypt's offer to co-finance the GERD and request to increase the bottom outlets respectively (T_{Et}-S_{Eg})	

than ever before. Under this extraordinary context, Ethiopia announced unilaterally the construction of the GERD without prior notification, with a massive size several times larger than the one investigated in all previous prefeasibility studies, and without completing all impact studies that befit the huge magnitude of the dam. As seen in the comments of the Ethiopian officials, this step has been taken reciprocating the way by which Egypt has implemented its previous national projects, mainly the AHD, the Peace Canal, and the Toshka project. For Egypt, these schemes have had no negative impacts on the upstream riparians, and thus they did not require prior consultation with them. In contrast, Ethiopia views the Egyptian projects as facts imposed on the ground to foreclose development upstream. Unlike the colonial and Cold War rounds, Ethiopia has been able to form a new cluster with Sudan and change the status quo in the basin.

The effective Tit for Tat strategy employed by Ethiopia has been clear along the round. Reciprocating the Egyptian popular and official visits to Addis Ababa in May 2011, Ethiopia's Zenawi initiated the establishment of the IPoE and the postponement of the CFA ratification. Two years later, when Ethiopia diverted the course of the Blue Nile unilaterally few days before the IPoE submitted its final report, Egypt's Morsi held a meeting during which the Egyptian politicians heavily offended the GERD, and then he made explicit threats against Ethiopia. Ethiopia responded forcefully through Desalegn's statement that "nothing" and "no one" would halt the project, in addition to the decision of the Ethiopian parliament to ratify the CFA without modifying the articles to which Egypt and Sudan have firmly objected since 2010. Realizing that confrontation would make all sides worse off, the new regime in Egypt following the June 30 Revolution decided to change its position into pacification and the search for an acceptable solution. Ethiopia welcomed Egypt's new position, and important steps were followed through issuing the Malabo statement in June 2014, concluding the DoP in March 2015, and signing the contracts of the joint studies in September 2016. Nevertheless, along this path, Ethiopia has insisted not to modify the dam design, in any degree, and to speed up construction without waiting for the completion of the required studies. Still, it is clear that Egypt have accepted the GERD as a fact on the ground for the purpose of hydropower generation and economic development; yet, as repeatedly declared, it will not accept the reduction of a cubic meter of its water share. Surprisingly, Abiy Ahmed, who took office in April 2018 as a result of 3-year anti-government protests, showed initial understanding over preserving Egypt's Nile share. Evidently, the changing contexts during the whole round have profoundly affected the balance of interests and capabilities between Egypt, Ethiopia and Sudan, driving the game equilibrium away from the sole utilization outcome to somewhere between the contention and the cooperation outcomes.

Chapter 7
Conclusion: Tug-of-War or Benefits for All?

7.1 Drawing a Comparative Analysis

History is a discipline widely cultivated among nations and races. It is eagerly sought after. The men in the street, the ordinary people, aspire to know it. Kings and leaders vie for it. Both the learned and the ignorant are able to understand it. For on the surface, history is no more than information about political events, dynasties, and occurrences of the remote past, elegantly presented and spiced with proverbs. It serves to entertain large, crowded gatherings and brings to us an understanding of human affairs. It shows how changing conditions affected (human affairs), how certain dynasties came to occupy an ever wider space in the world, and how they settled the earth until they heard the call and their time was up. The inner meaning of history, on the other hand, involves speculation and an attempt to get at the truth, subtle explanation of the causes and origins of existing things, and deep knowledge of the how and why of events. History, therefore, is firmly rooted in philosophy. It deserves to be accounted a branch of it.
Ibn Khaldun (1377/1969: 5)

This book has been inspired by the dramatic changes occurring in the Nile Basin over the recent years. The study assumed that the water-related events in the basin represent only the tip of the iceberg and that the answer to the above question requires diving deep into the history in order to examine the entire picture portraying the Nile story. Instead of looking at the macro-correlation between causes and consequences of conflict and cooperation in transboundary river basins, the book aimed to closely investigate the process and mechanisms through which the status quo in the Nile Basin has been changing. The research hypothesis states that development on the Nile represents a Prisoners' Dilemma game, whereby the equilibrium is driven by the changing contexts through their varying influences on the inherited heterogeneities among the different riparian countries.

Process tracing was used to analyze the interactions along the game and to collect evidence in order to prove the research hypothesis. The starting point of the process was the colonial age, where the change in equilibrium began to be apparent enough for analysis. The process then was divided into four main rounds, based on the profound changes taking place in the context; the colonial round, the Cold War round, the post-Cold War round and the post-2011 round. Tracing the interactions

M. M. Samaan, *The Nile Development Game*,
https://doi.org/10.1007/978-3-030-02665-3_7

along the four rounds confirmed the hypothesis and explored the mechanisms through which the changing contexts have driven the inherited heterogeneities; thus generating different outcomes and shifting the game equilibrium from one point to another.

In the final analysis, the accumulated influences of the context can preserve the game equilibrium at the sole utilization outcome (one side reaps the highest payoffs while the other gets the lowest) and can turn it into the cooperation or contention outcomes (both yield rewarding or punishing payoffs respectively). The political and natural events taking place in a given context may work in favor of the side which has a vital interest and possesses the capacity to utilize the shared watercourse, maintaining its position as the sole user of the basin resources either by providing the others with adequate alternatives or by preventing them from altering the status quo. However, the context may also push the riparian states towards or away from cooperation. On the one hand, its overall influence may lead the riparians to be interested in working together, but even trusting and in need of, each other. On the other hand, the context may render cooperation among them less fruitful, less needed, and less trusted. Finally, turning into contention requires that all sides are capable of benefiting from the river unilaterally and full of mistrust towards one another. Hence, the mechanism through which the context drives the game equilibrium is influencing the degree of heterogeneity among the riparian states over their interests, capabilities and beliefs.

The Nile's hydrologic, climatic and demographic characteristics gave birth to the equilibrium at the sole utilization outcome. Egypt remained for millennia the sole user of the Nile waters, given that it is the only riparian that has no alternative freshwater resources, while the others did not possess the capacity to alter the Nile flow. Yet, the opening of the Suez Canal in 1869 and the consequent political and economic instability in Egypt changed the game entirely. In 1882, Britain entered the scene by occupying Egypt to secure the new sea-route to India. London soon realized that the security of the Suez Canal required the stability of Egypt, which ultimately led to its domination over the entire Nile, either by direct colonization or by signing treaties. As Britain's interests extended beyond the security of the Suez Canal to the expansion of cotton cultivation across its colonies, it played the role of moderator to convince Cairo that upstream dams would be important for the prosperity of the Egyptian agriculture and that irrigation schemes in Sudan would not prejudice its water rights.

Lengthy negotiations were conducted between Egypt and Sudan during the 1910s and the 1920s over the Gezira scheme, the first large-scale irrigation project to be constructed upstream of Aswan. Generating a cooperation outcome was seriously challenged by the low-flood year of 1914 and the assassination of the British governor of Sudan in Cairo in 1924, but the effective role played by Britain facilitated the conclusion of the 1929 Agreement. On the one hand, this agreement formed the downstream cluster between Egypt and Sudan, while on the other, it generated a sole utilization outcome with the Equatorial colonies, since they were left with no benefits and with the obligation not to construct any upstream project without Egypt's prior approval. In the late 1940s, after the failure of negotiations over the Equatorial Nile Project, Egypt and Uganda found a common interest and a mutual need to construct the Owen Falls Dam, which led them to sign separately the 1949–1953 Agreements. Such deals strengthened Egypt's position on the Nile, while giving considerable benefits to Sudan and Uganda for irrigation and hydropower generation respectively. There was another attempt to generate cooperation over the Tana concession with Ethiopia; however it was hindered due to the weakness of Addis Ababa's central government and its mistrust of British and Italian ambitions. At the same time, the world economic depression of the late 1920s made Britain no longer interested in the project, while Ethiopia's poor credit rating prevented it from building the dam privately. Thus, during the colonial round, the global context pushed towards the generation of cluster cooperation and preserved the game equilibrium around the sole utilization outcome in favor of the downstream cluster.

By the 1950s and the early 1960s, the riparian states gained their independence one after another. One of the early decisions took by the new revolutionary regime, which abolished the monarchy in Egypt, was to build one massive dam at Aswan instead of the Century Storage Scheme that involved several upstream dams; over which it failed to reach a compromise with the other riparians during the preceding decades. Sudan also decided to build the Roseires Dam; yet neither was capable of financing such national projects. Both resorted to the IBRD, which required concluding a joint agreement over their plans. However, the generation of cooperation was halted for a while due to their incompatible visions over water allocation and cost sharing, in addition to their border dispute over the Hala'ib. Triangle. Besides, nationalizing the Suez Canal to build the AHD using its revenues resulted in the Tripartite Aggression of Britain, France and Israel against Egypt. Only the military takeover in Sudan by late 1958 made it possible to sign the 1959 Agreement with Egypt. Moreover, the termination of the first civil war in Southern Sudan in 1972 and the launching of the peace process between Egypt and Israel after the 1973 War allowed for the initiation of the Jonglei I project, but it was soon halted after the civil war was sparked again in 1983.

As for Egypt and Ethiopia, relations were strained in the context of their opposing positions on the global rivalry between the U.S. and the Soviet Union, as well as the Arab-Israeli conflict, the Ethiopian-Somali dispute and the Ethiopian-Eritrean war. For more than three decades, the contention outcome between Egypt and Ethiopia had been evolving, whereby Egypt was allocating the entire Nile flow with Sudan, building the AHD, and launching the Peace Canal project, while Ethiopia was conducting extensive surveys on the Blue Nile and constructing the Finchaa Dam and the Lake Tana regulator. Ironically, all these unilateral actions were undertaken with the support of the Soviet Union and the U.S. to Egypt and Ethiopia alternatively. After all, the continuing internal fighting and frequent famines in Ethiopia, In addition to the blocking of international funds by Egypt, hindered its capability to fully implement the Blue Nile scheme. In parallel, the great rains of 1961–1964 and the severe drought of 1983–1985 motivated the riparians to cooperate over the Hydromet survey and the Undugu group with the support of international developmental partners. Yet, general mistrust limited these initiatives to information sharing and technical cooperation, and therefore they did not evolve to joint planning and management. Hence, during the Cold War round, the game equilibrium slightly moved towards contention and cooperation outcomes; however it remained closer to sole utilization in favor of the downstream cluster of Egypt and Sudan.

In the early 1990s, the context was profoundly changed once more, since by the collapse of the Soviet Union, the world was no longer bipolar. Consequently, international organizations were able to catalyze bilateral and multilateral cooperation among the Nile riparians, as they were the major source of funds to rebuild economies emerging from decades of wars and destruction. In 1993, the Hydromet survey was folded into the TECOONILE, and in 1999 the NBI was launched to replace the latter and involve all the riparian states. The upstream riparians successfully formed a cluster through which they pushed for developing a basin-wide framework that they hoped to replace the existing agreements and reallocate the Nile waters. Along the 1990s, the downstream cluster was about to breakdown in light of the acute tension between Egypt and Sudan over the Gulf War, the old Hala'ib. Triangle dispute, and the assassination attempt against Mubarak. However, the latter sought to bypass the tension to maintain the unity of Sudan and the coherence of the downstream position on the Nile.

After a considerable portion of Egypt's foreign debts was cancelled in recognition of its role in the Gulf War, the government was encouraged to launch the Toshka reclamation project, benefiting from the rising of the water levels of Lake Nasser at the time. This fueled Ethiopia's mistrust, since it viewed the move as creating new facts on the ground while attempting to confine the upstream projects under the NBI umbrella to non-consumptive water uses. Egypt's point of view was completely different, believing that it has the full right to manage its water share as a rainless downstream riparian in the most efficient way. The emergence of China as an ideal alternative source of financial and technical assistance in addition to the relative economic and political stability enabled the upstream riparians to build their projects similarly outside the NBI. As a result, a race started over unilateral projects across the Nile Basin. Ethiopia was able to construct the Tekeze Dam and the Beles hydropower project, while Sudan built the Merowe Dam and the Atbara-Setit Dam and heightened the Roseires Dam. The contention outcome, therefore, was fully generated between the upstream and downstream riparians for the first time ever in the game. Furthermore, despite the unprecedented international support, the riparians failed to generate a cooperation outcome over the CFA. Obviously, each of the downstream and upstream clusters found it harmful to accept the other's vision on the agreement, and both neither needed nor trusted cooperation with one another, especially after they had carried out their water plans unilaterally. Ultimately, five upstream riparians signed individually the CFA in May 2010 (Burundi, the sixth riparian, joined them in February 2011), and the downstream riparians responded by suspending their participation in the NBI. Thus, during the post-Cold War round, the game equilibrium completely left the sole utilization and stabilized at the contention outcome.

By January 2011, the game context witnessed dramatic changes, however this time at the domestic level, through the referendum on the independence of South Sudan and the spark of the Egyptian Revolution. Sudan became more interested in cooperation with Ethiopia over large dams upstream, since these would regulate the water flow, control sediments, manage floods, and generate hydropower. Experiencing acute shortages of hard currency and energy after the loss of oil fields in the south, such benefits will help Sudan attract new investments in the agricultural sector and import inexpensive electricity. In Egypt, the political disturbance that lasted for years led it to be extremely busy with domestic affairs and reduced its capacity to deal effectively with external issues. Under these extraordinary conditions, Ethiopia announced the construction of the GERD with a maximized size, several times larger than the one investigated in the previous plans and prefeasibility studies.

Since April 2011 to date, the game equilibrium has been swinging between the cooperation and contention outcomes over the GERD. As highlighted by the panel of international experts formed by the three Eastern Nile countries, Ethiopia launched construction without carrying out the full impact assessment and dam safety studies that befit the massive size of the dam and the vitality of the Blue Nile (IPoE 2013: 37). Obviously, the suitable timing and the lack of trust influenced Ethiopia's unilateral step and its choice not to follow a participatory approach in designing the dam; neither with the downstream riparians as stipulated in the UNWC, nor with the local stakeholders as recommended by the WCD (2000) framework. Egypt, on the other hand, worried that the downstream communities and ecosystems would pay alone the enormous cost, since the dam is built on the Ethiopian-Sudanese border. In light of the accelerated construction and the loss of time in protracted negotiations, the Egyptian fears have increased over time, ranging from the potential "catastrophic" risks in case of any dam safety problem to large deficit in water supplies and salinity of coastal lands caused by inappropriate reservoir first filling and dam operation (MIT 2014: 6). What aggravated Egypt's anxiety has been the explicit Sudanese support for the project due to its direct benefits. For Ethiopia, despite Chinese assistance in building transmission lines and other complementary projects, the decision to take the step unilaterally and to speed up construction required that the GERD itself would be nationally funded, which exposes the country to a serious foreign currency shortage (IMF 2013; Reuters 2018d).

To avoid the punishing payoffs of the contention outcome, all the three countries have been seeking to reach a compromise, which led to the establishment of the IPoE with the initiative of Ethiopia in 2011, the signature of the DoP in March 2015, and the signature of the technical contracts on the IPoE recommended studies in September 2016. Apparently, the context has been playing a significant role in driving the game outcomes. Contention was more likely to be generated by the threat of Egypt's Morsi against Ethiopia under the pressure of popular anger about the latter's decision to divert the course of the Blue Nile few days before the submission of the IPoE report and few hours after the meeting of Morsi and Ethiopia's Desalegn.

However, the spark of the June 30 Revolution in Egypt and the strategy adopted by the new regime brought a relief by focusing on how to mitigate the negative impacts of the GERD rather than halting the project. Since 2014, the leaders of the three countries have made considerable efforts to solve sharp technical differences over the dam and strengthen ties between their countries. However, this has been challenged by concurrent tensions, such as the renewal of the Hala'ib. Triangle dispute between Egypt and Sudan, and Ethiopia's accusation against Egypt of backing protests and disturbances in Addis Ababa.

Moreover, the shadow casted by the future indicates that all scenarios remain possible in the Nile game. This depends on the way the riparians will manage their common risks, mainly the uncertainties of climate change, the rapid population growth, and the wide gaps in water, energy and food needs. If every riparian or cluster proceeds with its unilateral projects regardless of the harm may be caused to others, then contention is more likely to be the next destination of the game equilibrium.

7.2 Main Findings

Overall, the main findings of the study can be summarized as follows:

- Since the game is iterated, equilibrium is not static, and its dynamics is driven by the changing contexts that continuously influence the inherited heterogeneities among the riparian states over their interests, capabilities and beliefs.
- The Nile is a clear example in this respect. While the natural configuration of the basin had sustained equilibrium at the sole utilization outcome for millennia, this was ultimately transformed given the changing contexts during the past century and half. These include the British colonization following the opening of the Suez Canal, the Cold War, the longstanding civil wars and regional conflicts in Africa and the Middle East, the frequent regime transformations, and the growing Gulf and Chinese investments in the region, in addition to the extreme environmental crises.
- If there is a powerful riparian that heavily depends on the shared watercourse and is capable of controlling the game, it should be aware that sponsoring comprehensive development schemes in early stages that bring "benefits for all" is significantly more beneficial than waiting until the others possess the capacity to compete over the use of water. Evidently, when those, which have lacked the power to benefit from the river for long times, become capable of implementing their national water projects unilaterally, they will not prefer cooperation nor consider the interests of that powerful state; reflecting a sort of retaliation. Nevertheless, it is crucial that when the powerful side initiates programs of benefit sharing, these should generate satisfactory rewarding payoffs of cooperation.

- Cooperation (R-R) is more fruitful than both contention (P-P) and shifting between exploiting (T-S) and being exploited (S-T) by the other. However, mistrust remains the chief obstacle to the emergence of cooperation. For instance, Egyptians have never forgotten the use of the Nile water as a weapon in Emperor Tekle Haymanot's 1706 letter, King Theodore's ambitions in 1855–1868, the Fashoda incident by France in 1898, Lord Allenby's 1924 ultimatum, Britain's secret plans during the 1956 Suez Crisis, and Mengistu's threats to Sadat in 1978. For Egyptians, all these events have been in the back of their minds while interpreting the American Blue Nile survey during the Cold War, the individual signature of the CFA by the upstream riparians, and Zenawi's sudden announcement of the GERD with its tremendous size few weeks after the January-25 Revolution. At the same time, Ethiopians still remember the military expeditions by the Muhammad Ali Dynasty to secure the Nile sources during the nineteenth century and the British recognition of Ethiopia's sovereignty in the 1902 Treaty in return for the commitment not to alter the Nile flow to Egypt. These also have been directly linked by Ethiopians to Egypt's initiatives for non-consumptive water joint projects since the 1980s and its objection to the CFA or the GERD.
- It was supposed that all riparians commit to the UNWC with all its rights and obligations, but due to the long history of mistrust between the downstream and upstream riparians, each side sought to tailor the UNWC in the way that assures its own interests.
- Cooperation outcomes require that all sides provide concessions to reach a compromise. This explains why cooperation in the Nile game has been more facilitated in the formation of clusters, where it is possible to create compatible interests, interdependent capabilities and trusted beliefs. Sudan gave Egypt the right to build the Jabal Auliya Dam for water storage and to collect data across its territory in return for the approval of the Egyptian government that Sudan would build the Sennar Dam and irrigate the Gezira scheme with 4 BCM annually. Egypt allowed for the construction of the Owen Falls hydropower plant and shared the dam costs, while Uganda, in return, stored water for Egypt and accepted to host a resident engineer to oversee dam operation. More recently, Sudan is not involved in financing the GERD, despite its potential extensive benefits; however for Ethiopia, Khartoum's explicit support to the project has been a sufficient payback. In contrast, basin-wide cooperation is more likely to be hindered due to incompatibility of interests, independency of capabilities, and lack of trust among the entire riparian states, all of which make each side reluctant to accept the vision of the other. This has been evident in the CFA and the GERD stalemates.
- Although international institutions have made enormous efforts to attain such basin-wide cooperation, multilateralism could not replace unilateralism, particularly because they were mainly working on the compatibility of interests and were less effective in dealing with the other two pillars. Concerning dependence of capabilities, their roles were significantly reduced by the emergence of alternative sources of fund, which made the riparians no longer in need of cooperation

with one another. With respect to trusted beliefs, the Nile talks in general have not followed a participatory approach, given that the whole issue was left in the hands of the officials in each country, whereby none of the concerned parties, even international institutions, seriously enabled civil societies to participate in the process. Moreover, economic ties such as mutual trade between Egypt and the upstream riparians have not been strong enough to build interdependence that can create an atmosphere of trust or a ground of common interests. In other words, international institutions have neither been able to protect the Nile game from negative contextual influences nor to create corresponding positive ones.

7.3 Key-Message

The missing basin-wide cooperation has turned the game into "tug-of-war" over the Nile, in which the river has been the rope, and each of Egypt and Ethiopia has been tugging it in full capacity to assure maximum payoffs. Each has sought to form clusters from inside the basin and request the support of external actors in order to strengthen its side against the other. It is undeniable that the Nile is the lifeline of Egypt and a pillar of development for Ethiopia. However, continuing to play the tug-of-war over the river will extremely exhaust all the riparian states on the long run. More important, this will severely harm the basin itself; its natural and human capitals. Instead of paying more attention to the prosperity of its communities and ecosystems, the river has been suffering from being tugged in every direction by building massive structures, with long-term costs far exceeding their shortsighted benefits.

Interaction over the Nile is not a zero-sum game without a win-win scenario. Yet, it is a dilemma in which the decision to cooperate entails concessions and requires different sides to fully trust and vitally need each other. Each side may think that if the balance of power is in its favor today, it is better not to provide concessions and rather to seek maximum tempting payoffs. However, the changing contexts will turn over the balance tomorrow or any day after, and everything may be dramatically changed. On that day, the one that previously got the lowest payoff would act to retaliate, whereas the former will also wait to counter-retaliate. Ultimately, both will enter into closed circle of retaliation and counter-retaliation and will get exponential punishing payoffs. Unfortunately, the harsh punishment would be imposed on the river itself that may lead to unsustainable future for the basin's ecosystems, local communities, and more important the future generations that are supposed to find a healthy, peaceful and developed basin.

7.4 Policy Implications

Addressing the relevant policy makers in the Nile Basin, this study may conclude the following short messages:

- *To all the riparian states*: The policy of building mega water schemes should be revised in terms of long-term benefits and costs. Actually, such massive schemes generate relatively shortsighted economic benefits at the national level and leave the local communities and ecosystems to bear the heavy costs. Instead of proceeding with harmful competition over unilateral projects, it is more beneficial to follow a participatory approach involving all different stakeholders and to return to conducting joint technical studies with the support of independent international experts in order to develop balanced plans that take into consideration all socio-economic and environmental aspects. The Blue Nile canyons, the Sudd swamps, the Equatorial Lakes, and all other ecosystems in the Nile are substantial assets that should be saved for the sake of future generations, not only related to the basin, but even to the entire world, as Whittington (2004: 19) points out. Moreover, every state should be ready to make concessions for reaching compromises, in order to avoid causing harm to others according to the recommendations raised by international consultants.
- *To Egypt*: Since the Nile is Egypt's sole source of freshwater, Egyptian policy makers will always consider it a priority and a duty to maintain their country's interests in the basin. However, after more than a century of several rounds of water interactions, it is clear that the adoption of reactive project-by-project approach has not led to favorable outcomes. Instead, future national plans upper and lower the basin should be discussed during early stages with an adequate level of flexibility. For instance, Ethiopia's national plans comprise the construction of three large dams upstream of the GERD (Karadobi, Beko Abo and Mandaya) with a total capacity exceeding 125 BCM (Table 6.1) and the implementation of irrigation schemes with a total area of over one million hectares (Sect. 5.4.1). If all these projects are carried out unilaterally the same way the GERD has been built, downstream riparians will be significantly harmed. Thus, it is necessary that Egypt should promptly initiate technical discussions and hold scientific conferences over all such plans across the basin. International partners and investors should be invited to the discussions. By doing so, Egypt will send an unmistakable message that it is sponsoring development in the basin that brings benefits for all without causing significant harm downstream.
- *To Ethiopia*: It is doubtless that the Nile represents a major resource in Ethiopia's war against poverty, starvation and blackouts, and that exercising its right of utilizing the Nile waters has been hindered for long decades due to different circumstances. Yet, this does not justify constructing huge dams on the Nile without conducting adequate safety and impact assessment studies jointly with co-riparian states in advance, as stipulated in the UNWC. Despite speeding up construction before completing all required studies, Ethiopia has taken important steps in showing its good faith by establishing the IPoE, concluding the DoP, and signing the contracts of the joint studies. However, this should be complemented by further collaboration along the next steps, especially over the first filling and dam operation policies. More important, inviting Egypt and Sudan to study the other large dams prior to construction will benefit Ethiopia the most, since continuing to implement unilateral projects with national funds may harm the

Ethiopian economy on the long term (IMF 2013). In contrast, reaching a compromise with downstream riparians over such projects will help Ethiopia attract international investments. It is understandable that the process of joint planning proved fruitless in the past, but the case is completely different today after the experience of the GERD.

- *To Sudan*: It is completely true that Sudan is the "master of the middle" (Waterbury 2002: 128). More than any time before, Sudan is occupying a central position in the Nile game. It still maintains the 1959 Agreement with Egypt, has strengthened ties with Ethiopia, has played the role of moderator for cooperation over the GERD, and has resumed its full participation in the NBI. The more Sudan's position is balanced, the more the whole basin will be prosperous. Yet, at the domestic level, planning development schemes should not overlook the livelihoods and historical assets of local communities.
- *To international institutions*: In spite of the enormous efforts made in the Nile since the 1960s to date, basin-wide cooperation has not been established yet. Concerning the NBI, the focus has been given to making the riparians' interests more compatible over water issues, but less attention was devoted to boosting interdependence and building trust among the riparians over broader economic and political issues. Promoting regional economic integration and strengthening ties between civil societies and scientific networks across the basin may create a positive context that pushes state-actors towards cooperation, rather than leaving them to negative contextual influences of political and environmental crises. Regarding the CFA, relevant international bodies should have made greater efforts to convince the riparians to adopt the UNWC as it is. Unfortunately, the downstream and upstream riparians were left to exert pressure on one another so that each would accept the tailored version adopted by the other. The same is currently done with respect to the GERD, as international organizations remain silent on the accelerated construction of such a huge dam on the most vital tributary in the Nile without completing the required joint studies that would assure eliminating potential risks of the project. Seeking peace and prosperity in the Nile is not an aim related only to the peoples of the basin, but to the entire region and even the world. For instance, the EU is currently facing serious challenges in absorbing the continuous waves of immigrants from the surrounding hotspots suffering from political and economic instability. However, potential environmental crises and regional fragmentation due to water disputes and overlooking biophysical and socio-economic aspects in implementing mega schemes could result in additional waves.

7.5 Theoretical Implications

This study has sought to contribute to the progressive debates among the three main branches of hydropolitics. It has questioned the neo-realist notion that interactions over scarce resources represent a zero-sum game, showing that win-win outcomes were achieved between several pairs of riparians in the Nile game. At the same time, the study demonstrated that the dilemma embedded in the game has hindered basin-wide cooperation in the Nile. It also illustrated that the bilateral cooperation outcomes were directed to the formation of clusters that generated outcomes of sole utilization and contention.

Regarding the critical branch, the study endorsed the notion that conflict and cooperation coexist, since, in every round, the sole utilization, cooperation and contention outcomes have been generated simultaneously. In addition, following the recommendation of Hasenclever et al. (2000: 10, 11, 25), the study shed light on the significance and usefulness of supplementing the "minimalist cognitivism" with the rationalist approach of Axelrod and Keohane (1985), which was implied in Goldstein and Keohane (1993), Keohane and Martin (1995), Keohane and Ostrom (1995), and Keohane (1998, 2010). It was shown that the state-centric model, founded on self-interests, has to be complemented with the consideration of the capabilities and beliefs of states, the influences of political and environmental contexts at the multiple levels, and the role of non-state actors. Such an integrative approach has helped in revisiting the theory of "cooperation based upon reciprocity" by Axelrod (1984, 2000). Tracing the Nile game in this study highlighted two crucial factors restricting the effectiveness of the reciprocal strategy (Tit for Tat); the lack of power in reciprocating defection and the lack of trust in reciprocating cooperation. It was also shown how the changing contexts affect the degree of heterogeneity of interests, capabilities and beliefs; thus shifting equilibrium from one point to another. Moreover, the role of non-state actors cannot be overlooked, such as the Chinese/Italian construction companies, the Gulf agricultural investments, the MIT expert group, and the International Rivers Network. However, the study's findings question the critical notion that international norms can drive states to cooperation. This is evident in how the Nile riparians have dealt with the UNWC principles of equitable use, causing no significant harm, and prior consultation.

Overall, the analytic framework, constructed by the study, proved comprehensive and flexible in tracing the hydropolitical interactions on international rivers. Although the focus was on the Nile case, the clear structure of scenarios and payoffs, the possibility of cluster formation, and the involvement of domestic, regional and global actors ensure the applicability of the framework on other river basins and even on other IR topics. Thus, the study agrees with the KISS principle adopted by Axelrod (2000), "Keep it simple, stupid," while applying the iterated PD game.

7.6 Methodological Implications

Every method has its specific limitations, and being aware of these enables the study to attain its goal. While process tracing allows for in-depth analysis within a single case study, the domain of the independent and intervening variables and the range of the dependent variable remain limited to the events and observations of this single case. Thus, the significant limitation of using process tracing in this study lies in its power of generalization. In the Nile case, the domain of the inherited heterogeneities among the riparians over interests, capabilities and beliefs (independent variables) is restricted by the special hydrological configuration in the basin; that is rainless downstream countries with almost no alternative freshwater resources vis-a-vis rich upstream riparians of unevenly distributed water resources. The range of the equilibrium dynamics is bounded between maintaining sole utilization by Egypt and shifting to contention or cooperation with the other riparian states.

Nevertheless, three important factors have been taken into consideration while designing the case study. First, since the scope of this book is to look at the role of the context in driving the equilibrium in the Nile iterated game, it has been a priority to assure a wide domain of various contextual influences along the game rounds (the intervening variable). This paved the way for analyzing how the different intervening actors and linked issues affect the generation of outcomes at the multiple levels. Second, before discussing the case study, Chap. 3 (the analytic framework) was entirely devoted to study various configurations of transboundary river basins, distinguishing symmetric and asymmetric balance of power among the riparian states. This step has a twofold benefit: to confirm the validity of the theoretical model in investigating other international rivers with different configurations and to specify the category to which the Nile case belongs in order to define the generalization limits of the study findings. Third, by dividing the Nile case into separate four rounds, the study was able to draw a comparative analysis on the process and mechanisms through which the changing contexts have influenced the equilibrium dynamics.

Above all, one of the most important objectives of the book is to come up with specific findings and recommendations that can guide relevant policy makers in achieving a better future for the basin communities and ecosystems.

7.7 Future Research

One of the research main findings is that the attempts to change the status quo in the Nile into basin-wide cooperation have been unsuccessful to date because relevant institutions were not able to protect the Nile from the negative contextual influences and create other positive ones. Because the contexts are always in change and their influences on the inherited heterogeneities are random, the basin should not be left without effective institutions. The argument that the role of institutions is of such a great importance in catalyzing cooperation among riparians has to be examined more closely. How should they learn the lessons of the past?

The study points out that driving the riparians towards sustainable cooperation would not be attained through focusing only on state actors. Greater attention should be paid to the mechanisms of connecting the civil societies, empowering the scientific networks, and strengthening the economic ties among the riparian states. Thus, it should be investigated how institutions at the multiple levels can protect the riparians from the random effects of the changing contexts surrounding the game and create new intervening contexts that work positively in favor of communities and ecosystems for the sake of current and upcoming generations. Such a research project may require conducting further single case studies in addition to employing the Qualitative Comparative Analysis (QCA) method to compare multiple cases of transboundary river basins where institutions have been developed in different degrees.

Maps for the Natural and Demographic Characteristics of the Nile Basin

M. M. Samaan, *The Nile Development Game*,
https://doi.org/10.1007/978-3-030-02665-3

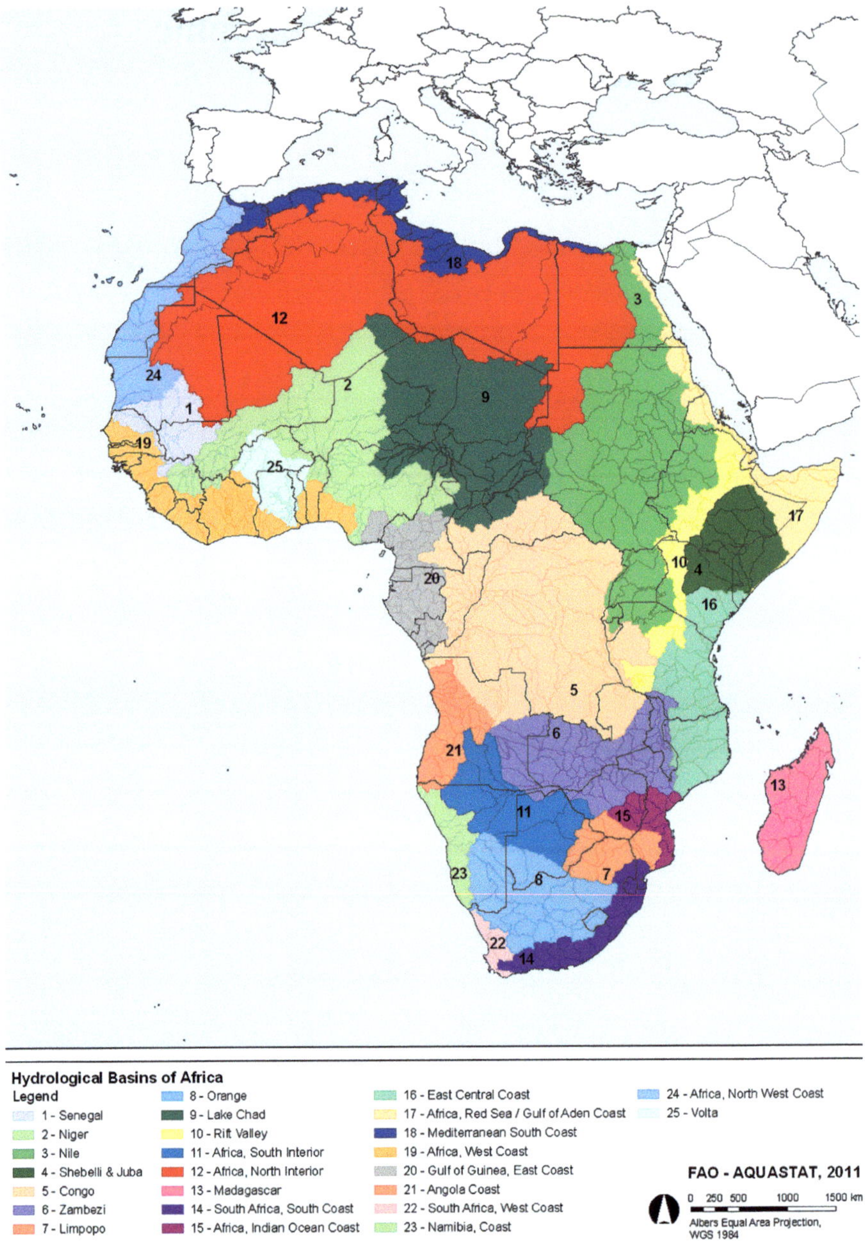

Fig. A.1 Transboundary river basins in Africa. (Source: FAO, Aquastat Database, Regional thematic maps, available at: http://www.fao.org/nr/water/aquastat/maps/AfricaBasins.png (accessed on July 11, 2018))

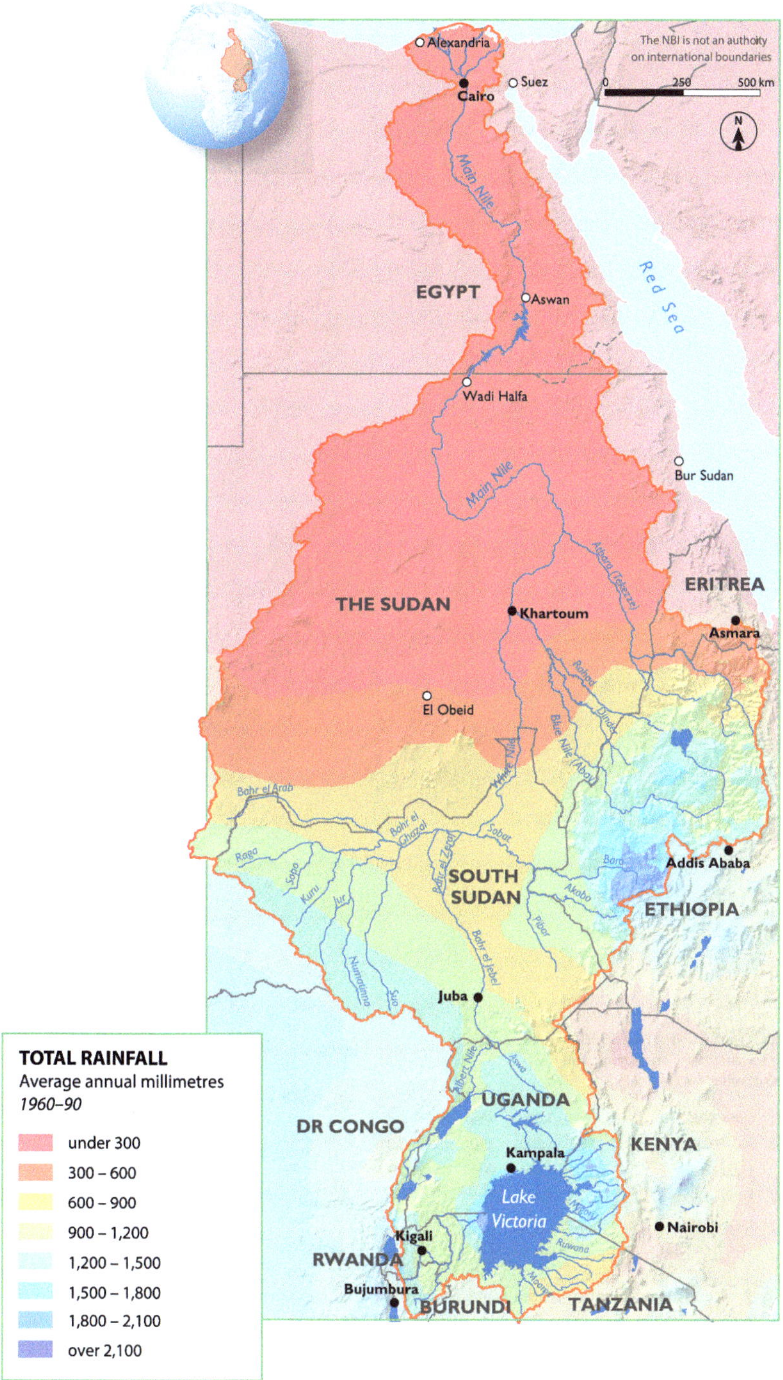

Fig. A.2 Rainfall precipitation across the Nile Basin. (Source: Prepared by NBI, the State of Basin Report (2012a: 30) after the data obtained from the Climatic Research Unit)

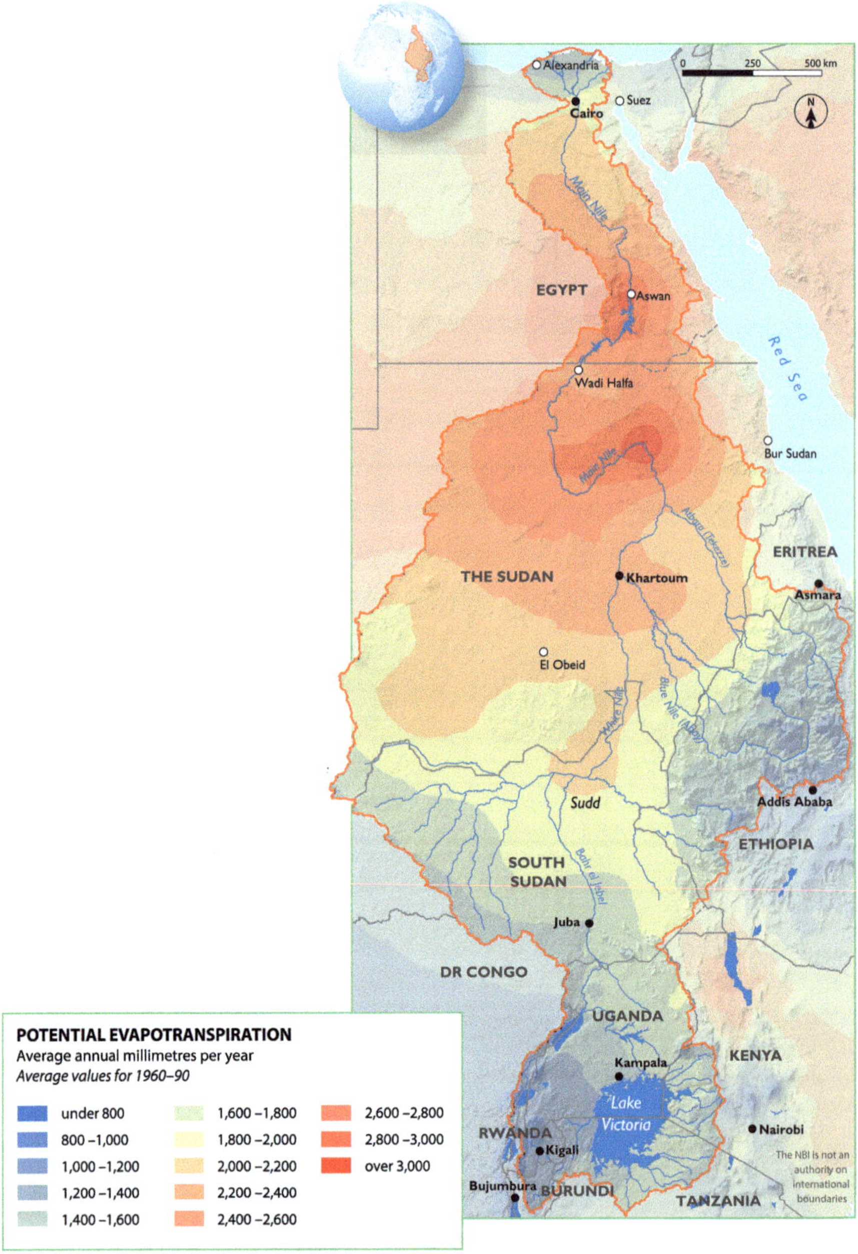

Fig. A.3 Evapotranspiration rates across the Nile Basin. (Source: Prepared by NBI, the State of Basin Report (2012a: 35) after the data obtained from FAO)

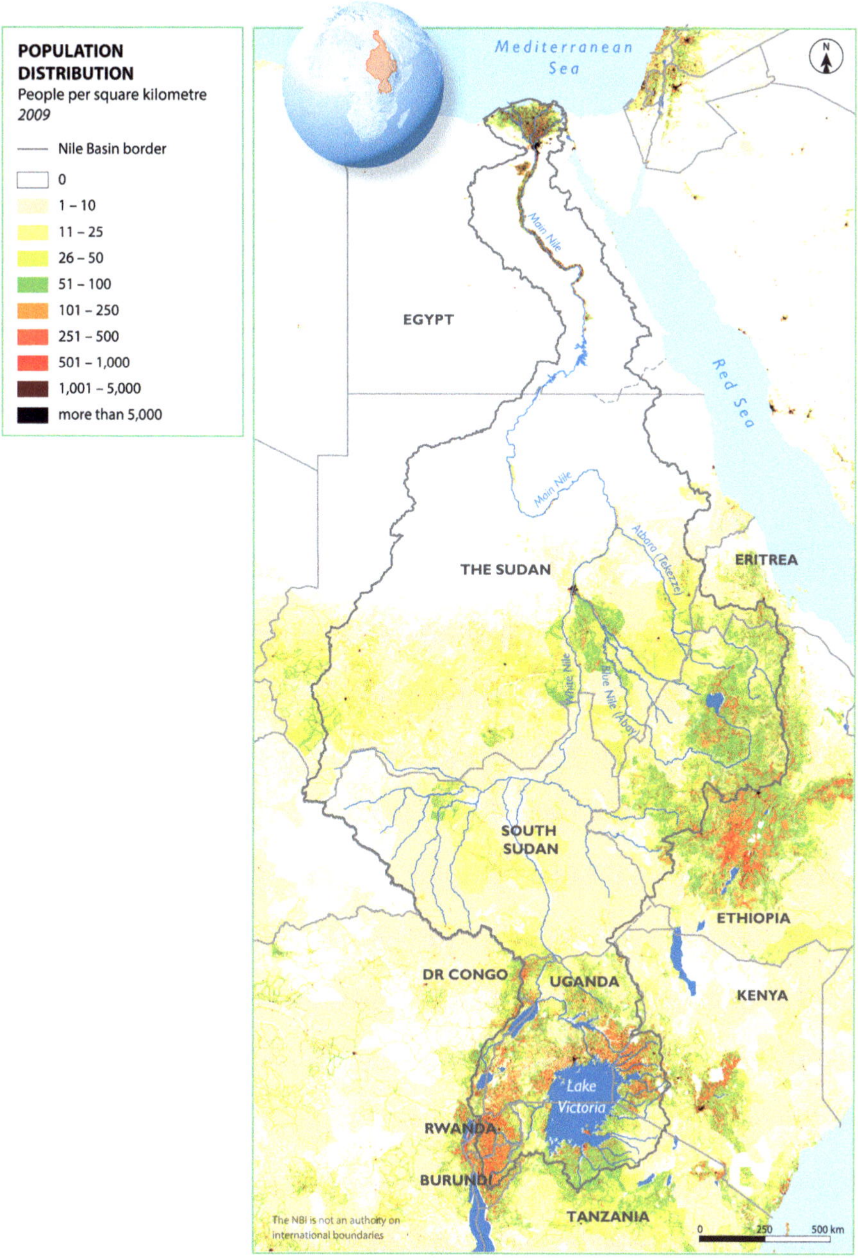

Fig. A.4 Population distribution across the Nile Basin. (Source: Prepared by NBI, the State of Basin Report (2012a: 103) after the data obtained from LandScan Global Population Database in 2009)

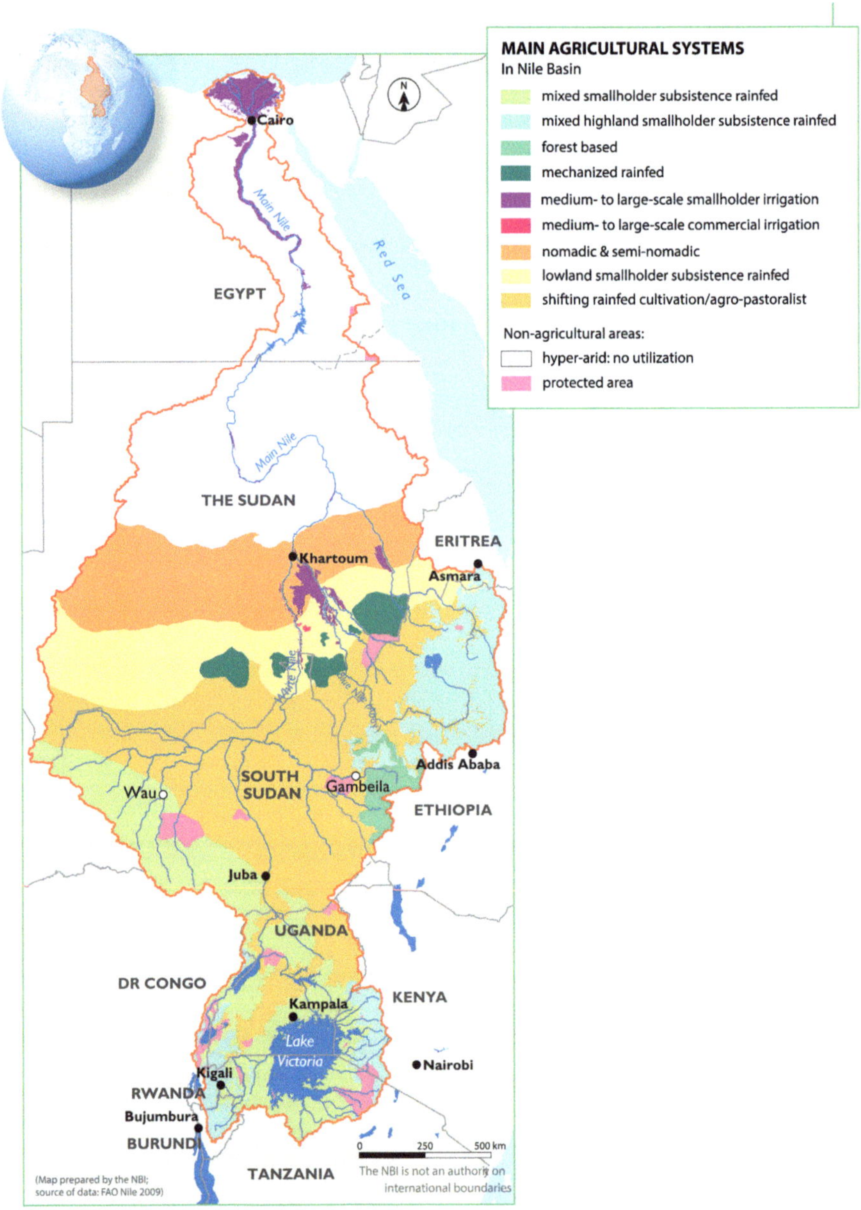

Fig. A.5 Agriculture in the Nile Basin. (Source: Prepared by NBI, the State of Basin Report (2012a: 124) after the data obtained from FAO)

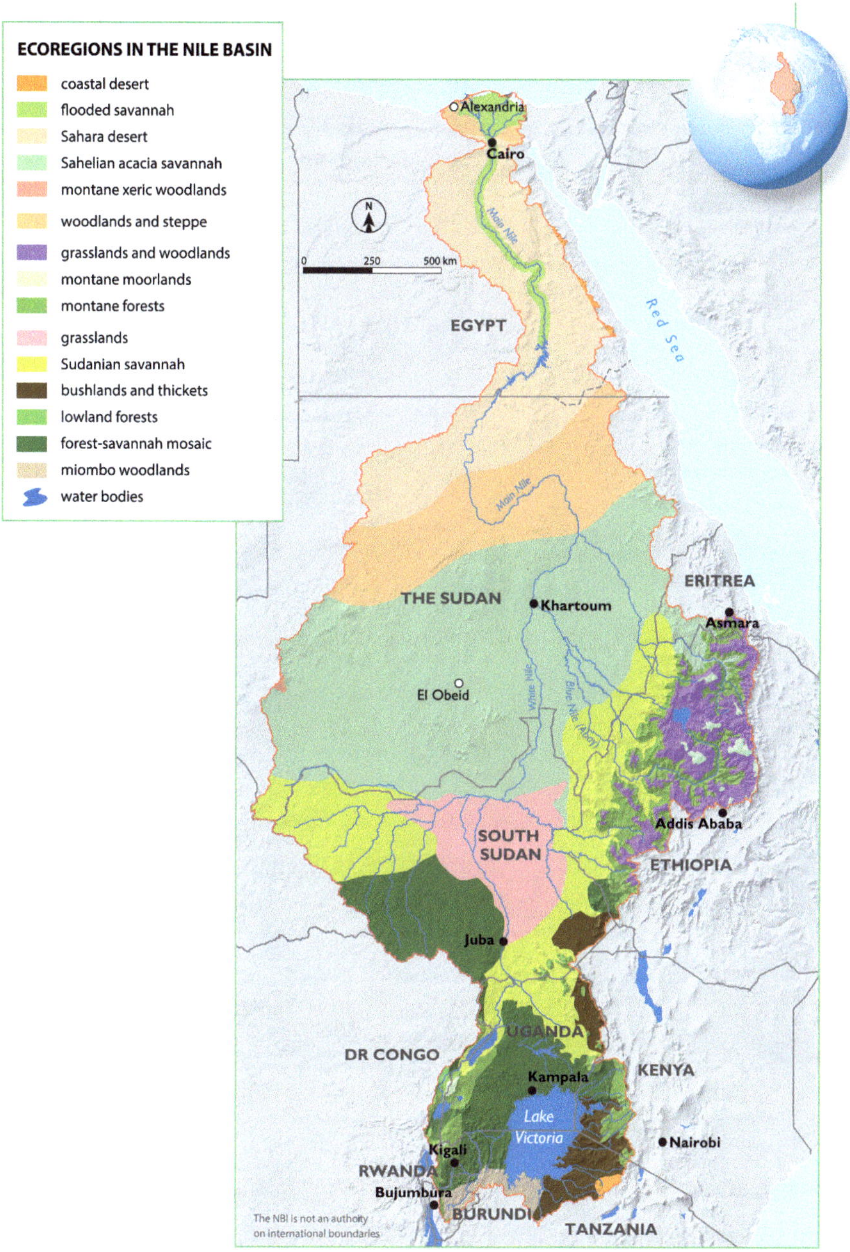

Fig. A.6 Ecoregions in the Nile Basin. (Source: Prepared by NBI, the State of Basin Report (2012a: 68) after the data obtained from the Terrestrial Ecoregions GIS Database)

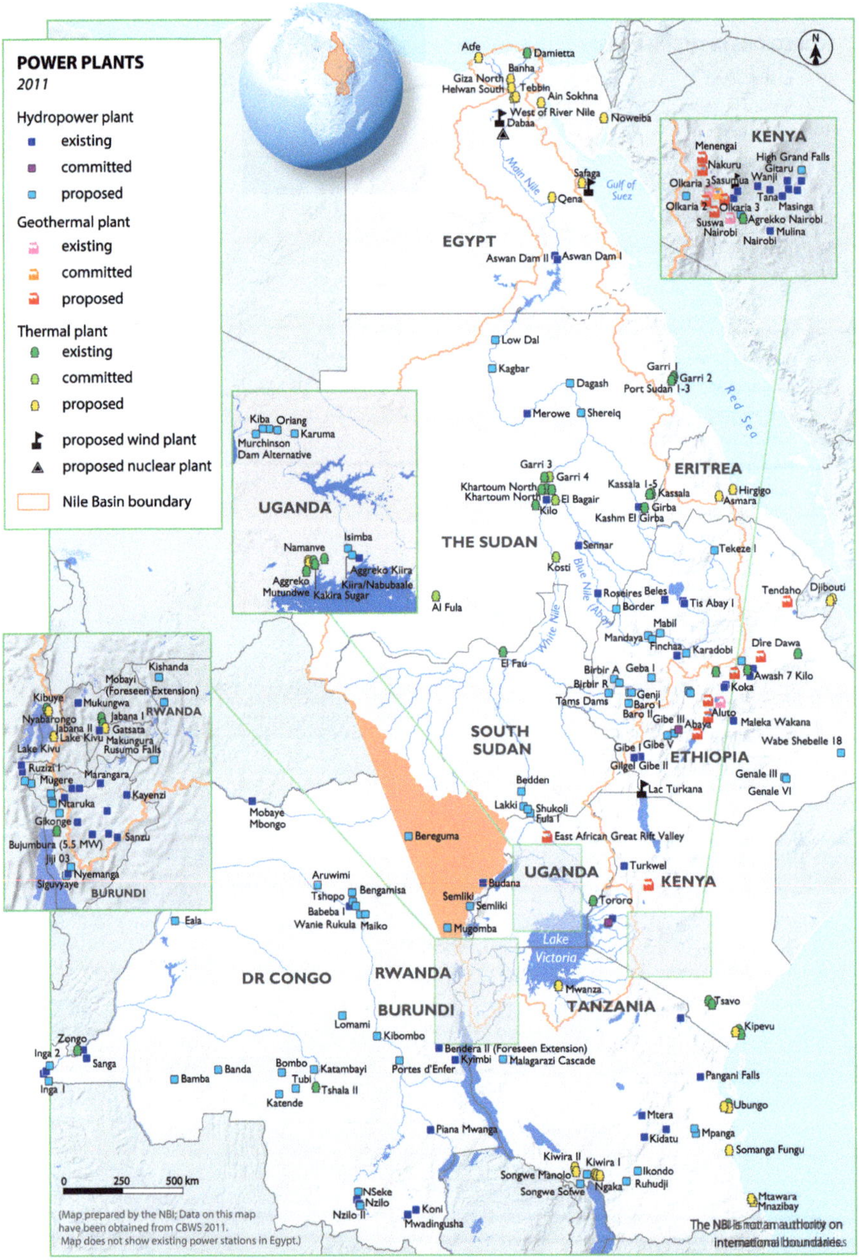

Fig. A.7 Existing and proposed power plants according to the EAPP and EAC 2011 master plan. (Source: Prepared by NBI, the State of Basin Report (2012a: 181) after the data obtained from the EAPP and EAC 2011 master plan)

References

Abdel-Alim, T. *Judiciary on the chopping block.* May 1, 2013. http://english.ahram.org.eg/NewsPrint/70512.aspx (accessed December 7, 2016).

Abdelaziz, K. *Sudanese police clash with students in Khartoum.* January 30, 2011. http://uk.reuters.com/article/uk-sudan-protests-demonstrators-idUKTRE70T0LE20110130 (accessed November 28, 2016).

—. *One person killed in Sudanese protests over fuel subsidies.* September 24, 2013. http://uk.reuters.com/article/uk-sudan-protest-idUKBRE98N1B320130924 (accessed November 28, 2016).

Abdelaziz, K., and A. Dziadosz. *Sudan police, protesters clash over austerity cuts.* June 20, 2012. http://uk.reuters.com/article/uk-sudan-protests-idUKBRE85J0ZL20120620 (accessed November 28, 2016).

Abdel-Hadi, H. *The activists blame Morsi for the crisis of the Renaissance Dam.* May 29, 2013. http://www.vetogate.com/362712 (accessed December 6, 2016).

Abdelwahab, H.B. *Influence (supremacy) of religion of Sudan's foreign policy decision-making.* Bloomington IN, USA: Authorhouse, 2012.

Abir, M. *Oil, power and politics: Conflict in Arabia, the Red Sea and the Gulf.* London, UK: Frank Cass, 1974.

Abou-Amra, M., and M. Abnoudi. *Morsi visited Ethiopia, then the course of the Blue Nile was diverted.* May 29, 2013. http://www.elwatannews.com/news/details/190371 (accessed December 6, 2016).

Aboul-Gheit, A. *My testimony: The Egyptian foreign policy, 2004–2011 (in Arabic).* Cairo, Egypt: Nahdet Misr Publishing Group, 2014.

Abouyoub, Y. "The Sudan: Africa's civilizational fault line? U.S. foreign policy and the National Islamic Front". In *Towards the dignity of difference? Neither "end of history" nor "clash of civilizations"*, edited by M. Mahdavi and W.A. Knight, 317–336. New York NY, USA: Routledge, 2016.

Abul-Magd, Z.A. *Empire and its discontents: Modernity and subaltern revolt in Upper Egypt, 1700–1920.* Ann Arbor MI, USA: Proquest, 2008.

Abusamhadaneh, K. *Transboundary water management in the Western part of Jordan: Challenges and solutions (M.A. thesis).* Braunschweig, Germany: Institute of Social Sciences, TU-Braunschweig, 2012.

Africa Confidential. *Sudan: Talk in New York, war in the Horn – Volume 37, Issue 8.* April 12, 1996. http://www.africa-confidential.com/article-preview/id/11538/Talk_in_New_York%2c_war_in_the_Horn (accessed October 23, 2016).

M. M. Samaan, *The Nile Development Game*,
https://doi.org/10.1007/978-3-030-02665-3

Ahram. *Islamist Sudan welcomes election of Egypt's Morsi.* June 25, 2012. http://english.ahram.org.eg/NewsPrint/46095.aspx (accessed November 30, 2016).

—. *Ethiopian dam won't affect Nile water share: Egypt presidency.* May 28, 2013a. http://english.ahram.org.eg/News/72569.aspx (accessed December 5, 2016).

—. *Sudan, Egypt may call on Arab League over Nile dam.* May 28, 2013b. http://english.ahram.org.eg/NewsContent/1/0/72568/Egypt/0/Sudan,-Egypt-may-call-on-Arab-League-over-Nile-dam.aspx (accessed December 5, 2016).

—. *Egypt's live TV coverage of Nile crisis meeting draws fire.* June 4, 2013c. http://english.ahram.org.eg/NewsContent/1/64/73174/Egypt/Politics-/Egypts-live-TV-coverage-of-Nile-crisis-meeting-dra.aspx (accessed July 8, 2015).

—. "Egypt's Coptic pope holds meeting on Nile row before meeting Ethiopian counterpart." *Ahram Online.* June 11, 2013d. http://english.ahram.org.eg/NewsContent/1/64/73746/Egypt/Politics-/AboutUs.aspx (accessed December 16, 2016).

—. "Visit by Ethiopia's Orthodox Church head to Cairo postponed." *Ahram Online.* June 13, 2013e. http://english.ahram.org.eg/NewsContent/1/64/73887/Egypt/Politics-/Visit-by-Ethiopias-Orthodox-Church-head-to-Cairo-p.aspx (accessed December 16, 2016).

—. *Jumping, not walking, out of Egypt's crisis: El-Sisi interview, part 2.* May 7, 2014a. http://english.ahram.org.eg/NewsContent/1/64/100663/Egypt/Politics-/Jumping,-not-walking,-out-of-Egypts-crisis-ElSisi-.aspx (accessed December 11, 2016).

—. *El-Sisi wins Egypt's presidential race with 96.91%.* June 3, 2014b. http://english.ahram.org.eg/NewsContent/1/64/102841/Egypt/Politics-/ElSisi-wins-Egypts-presidential-race-with-.aspx (accessed December 11, 2016).

—. *Egypt and Ethiopia agree that tripartite dam committee will resume its work.* June 27, 2014c. http://english.ahram.org.eg/NewsContent/1/64/104937/Egypt/Politics-/Egypt-and-Ethiopia-agree-that-tripartite-dam-commi.aspx (accessed July 8, 2015).

—. *Egypt, Ethiopia, Sudan agree on impact studies on Nile dam.* August 26, 2014d. http://english.ahram.org.eg/NewsContent/1/64/109313/Egypt/Politics-/Egypt,-Ethiopia,-Sudan-agree-on-impact-studies-on-.aspx (accessed December 11, 2016).

—. *Egypt's irrigation minister visits Ethiopia's Grand Renaissance Dam.* September 22, 2014e. http://english.ahram.org.eg/NewsContent/1/64/111397/Egypt/Politics-/Egypts-irrigation-minister-visits-Ethiopias-Grand-.aspx (accessed December 11, 2016).

—. *Egypt garners pledges of financial and political support on 1st day of Sharm conference.* March 13, 2015a. http://english.ahram.org.eg/NewsContent/3/162/125127/Business/EEDC-/Live-updates-Egypts-economic-conference-opens-in-S.aspx (accessed December 12, 2016).

—. *A new era of cooperation: Sisi to Ethiopian parliament.* March 25, 2015b. http://english.ahram.org.eg/NewsContent/1/64/126072/Egypt/Politics-/A-new-era-of-cooperation-Sisi-to-Ethiopian-parliam.aspx (accessed December 13, 2016).

—. *Questions about the CFA.* April 22, 2015c. http://weekly.ahram.org.eg/Print/11030.aspx (accessed July 8, 2015).

—. "Egypt's Sisi tells Egyptians 'not to worry' about Ethiopian dam." *Ahram Online.* December 30, 2015d. http://english.ahram.org.eg/NewsContent/1/64/177773/Egypt/Politics-/Egypts-Sisi-tells-Egyptians-not-to-worry-about-Eth.aspx (accessed December 17, 2016).

—. *Egypt, Ethiopia, Sudan agree on measures to strengthen trilateral ties.* February 20, 2016a. http://english.ahram.org.eg/NewsContent/1/64/188088/Egypt/Politics-/Egypt,-Ethiopia,-Sudan-agree-on-measures-to-streng.aspx (accessed December 13, 2016).

—. *Egypt, Ethiopia, Sudan sign final contracts on Nile dam studies.* September 20, 2016b. http://english.ahram.org.eg/NewsContent/1/64/244290/Egypt/Politics-/Egypt,-Ethiopia,-Sudan-sign-final-contracts-on-Nil.aspx (accessed December 15, 2016).

—. *'We do not interfere in other countries' domestic affairs': Egypt ambassador told Ethiopia.* October 9, 2016c. http://english.ahram.org.eg/NewsContentP/1/245474/Egypt/We-do-not-interfere-in-other-countries-domestic-af.aspx (accessed December 15, 2016).

—. *Egypt's red lines on water.* November 23–29, 2017. http://weekly.ahram.org.eg/News/23064.aspx (accessed July 9, 2018).

—. *'The interests of Egypt, Sudan and Ethiopia are one,' President Sisi says after tripartite summit in Addis Ababa.* January 29, 2018a. http://english.ahram.org.eg/NewsContent/1/64/288980/Egypt/Politics-/The-interests-of-Egypt,-Sudan-and-Ethiopia-are-one.aspx (accessed July 10, 2018).

—. *Egypt, Sudan and Ethiopia reach agreement on next steps for dam technical studies.* May 16, 2018b. http://english.ahram.org.eg/NewsContent/1/64/299614/Egypt/Politics-/Egypt,-Sudan-and-Ethiopia-reach-agreement-on-next-.aspx (accessed July 13, 2018).

AIPC. *Red shadow on the Nile – To the point: News in depth – Volume 8, Issues 18–35.* Sandton, South Africa: African International Publishing Company, 1979.

Alaa el-Din, M. *Egypt does not view Ethiopian dam as threat, says minister.* January 14, 2016. http://english.ahram.org.eg/NewsContent/1/64/181008/Egypt/Politics-/Egypt-does-not-view-Ethiopian-dam-as-a-threat,-say.aspx (accessed December 15, 2016).

Al-Arabiya. *New Nile agreement a wake-up call for Egypt.* May 21, 2010. https://www.alarabiya.net/articles/2010/05/21/109210.html (accessed November 24, 2016).

Albanese, R., and D.D. Van Fleet. "Rational behavior in groups: The free-riding tendency." *Academy of Management Review* 10, no. 2 (1985): 244–255.

Al-Haj, Al-Meqdad. *Al-Bashir: We Support the Construction of the Renaissance Dam.* December 5, 2013. http://alintibaha.net/portal/home1/%D8%A2%D8%AE%D8%B1-%D8%A7%D9%84%D8%A7%D8%AE%D8%A8%D8%A7%D8%B1-1/42871-%D8%A7%D9%84%D8%A8%D8%B4%D9%80%D9%8A%D9%80%D8%B1-%D9%86%D9%80%D8%A4%D-9%8A%D9%80%D8%AF-%D9%82%D9%80%D9%8A%D9%80%D8%A7%D9%85-%D8%B3%D9%80%D9%80%D9%80%D8 (accessed March 19, 2014).

Aljazeera. *Timeline: Egypt's revolution – A chronicle of the revolution that ended the three-decade-long presidency of Hosni Mubarak.* February 14, 2011a. http://www.aljazeera.com/news/middleeast/2011/01/201112515334871490.html (accessed November 29, 2016).

—. *Egypt and Ethiopia to review the Nile River dam.* September 17, 2011b. http://www.aljazeera.com/news/middleeast/2011/09/2011917132445980153.html (accessed July 8, 2015).

Allam, M.N. *The crisis of the Ethiopian Renaissance Dam: A political issue or a technical problem? (in Arabic).* Cairo, Egypt: Mahrousa Publishing Center, 2014.

Allan, J.A. *The Middle East water question: Hydropolitics and the global economy.* London, UK: Tauris Publishers, 2001.

—. "Nile Basin asymmetries: A closed fresh water resource, soil water potential, the political economy and Nile transboundary hydropolitics." In *The Nile: Origin, environments, limnology and human use*, edited by H.J. Dumont, 749–770. New York NY, USA: Springer, 2009.

Al-Monitor. *Morsi supporters attack opposition demonstrators.* December 5, 2012. http://www.al-monitor.com/pulse/originals/2012/al-monitor/morsi-supporters-attack.html (accessed December 6, 2016).

Aman, A. *Ethiopian dam talks jammed with no compromise in sight.* April 16, 2018. https://www.al-monitor.com/pulse/originals/2018/04/meeting-sudan-ethiopia-egypt-renaissance-dam-ends-impasse.html (accessed July 9, 2018).

Amare, G. *Contentious issues in the negotiations process of the Cooperative Framework Agreement on the Nile – Presented paper.* Addis Ababa, Ethiopia: The National Consultative Workshop on Nile Cooperation, 2009.

Ambrosio, T. *Ethnic identity groups and U.S. foreign policy.* London, UK: Praeger, 2002.

Amin, I. *Idi Amin speaks: An Annotated selection of His speeches.* Edited by B. Turyahikayo-Rugyema. Madison WI, USA: African Studies Program, University of Wisconsin-Madison, 1998.

Anisfeld, S.C. *Water resources.* Washington, DC: Island Press, 2010.

Antar, A. "Irrigation Minister: No worries about the one and half million feddans as we learned the lesson from Toshka." *Al-Watan News.* December 31, 2015. http://www.elwatannews.com/news/details/889759 (accessed December 17, 2016).

Arnold, M., and M. LeRiche. *The South Sudan: From revolution to independence.* New York NY, USA: Oxford University Press, 2013.

Arsano, Y. "Institutional development and water management in the Ethiopian Nile Basin." In *The River Nile in the post-colonial age: Conflict and cooperation among the Nile Basin countries*, edited by T. Tvedt, 161–178. New York NY, USA: I.B. Tauris, 2010.

Ashra, N. *After the solemnity of 40 years since the Sadat era, Ethiopia dare to Egypt during the reign of Morsi.* June 8, 2013. http://alwafd.org/search?q=%D8%A3%D8%AB%D9%8A%D9%88%D8%A8%D9%8A%D8%A7%20%D8%AA%D8%AA%D8%AC%D8%B1%D8%A3%20%D8%B9%D9%84%D9%89%20%D9%85%D8%B5%D8%B1%20%D9%81%D9%8A%20%D8%B9%D9%87%D8%AF%20%D9%85%D8%B1%D8%B3%D9%8A (accessed December 6, 2016).

Asser, M. *Obstacles to Arab-Israeli peace: Water.* September 2, 2010. http://www.bbc.com/news/world-middle-east-11101797 (accessed January 3, 2017).

AUB. *Middle East Forum (Volume 37).* Beirut, Lebanon: The Alumni Association of the American University of Beirut, 1961.

Awad, M. *Clashes in Cairo after Mursi seizes new powers.* November 23, 2012. http://www.reuters.com/article/us-egypt-president-idUSBRE8AM0DO20121123 (accessed December 6, 2016).

Awad, M., and D. Zayed. *Egypt police prepare for "Day of Wrath" protest.* January 25, 2011. http://www.reuters.com/article/egypt-protest-idAFLDE70O0GM20110125 (accessed November 29, 2016).

Awadalla, N., and E. Knecht. *Egypt FM to head to Ethiopia after Nile dam talks stall.* December 24, 2017. https://www.reuters.com/article/us-egypt-ethiopia-dam/egypt-fm-to-head-to-ethiopia-after-nile-dam-talks-stall-idUSKBN1EI09N (accessed July 9, 2018).

Awulachew, S.B., V. Smakhtin, D. Molden, and D. Peden. *The Nile River Basin: Water, agriculture, governance and livelihoods.* London, UK: Routledge, 2012.

Axelrod, R. "Effective choice in the Prisoner's Dilemma." *Conflict Resolution* 24 (1980a): 3–25.

—. "More effective choice in the Prisoner's Dilemma." *Conflict Resolution* 24 (1980b): 379–403.

—. "The emergence of cooperation among egoists." *The American Political Science Review* 75, no. 2 (1981): 306–318.

—. *The evolution of cooperation.* New York, USA: Basic Books Inc. Publishers, 1984.

—. "On six advances in cooperation theory." *Analyse & Kritik* 22, no. 1 (2000): 130–151.

Axelrod, R., and D. Dion. "The further evolution of cooperation." *Science* 242, no. 4884 (1988): 1385–1390.

Axelrod, R., and W.D. Hamilton. "The evolution of cooperation." *Science* 211, no. 4489 (1981): 1390–1396.

Axelrod, R., and R.O. Keohane. "Achieving cooperation under anarchy: Strategies and institutions." *World Politics* 38, no. 1 (1985): 226–254.

Azzam, H. *The Arab World facing the challenge of the new millennium.* London, UK: I.B. Tauris, 2002.

Badawi, A. *The position of Egypt's presidents concerning the Hala'ib Triangle issue (in Arabic).* April 18, 2016. http://www.elwatannews.com/news/details/1106462 (accessed December 15, 2016).

Baker, M.M. *Mubarak: Toshka project opens way towards new civilization in Egypt, printed in the Excutive Intelligence Review.* December 1997. http://american_almanac.tripod.com/toshka.htm (accessed October 31, 2016).

Batisha, A.F. "Sustainability assessment of Grand Renaissance Dam: Regarding landslides in Ethiopia." *International Journal of Sustainability* 2, no. 1.3 (June 2013): 25–38.

Bazilian, M., H. Rogner, M. Howells, S. Hermann, D. Arent, D. Gielen, P. Steduto, A. Mueller, P. Komor, R.S.J. Tol, and K.K. Yumkella. "Considering the energy, water and food nexus: Towards an integrated modelling approach." *Energy Policy* 39, no. 12 (2011): 7896–7906.

BBC. *Egypt protests: Three killed in 'day of revolt'.* January 26, 2011. http://www.bbc.com/news/world-africa-12272836 (accessed November 29, 2016).

—. *Egyptian politicians caught in on-air Ethiopia dam gaffe.* June 4, 2013a. http://www.bbc.com/news/world-africa-22771563 (accessed July 8, 2015).
—. *Egyptian warning over Ethiopia Nile dam.* June 10, 2013b. http://www.bbc.com/news/world-africa-22850124 (accessed July 8, 2015).
—. *Egypt and Ethiopia vow to defuse Blue Nile dam row.* June 18, 2013c. http://www.bbc.com/news/world-africa-22951276 (accessed December 7, 2016).
—. *Profile: Egypt's Tamarod protest movement.* July 1, 2013d. http://www.bbc.com/news/world-middle-east-23131953 (accessed December 6, 2016).
—. *Egypt crisis: Morsi supporters gather for protests.* July 5, 2013e. http://www.bbc.com/news/world-middle-east-23192472 (accessed December 8, 2016).
—. *Ethiopia PM Hailemariam Desalegn in surprise resignation.* February 15, 2018a. https://www.bbc.com/news/world-africa-43073285 (accessed July 19, 2018).
—. *Abiy Ahmed: Ethiopia's prime minister.* July 9, 2018b. https://www.bbc.com/news/world-africa-43567007 (accessed July 10, 2018).
Bennett, A., and C. Elman. "Qualitative research: Recent developments in case study methods." *Annual Review of Political Science* 9 (2006): 455–476.
Bennett, A., and A.L. George. *Process tracing in case study research – MacArthur Foundation Workshop on Case Study Methods.* Cambridge MA, USA: Belfer Center for Science and International Affairs (BCSIA), Harvard University, 1997.
Berhane, D. *Nile, Ethiopia: Panic over the river 'diversion' an Egyptian hoax.* June 15, 2013. http://hornaffairs.com/en/2013/06/15/nile-ethiopia-panic-river-diversion-egyptian-hoax/ (accessed December 5, 2016).
—. *Ethiopia rejects Egypt's offer to finance the Renaissance dam.* April 8, 2014. http://hornaffairs.com/en/2014/04/08/ethiopia-reject-egypt-finance-renaissance-dam/ (accessed December 9, 2016).
Beyene, M. *How efficient is the Grand Ethiopian Renaissance Dam?* July 14, 2011. http://www.internationalrivers.org/files/attached-files/ethiopiadamefficiency.pdf (accessed July 8, 2015).
Bishku, M.B. "Israel and Ethiopia: From a special to a pragmatic relationship." *Conflict Quarterly* 14, no. 2 (1994): 39–62.
Biswas, A.K. "Management of international waters: Opportunities and constraints." *Water Resources Development* 15, no. 4 (1999): 429–441.
Blackmore, D., and D. Whittington. *Opportunities for cooperative water resources development on the Eastern Nile: Risks and rewards – An independent report of the Scoping Study Team for the Eastern Nile Council of Ministers.* Edited by planning.unc.edu/people/faculty/dalewhittington/BlackmoreWhittingtonENScopingStudyFinal2009.pdf. The Nile Basin Initiative, 2008.
Blair, A. *Britain and the World since 1945.* London, UK: Routledge, 2015.
Bloomberg. *Ethiopia rejects Egypt proposal on Nile as dam talks falter.* January 8, 2014. http://www.bloomberg.com/news/articles/2014-01-08/ethiopia-rejects-egyptian-proposal-on-nile-as-dam-talks-falter (accessed July 8, 2015).
Booth, C. *The ancient Egyptians for dummies.* Chichester, UK: John Willey & Sons, Ltd., 2007.
British and Foreign State Papers. *Correspondence Respecting Abyssinia (No. 212) in Volume LII (1861–1862).* London, UK: The British Foreign Office, 1868.
Brochmann, M., and P.R. Hensel. "The effectiveness of negotiations over international river claims." *International Studies Quarterly* 55, no. 3 (2011): 859–882.
Brown, P.H., D. Tullos, B. Tilt, D. Magee, and A.T. Wolf. "Modeling the costs and benefits of dam construction from a multi-disciplinary perspective." *Environmental Management* 90 (2009): 5303–5311.
Brunnee, J., and S.J. Toope. "The changing Nile Basin regime: Does law matter?" *Harvard International Law Journal* 43, no. 1 (2002): 105–159.
Budge, E.A.W. *The Egyptian Sudan: Its history and monuments.* New York NY, USA: Cosimo Classics, 1907/2010.
Burchill, S., and A. Linklater. "Introduction (Chapter 1)." In *Theories of International Relations*, edited by S. Burchill, et al., 1–28. New York NY, USA: Palgrave Macmillan, 2005.

Butcher, A.D. *The Jonglei Canal scheme.* Cairo, Egypt, 1936.

—. *The Sadd Hydraulics.* Cairo, Egypt, 1938.

Buzan, B. *People, states & fear: An agenda for International Security studies in the Post-Cold War era.* Second edition. Colchester, UK: ECPR Press, 1991/2007.

Carlson, A. "Who owns the Nile? Egypt, Sudan, and Ethiopia's history-changing dam." *Origins: Current Events in Historical Perspective* (Ohio State University) 6, no. 6 (2013).

Carol, S. *From Jerusalem to the Lion of Judah and beyond: Israel's foreign policy in East Africa.* Bloomington IN, USA: iUniverse, 2012.

Cascao, A.E. "Ethiopia–Challenges to Egyptian hegemony in the Nile Basin." *Water Policy* 10 (2008): 13–28.

—. "Changing power relations in the Nile River Basin: Unilateralism vs. cooperation." *Water Alternatives* 2 (2009): 245–268.

Cascao, A.E., and A. Nicol. "GERD: New norms of cooperation in the Nile Basin?" *Water International* 41, no. 4 (2016): 550–573.

Cascao, A.E., and M. Zeitoun. "Power, hegemony and critical hydropolitics." In *Transboundary water management: Principles and practice*, edited by A. Earle, A. Jaegerskog and J. Oejendal, 27–42. London, UK: Earthscan Ltd., 2010.

Caso, F. *Global issues: Freshwater supply.* New York NY, USA: Facts On File, 2010.

Chen, H., and A. Swain. "The Grand Ethiopian Renaissance Dam: Evaluating its sustainability standard and geopolitical significance." *Energy Development Frontier* 3, no. 1 (2014): 11–19.

Chimp Reports. *Ethiopia warns Egypt as dam row deepens.* February 13, 2014. http://www.chimpreports.com/17020-ethiopia-warns-egypt-as-dam-row-deepens/ (accessed December 8, 2016).

Chwaszcza, Christine. "Game Theory (Chapter 8)." In *Approaches and methodologies in the social sciences: A pluralist perspective*, edited by Donatella della Porta and Michael Keating, 139–161. Cambridge University Press, 2008.

Cohen, I. *Theodor Herzl, founder of political Zionism.* New York NY, USA: Thomas Yoseloff, 1959.

Coker, C. *NATO, the Warsaw Pact and Africa – RUSI Defense Studies Series.* London, UK: MacMillan Press, 1985.

Collier, D. "Understanding process tracing." *Political Science and Politics* 44, no. 4 (2011): 823–830.

Collins, R.O. *European in Africa.* New York NY, USA: Alfred A. Knopf, 1971.

—. *The waters of the Nile: Hydropolitics and the Jonglei Canal, 1900–1988.* Oxford, UK: Clarendon Press, 1990.

—. "History, hydropolitics and the Nile: Myth or reality?" In *The Nile, sharing a scarce resource: A historical and technical review of water management and of economic and legal issues*, edited by P.P. Howell and J.A. Allan, 109–136. Cambridge, UK: Cambridge University Press, 1994.

—. "In search for the Nile waters, 1900–2000." In *The Nile: Histories, cultures, myths*, edited by E. Haggai and I. Gershoni, 245–268. London, UK: Lynne Rienner Publishers, 2000.

—. "Negotiations and exploitation of the Nile waters at the end of the millennium." *Water International* 31, no. 1 (2006): 116–126.

Cornes, R., and T. Sandler. *The Theory of Externalities: Public goods and club goods.* Cambridge, UK: Cambridge University Press, 1986.

Cox, R.W. "Social forces, states and world orders: Beyond International Relations Theory." *Millennium* 10, no. 2 (1981): 126–155.

CQ Researcher. *Issues in Peace and Conflict: Selections from CQ Researcher.* London, UK: Sage, 2011.

Daily Mail. *Egypt stresses Nile water rights in Ethiopia dam project.* March 25, 2015. http://www.dailymail.co.uk/wires/afp/article-3011635/Egypt-stresses-Nile-water-rights-Ethiopia-dam-project.html (accessed December 13, 2016).

—. *Nile dam won't harm Egypt, says new Ethiopian leader.* May 3, 2018. http://www.dailymail.co.uk/wires/afp/article-5687947/Nile-dam-wont-harm-Egypt-says-new-Ethiopian-leader.html (accessed July 13, 2018).

Daily News Egypt. *International firms to conduct studies on Ethiopian Renaissance Dam.* April 10, 2015a. http://www.dailynewsegypt.com/2015/04/10/international-firms-to-conduct-studies-on-ethiopian-renaissance-dam/ (accessed December 14, 2016).

—. "Ethiopia announces diversion of Blue Nile's course to run through Renaissance Dam." *Daily News Egypt.* December 26, 2015b. http://www.dailynewsegypt.com/2015/12/26/ethiopia-announces-diversion-of-blue-niles-course-to-run-through-renaissance-dam/ (accessed December 16, 2016).

Dawisha, K. *Soviet foreign policy towards Egypt.* London, UK: MacMillan Press, 1979.

Dawoud, K. *Bolstering a joint determination.* April 6–12, 2000. http://weekly.ahram.org.eg/Archive/2000/476/sum2.htm (accessed October 24, 2016).

De Chatel, F. *Water Sheikhs and dam builders: Stories of people and water in the Middle East.* New Brunswick NJ, USA: Transaction Publishers, 2007.

De Stefano, L., P. Edwards, L. de Silva, and A.T. Wolf. "Tracking cooperation and conflict in international basins: Historic and recent trends." *Water Policy* 12 (2010): 871–884.

Degefu, G.T. *The Nile: Historical, legal and developmental perspectives – A warning for the 21st century.* Bloomington: Trafford, 2003.

Dellapenna, J.W. "The Nile as a legal and political structure." In *The scarcity of water: Emerging legal and policy responses*, edited by E.H.P. Brans, E.J. De Haan, A. Nollkaemper and J. Rinzema, 121–134. London, UK: Kluwer Law International Ltd., 1997.

Delli Priscoli, J., and A.T. Wolf. *Managing and transforming water conflicts.* Cambridge, UK: Cambridge University Press, 2009.

Dent, D., and S. Kaufmann. "Green water credits for rainwater management by farmers." *Global Water News 4*, 2006: 4–5.

Dilke, C.W. "Egypt – Now, and after?" In *The Speaker: A review of politics, letters, science, and the arts (Volume IV, from July 4 to December 26, 1891)*, 730–733. London, UK: The Speaker, 1891.

Dinar, A., and G.S. Nigatu. "Distributional considerations of international water resources under externality: The case of Ethiopia, Sudan and Egypt on the Blue Nile." *Water Resources and Economics* 2–3 (2013): 1–16.

Dinar, S. "Assessing side-payment and cost-sharing patterns in international water agreements: The geographic and economic connection." *Political Geography* 25, no. 4 (2006): 412–437.

DIU. *President Al-Bashir witnesses signing of DIUDB Dam heightening contract.* April 27, 2008. http://diu.gov.sd/en/home/pages/show/8 (accessed November 12, 2016).

—. "Contract signed to implement Upper Atbara and Setet Dams." *The Dams Implementation Unit, the Government of Sudan.* May 25, 2010. http://merowedam.gov.sd/en/index.php/home/show/184 (accessed November 12, 2016).

—. "Council of Ministers reviews report on completion of studies for future dams on the Nile." *The Dams Implementation Unit, the Government of Sudan.* August 2011. http://diu.gov.sd/assets/uploads1/files/mag/88.pdf (accessed November 12, 2016).

—. *Merowe Dam project: Previous studies.* 2016a. http://www.merowedam.gov.sd/en/studies.html (accessed November 11, 2016).

—. *Formation decrees.* 2016b. http://www.merowedam.gov.sd/en/establishment.html (accessed November 12, 2016).

—. *About DIU.* 2016c. http://diu.gov.sd/en/home/pages/url/6 (accessed November 12, 2016).

Donnelly, J. "Realism (Chapter 2)." In *Theories of International Relations*, edited by S. Burchill, et al., 29–54. New York NY, USA: Palgrave Macmillan, 2005.

Duffy, G. *Amazon river 'longer than Nile'.* June 16, 2007. Retrieved January 8, 2017, from BBC: http://news.bbc.co.uk/2/hi/6759291.stm

Dumont, H.J. "A description of the Nile Basin." In *The Nile: Origin, environments, limnology and human use*, edited by H.J. Dumont, 1–21. New York NY, USA: Springer, 2009.

Dzurek, D.J., C. Schofield, and S. Furness. *Parting the Red Sea: Boundaries, offshore resources and transit.* Durham, UK: International Boundaries Research Unit, Maritime Briefing, 2001.

E.A., interview by M.M. Samaan. *Face-to-face interview with a late president of the Nile Basin Discourse during the Stockholm World Water Week* (August 31, 2016).

EAPP & EAC (Eastern Africa Power Pool and East African Community). "Final master plan report: Regional power system master plan and grid code study, Volume I: Generation supply study and planning criteria (WBS 1200)." 2011a. www.eac.int/energy/index.php?option=com_docman&task=doc_download&gid=89&Itemid=70 (accessed July 8, 2015).

—. "Final master plan report: Regional power system master plan and grid code study, Volume I: Supply-demand analysis and project identification (WBS 1300)." 2011b. www.eac.int/energy/index.php?option=com_docman&task=doc_download&gid=89&Itemid=70 (accessed July 8, 2015).

—. "Final master plan report: Regional power system master plan and grid code study, Appendix B-II: Summary information on hydro options (WBS 1200)." 2011c. www.eac.int/energy/index.php?option=com_docman&task=doc_download&gid=112&Itemid=70 (accessed July 8, 2015).

EDF & SW. *Strategic environmental and social assessment, M5 and M6, Pre-feasibility studies and regional investment program – Eastern Nile Power Trade Program Study, the Eastern Nile Technical Regional Office, NBI (unpublished report).* Electricité de France (EDF) and Scott Wilson of UK (SW), 2007.

Ehteshami, A., and E.C. Murphy. *The international politics of the Red Sea.* London, UK: Routledge, 2011.

El-Atawy, M.H. "Nilopolitics: A hydrological regime, 1870–1990." *Cairo Papers in Social Science, The American University in Cairo Press* 19, no. 1 (1996).

El-Bey, D. *Back to African roots: Popular diplomacy managed to close gaps in Egypt's relationship with Ethiopia over the Nile.* May 5–11, 2011. http://weekly.ahram.org.eg/archive/2011/1046/eg3.htm (accessed December 3, 2016).

—. *Ethiopian Renaissance Dam: No breakthrough in negotiations.* July 5, 2018. http://english.ahram.org.eg/NewsContent/1/64/306276/Egypt/Politics-/Ethiopian-Renaissance-Dam-No-breakthrough-in-negot.aspx (accessed July 13, 2018).

Elemam, H.E.R. "Egypt and collective action mechanisms in the Nile Basin." In *The River Nile in the post-colonial age: Conflict and cooperation among the Nile Basin countries*, edited by T. Tvedt, 217–236. New York NY, USA: I.B. Tauris, 2010.

Elhance, A.P. *Hydropolitics in the 3rd world: Conflict and cooperation in international river basins.* Washington, DC: The United States Institute of Peace, 1999.

El-Shazly, S. *Tphe crossing of the Suez.* San Francisco CA, USA: American Mideast Research, 1980.

Elster, J. *Nuts and bolts for the social sciences.* New York NY, USA: Cambridge University Press, 1989.

EMFA (The Egyptian Foreign Ministry). *The outcomes of the extraordinary meeting of the Ministers of Water of the Nile Basin countries Sharm El Sheikh.* April 21, 2010. http://www.mfa.gov.eg/ENGLISH/MEDIACENTER/STATMENTS/Pages/StatementDetails.aspx?Source=6781921f-3993-444a-859e-ee26ce851de8&newsID=b750e94c-aa0f-4efd-9998-46d07fa0b515 (accessed July 8, 2015).

—. *Egypt's perspective towards the Ethiopian Grand Renaissance Dam Project (GERDP).* 2014a. http://www.mfa.gov.eg/SiteCollectionDocuments/Egyptrenaissance%20dam111.doc (accessed July 8, 2015).

—. *Egypt explains its stance on the Renaissance Dam Project to the outside world.* March 25, 2014b. http://www.mfa.gov.eg/english/ministry/news/pages/NewsDetails.aspx?Source=6781921f-3993-444a-859e-ee26ce851de8&newsID=9c36a30f-7c02-4946-9f74-ce861a999e9a (accessed December 9, 2016).

—. *Joint statement by the Arab Republic of Egypt and the Federal Democratic Republic of Ethiopia.* June 28, 2014c. http://www.mfa.gov.eg/english/embassies/egyptian_embassy_

albania/mediacenter/pages/NewsDetails.aspx?Source=6781921f-3993-444a-859e-ee26ce851de8&newsID=942cac21-0b94-4eeb-a0ae-a1f678c7faea (accessed July 8, 2015).

EMWRI (The Egyptian Ministery of Water resources and Irrigation). *New Valley development project at Toshka region (in Arabic).* 2013. http://www.mwri.gov.eg/project/toshka.aspx (accessed October 30, 2016).

ENA (The Ethiopian News Agency). *Council of Ministers approves regulation establishing Council on Grand Dam.* April 16, 2011. https://web.archive.org/web/20110420135619/http://www.ena.gov.et/EnglishNews/2011/Apr/16Apr11/138447.htm (accessed December 1, 2016).

—. *Hailemariam urges for more cooperation with Egypt.* March 14, 2015a. http://www.ena.gov.et/en/index.php/politics/item/511-hailemariam-urges-for-more-cooperation-with-egypt (accessed July 9, 2015).

—. *El-Sisi's visit displays growing trust between Ethiopia, Egypt.* March 24, 2015b. http://www.ena.gov.et/en/index.php/politics/item/563-el-sisi-s-visit-displays-growing-trust-between-ethiopia-egypt (accessed July 8, 2015).

ENTRO. "Nile-Flow, a Quarterly newsletter of the Eastern Nile Technical Regional Office – Volume 3, Issue 1." *The Nile Basin Initiative.* March 2012. http://entroportal.nilebasin.org/Documents/nileFlowMarch2012.pdf (accessed November 15, 2016).

EPI (Earth Policy Institute, Rutgers University). *Food and agriculture: Wheat production, consumption and imports in Egypt according to the U.S. Department of Agriculture, Production, Supply and Distribution.* 2011. http://www.earth-policy.org/data_center/C24 (accessed October 26, 2016).

ESIS (The Egyptian State Information Service). *Sinai development projects.* 2006. http://www.sis.gov.eg/newVR/sinia/html/esinia10.htm (accessed October 2, 2016).

—. *Water resources.* April 17, 2013a. http://www.sis.gov.eg/section/341/1345?lang=en-us (accessed October 30, 2016).

—. *Ethiopia dam crisis can be solved.* November 20, 2013b. http://www.sis.gov.eg/En/Templates/Articles/tmpArticleNews.aspx?ArtID=71263#.UyvyU4U31OM (accessed July 8, 2015).

—. *Irrigation Minister: Ethiopia rejects any solutions to problem of Renaissance Dam.* February 12, 2014a. http://www.us.sis.gov.eg/En/Templates/Articles/tmpArticleNews.aspx?ArtID=76018 (accessed July 8, 2015).

—. *Irrigation Minister: Steps to Be Taken to Solve Renaissance Dam Crisis.* March 2, 2014b. http://www.sis.gov.eg/En/Templates/Articles/tmpArticleNews.aspx?ArtID=76391#.Uyydx4U31OM (accessed March 22, 2014).

—. *Egyptian-Ethiopian negotiations on Renaissance Dam.* March 26, 2018. http://www.sis.gov.eg/section/7302/8342?lang=en-us (accessed July 9, 2018).

Ethiopian News. *Ethiopia: Tana Beles, a multi-purpose hydro-power plant, inaugurated.* May 14, 2010. http://www.ethiopian-news.com/ethiopia-tana-beles-a-multi-purpose-hydro-power-plant-inaugurated/ (accessed November 8, 2016).

Ethiopia's Embassy in Brazilia. *The state visit of President El SiSi of the Arab Republic of Egypt to Ethiopia: The President addresses the Joint Session of Parliament.* March 27, 2015. http://ethiopianembassy.org.br/weeks/see/28 (accessed December 13, 2016).

Ethiopia's Embassy in Brussels. *PM Abiy Ahmed visits Djibouti on his first foreign trip.* April 30, 2018. https://ethiopianembassy.be/en/2018/04/30/pm-abiy-ahmed-visits-djibouti-on-his-first-foreign-trip/ (accessed July 13, 2018).

EUED. *Bilateral cooperation.* 2016. http://www.addisababa.mofa.go.ug/data/smenu/54/Bilateral-Cooperation.html (accessed October 19, 2016).

Ezzidin, T. *Egypt's ambassador to Ethiopia discusses Egypt's stance on the Oromo people.* October 9, 2016. http://www.dailynewsegypt.com/2016/10/09/557327/ (accessed December 15, 2016).

Fagan, B. *The attacking ocean: The past, present, and future of the rising sea levels.* New York NY, USA: Bloomsbury, 2013.

Fahim, H.M. *Dams, people and development: The Aswan High Dam case.* New York NY, USA: Pergamon Policy Studies, 1981.

Fahmy, N. "People's diplomacy." *Ahram Weekly.* January 1, 2015. http://weekly.ahram.org.eg/News/10029/21/-People%E2%80%99s-diplomacy.aspx (accessed December 16, 2016).

FAO. *Aquastat database.* 2016a. http://www.fao.org/nr/water/aquastat/data/query/index.html?lang=en (accessed October 28, 2016).

—. *Aqustat database: Ethiopia – Regional report.* 2016b. http://www.fao.org/nr/water/aquastat/countries_regions/eth/print1.stm (accessed November 5, 2016).

—. *Aquastat database: Sudan – Regional report.* 2016c. http://www.fao.org/nr/water/aquastat/countries_regions/SDN/index.stm (accessed November 13, 2016).

Fayed, S., and Y. Saleh. *Millions flood Egypt's streets to demand Mursi quit.* June 30, 2013. http://www.reuters.com/article/us-egypt-protests-idUSBRE95Q0NO20130630 (accessed December 7, 2016).

FDRE-GoE (The Ethiopian Government Portal). *Policies and strategies.* n.d. http://www.ethiopia.gov.et/policies-and-strategies1 (accessed November 3, 2016).

FDRE-MFA (The Ethiopian Ministry of Foreign Affairs). *A week in the Horn of Africa.* June 14, 2013. http://www.mfa.gov.et/weekHornAfrica/morewha.php?wi=1024 (accessed July 8, 2015).

—. *Prime Minister Hailemariam on the Grand Ethiopian Renaissance Dam Project.* February 11, 2014. http://www.mfa.gov.et/news/more.php?newsid=3009 (accessed March 17, 2014).

FDRE-MoWR (The Ethiopian Ministry of Water Resources). "Ethiopian Water Sector Policy." *The Federal Democratic Republic of Ethiopia, Ministry of Water Resources.* 2001. http://www.mowr.gov.et/attachmentfiles/Downloads/Water%20policy.doc (accessed November 3, 2016).

—. *Border and Mendaya projects.* September 20, 2007. http://www.mowr.gov.et/index.php?pagenum=4.3&pagehgt=1000px&ContentID=55 (accessed July 8, 2015).

Finklestone, J. "War and deception (Chapter 9)." In *Anwar Sadat: Visionary who dared,* by J. Finklestone. London, UK: Frank Cass, 1996.

Fitzpatrick, M., T. Parkinson, and N. Ray. *East Africa.* Oakland CA, USA: Lonely Planet, 2006.

Foster, V., W. Butterfield, C. Chen, and N. Pushak. *Building bridges: China's growing role as infrastructure financier for Sub-Saharan Africa.* Washington DC, USA: The International Bank for Reconstruction and Development/The World Bank, 2009.

French, P. "Deltaic and estuarine environments." In *Environmental sedimentology,* edited by C. Perry and K. Taylor, 223–262. Madlen MA, USA: Blackwell Publishing, 2007.

Fueter, E. *World history, 1815–1920.* Edited by S.B. Fay. New York NY, USA: Harcurt, Brace, 1922.

Funk, K., and S. Fake. *Scramble for Africa: Darfur-intervention and the USA.* Montreal QC, Canada: Black Rose Books, 2009.

Gabra, G. *The A to Z of the Coptic Church.* Lanham MD, USA: The Scarecrow Press, 2009.

Garstin, W. *Report upon the basin of the Upper Nile: With proposals for the improvement of that river.* Cairo, Egypt: The Egyptian Ministry of Public Works – National Printing Department, 1904.

George, A.L., and A. Bennett. *Case studies and theory development in the Social Sciences.* London, UK: MIT Press, 2005.

George, J., and D. Campbell. "Patterns of dissent and the celebration of difference: Critical Social Theory and International Relations." *International Studies Quarterly* 34, no. 3 (1990): 269–293.

Gilkes, P. "Ethiopia: Recent history." In *Africa South of the Sahara 2003 (32nd edition),* edited by K. Murison, 382–389. London, UK: Europa Publications, Taylor and Francis Group, 2003.

Giordano, M., A. Drieschova, J.A. Duncan, Y. Sayama, L. De Stefano, and A.T. Wolf. "A review of the evolution and state of transboundary freshwater treaties." *International Environmental Agreements* 14 (2014): 245–264.

Glantz, M.H. "Aral Sea Basin: A sea dies, a sea also rises." *AMBIO: A Journal of the Human Environment* 36, no. 4 (2007): 323–327.

Gleick, P.H. "Water and conflict: Freshwater resources and international security." *International Security* 18, no. 1 (1993): 79–112.

Goldstein, J., and R.O. Keohane. "Ideas and foreign policy: An analytical framework." In *Ideas and foreign policy: Beliefs, institutions, and political change*, edited by J. Goldstein and R.O. Keohane, 3–30. Ithaca NY, USA: Cornell University Press, 1993.

Great Britain War Office. *Record of the expedition to Abyssinia (Volume I), completed by order of of the Secretary of State for War by Major T.J. Holland and Captain H. Hozier.* London: W. Clowes & Sons and Harrison & Sons, 1870.

Greenville News. *Administration plans: More aid for Africa, p. 28.* March 29, 1970. https://www.newspapers.com/newspage/189185681/ (accessed September 30, 2016).

Griffiths, R.J. *U.S. security cooperation with Africa: Political and policy challenges.* New York NY, USA: Routledge, 2016.

Gupta, B.B. *Arab-Israeli relations.* New Delhi, India: Ashish Publishing House, 1978.

GWP. *What is IWRM?* March 25, 2010. http://www.gwp.org/The-Challenge/What-is-IWRM/ (accessed March 9, 2016).

Haas, P.M., R.O. Keohane, and M.A. Levy. *Institutions for the Earth: Sources of effective environmental protection.* London, UK: MIT Press, 1993.

Haggai, E. *The Cross and the river: Ethiopia, Egypt, and the Nile.* Boulder CO, USA: Lynne Rienner Publishers, 2002.

Hailu, A. *GERD – No financial, technical constraints, and tripartite discussion terminated owing to Egyptian proposal against Ethiopia's interest: Official.* January 17 and 20, 2014. http://www.ethpress.gov.et/herald/index.php/herald/editorial/5659-gerd-no-financial-technical-constraints and http://www.ethpress.gov.et/herald/index.php/herald/news/5628-tripartite-discussion-terminated-owing-to-egyptian-proposal-against-ethiopia-s-inter (accessed March 21, 2014).

Hammond, M. *The Grand Ethiopian Renaissance Dam and the Blue Nile: Implications for transboundary water governance, Discussion paper no. 1307.* University of Exeter, 2013.

Hamner, J.H., and A.T. Wolf. "Patterns in international water resource treaties: The Transboundary Freshwater Dispute Database." *Colorado Journal of International Environmental Law and Policy* The 1997 Year Book (1998).

Hanna, R.L. "Transboundary resources and the political economy of large-scale land investments in the Nile: Sudan, hydropolitics and Arab food security." In *Land and hydropolitics in the Nile River Basin: Challenges and new investments*, edited by E. Sandstroem, A. Jaegerskog and T. Oestigaard, 117–149. New York NY, USA: Routledge, 2016.

—. *Transnational state-capital alliances & multi-level hydropolitics of land-water investments in Egypt and the Nile Basin (PhD thesis).* London, UK: Institute of Development Studies, University of Sussex, forthcoming.

Hardin, G. "The tragedy of the commons." *Science* Volume 162, No. 3859 (1968): 1243–1248.

—. "Political requirements for preserving our common heritage." In *Wildlife and America*, edited by H.P. Brokaw, 310–317. Washington DC, USA: Council on Environmental Quality, 1978.

Hargreaves Heap, S., Lyons, B., Hollis, M., Sugden, R., & Weale, A. *The theory of choice: A critical guide.* Wiley-Blackwell, 1992.

Hasenclever, A., P. Mayer, and V. Rittberger. *Theories of international regimes.* Cambridge, UK: Cambridge University Press, 1997.

—. "Integrating theories of international regimes." *Review of International Studies* 26, no. 1 (2000): 3–33.

Heyns, P. "Water resources management in Southern Africa." In *International waters in Southern Africa*, edited by M. Nakayama, 5–37. Tokyo, Japan: United Nations University Press, 2003.

Hoffman, M. *Critical theory and the inter-paradigm debate.* Vol. 16, in *The study of International Relations*, edited by H.C. Dyer and L. Mangasarian, 60–86. London, UK, Palgrave Macmillan, 1989.

Holsti, K.J. *The dividing discipline: Hegemony and diversity in international theory.* Boston MA, USA: Allen & Unwin, 1985.

Homer-Dixon, T.F. *Environment, scarcity, and violence.* Princeton NJ, USA: Princeton University Press, 1999.

Hooper, V., and M. McWilliams. "Case studies of transboundary water management: Fifteen initiatives from various parts of the world." In *Transboundary water management: Principles and practice*, edited by A. Earle, A. Jaegerskog and J. Oejendal, 187–236. London, UK: Earthscan Ltd., 2010.

Howell, P.P. "East Africa's water requirements: The Equatorial Nile Project and the Nile Waters Agreement of 1929, a brief historical review." In *The Nile, sharing a scarce resource: A historical and technical review of water management and of economic and legal issues*, edited by P.P. Howell and J.A. Allan, 81–108. Cambridge, UK: Cambridge University Press, 1994.

Howell, P.P., and J.A. Allan. *The Nile, sharing a scarce resource: A historical and technical review of water management and of economic and legal issues.* Cambridge, UK: Cambridge University Press, 1994.

Howell, P.P., M. Lock, and S. Cobb. *The Jonglei Canal: Impact and opportunity.* Cambridge, UK: Cambridge University Press, 1988.

Hozier, H. *The British expedition to Abyssinia.* London, UK: Macmillan and Co, 1869.

Human Rights Watch. *Evil days: 30 years of war and famine in Ethiopia – An Africa Watch report.* New York NY, USA: Human Rights Watch, 1991.

—. *Egypt: Mass attacks on churches.* August 21, 2013. https://www.hrw.org/news/2013/08/21/egypt-mass-attacks-churches (accessed December 10, 2016).

Hurst, H.E., R.P. Black, and Y.M. Simaika. *The Nile Basin: Future Conservation of the Nile (Volume VII).* Cairo, Egypt: The Egyptian Ministry of Irrigation, 1946.

Hussein, W. *Egyptian-Ethiopian disputes stall Renaissance Dam.* August 7, 2015. http://www.al-monitor.com/pulse/originals/2015/08/egypt-ethiopia-sudan-renaissance-dam-disputes-brl-deltares.html (accessed December 14, 2016).

—. *Is Egypt-Sudan border dispute new thorn in Renaissance Dam negotiations?* May 8, 2016. http://www.al-monitor.com/pulse/en/originals/2016/05/egypt-sudan-dispute-halayeb-renaissance-dam-ethiopia-water.html (accessed December 15, 2016).

Ibn Khaldun. *The Muqaddimah: An introduction to history (translated by Franz Rosenthal).* Edited by N.J. Dawood. New York: Bollingen Series, Princeton University Press, 1377/1969.

Ibrahim, A.M. "Development of the Nile River system." In *The Nile Valley countries, continuity and change – Volume 1, Issue 12 of the Sudanese Library Series*, edited by M.O. Beshir. Khartoum, Sudan: Institute of African and Asian Studies, University of Khartoum, 1984.

Ibrahim, A.M. "The Nile Basin Cooperative Framework Agreement: The beginning of the end of Egyptian hydro-political hegemony." *Missouri Environmental Law and Policy Review* 18, no. 2 (2011): 282–313.

Ibrahim, S. "Domestic developments in Egypt." In *The Middle East: Ten years after Camp David*, edited by W.B. Quandt, 19–62. Washington DC, USA: The Brookings Institution, 1988.

IMF. *Ethiopia: Poverty Reduction Strategy Paper – Annual progress report (2002/03), Country report no. 04/37.* Washington DC, USA: The International Monetary Fund and the Federal Democratic Republic of Ethiopia, Ministry of Finance and Economic Development, 2004.

—. "The Federal Democratic Republic of Ethiopia: 2013 article IV consultation." *The International Monetary Fund.* October 2013. http://www.imf.org/external/pubs/ft/scr/2013/cr13308.pdf (accessed July 8, 2015).

IMFA (The Israeli Ministry of Foreign Affairs). *PM Netanyahu on official visit to Ethiopia.* July 7, 2016. http://mfa.gov.il/MFA/PressRoom/2016/Pages/PM-Netanyahu-on-official-visit-to-Ethiopia-7-July-2016.aspx (accessed December 21, 2016).

International Rivers. "Field visit report: GERD project." *International Rivers Network.* August 2012a. http://www.internationalrivers.org/files/attached-files/grandren_ethiopia_2013.pdf (accessed July 8, 2015).

—. "The new great walls, A guide to China's overseas dam industry." *International Rivers.* 2012b. https://www.internationalrivers.org/sites/default/files/attached-files/intlrivers_newgreatwalls_2012_2.pdf (accessed November 11, 2016).

—. *GERD panel of experts report: Big questions remain.* March 31, 2014a. http://www.internationalrivers.org/gerd-panel-of-experts-report-big-questions-remain (accessed July 8, 2015).

—. *No room for debate on Grand Ethiopian Renaissance Dam?* April 17, 2014b. https://www.internationalrivers.org/blogs/229/no-room-for-debate-on-grand-ethiopian-renaissance-dam (accessed December 9, 2016).
—. *Tekeze Dam, Ethiopia.* 2016a. https://www.internationalrivers.org/resources/tekeze-dam-ethiopia-3607 (accessed November 7, 2016).
—. "Kajbar Dam." *International Rivers.* 2016b. https://www.internationalrivers.org/campaigns/kajbar-dam-sudan (accessed November 12, 2016).
IPCC. *Fourth assessment report, Working group II: impacts, adaptation and vulnerability.* Edited by R.L. Pachauri and A. Reisinger. Geneva, Switzerland: The Intergovernmental Panel on Climage Change (IPCC), 2007.
IPoE. "International Panel of Experts (IPoE) on Grand Ethiopian Renaissance Dam Project (GERDP): Final report." 2013.
IWLP. *United Nations General Assembly Press Release GA/9248.* May 21, 1997. http://www.internationalwaterlaw.org/documents/intldocs/convention_press.html (accessed November 20, 2016).
Iyob, B. *Resilience and adaptability of transboundary rivers: The principle of equitable distribution of benefits and the institutional capacity of the Nile Basin – A PhD Thesis.* Corvallis, OR: Oregon State University, 2010.
J.N., interview by M.M. Samaan. *Face-to-face interview with the then Executive Director of the NBI Secretariat during the Stockholm World Water Week* (August 28, 2016).
Jacobs, I. "A community in the Orange: The development of a multi-level water governance framework in the Orange-Senqu River Basin in Southern Africa." *International Environmental Agreement* 12, no. 2 (2012): 187–210.
Jeuland, M., and D. Whittington. "Water resources planning under climate change: A "real options" application to investment planning in the Blue Nile, Discussion paper series." *Environment for Development* 13–05 (March 2013).
Jhazbhay, I. "South Africa-North African relations: Revisiting the bridging of a continent." In *South African foreign policy review – Volume 1*, edited by C. Landsberg and J.A. Van Wyk, 239–256. Pretoria, South Africa: Africa Institute of South Africa, 2012.
JIT. *The Equatorial Nile Project and its effects in the Anglo-Egyptian Sudan (4 volumes).* The Jonglei Investigation Team, 1954.
Johnston, R. "Availability of water for agriculture in the Nile Basin." In *The Nile River Basin: Water, agriculture, governance and livelihoods*, edited by S.B. Awulachew, V. Smakhtin, D. Molden and D. Peden, 61–83. New York NY, USA: Routledge, 2012.
Johnston, R., and M. McCartney. *Inventory of water storage types in the Blue Nile and Volta river basins – Working paper no. 140.* Colombo, Sri Lanka: International Water Management Institute, 2010.
Kantoush, S.A. "The downstream iompacts of Ethiopia's cascade dams in the Upper Blue Nile on Egypt." *Sustainable Building International Conference 2013*, November 2013.
Katz, S.H. *Women and gender in early Jewish and Palestinian nationalism.* Miami FL, USA: University Press of Florida, 2003.
Keith, B., J. Enos, C.B. Garlick, G. Simmons, D. Copeland, and M. Cortizo. *Limits to population growth and water resource adequacy in the Nile River Basin, 1994–2100.* West Point NY, USA: Center for Nation Reconstruction and Capacity Development, United States Military Academy, 2013.
Keith, B., K. Epp, M. Houghton, J. Lee, and R. Mayville. *Water as a conflict driver: Estimating the effects of climate change and hydroelectric dam diversion on Nile River stream flow during the 21st century.* West Point NY, USA: Center for Nation Reconstruction and Capacity Development, United States Military Academy, 2014.
Kendie, D. *Egypt and the hydro-politics of the Blue Nile River.* August 8, 1999. http://chora.virtualave.net/kendie-nile.htm (accessed October 9, 2016).
Keneally, T. *Three famines: Starvation and politics.* New York NY, USA: The Serpentine Publishing, Public Affairs, 2011.

Keohane, R.O. *After hegemony: Cooperation and discord in the world political economy.* Princeton NJ, USA: Princeton University Press, 1984/2005.

—. "International institutions: Can interdependence work?" *Foreign Policy* 110 (1998): 82–96.

—. "Beyond the tragedy of the commons." *Perspectives on Politics* 8, no. 2 (2010): 577–580.

Keohane, R.O., and L.L. Martin. "The promise of the institutionalist theory." *International Security* 20, no. 1 (1995): 39–51.

Keohane, R.O., and J. Nye. *Transnational relations and world politics.* Cambridge MA, USA: Harvard University Press, 1972.

Keohane, R.O., and E. Ostrom. *Local commons and global interdependence.* London, UK: Sage Publications, 1995.

Kholaif, D. *Egypt's farmland offers minimal incentives to Arab investors.* March 28, 2015. http://www.dailynewsegypt.com/2015/03/28/egypts-farmland-offers-minimal-incentives-to-arab-investors/ (accessed October 29, 2016).

Kilgour, D.M., and Y. Wolinsky-Nahmias. "Game Theory and international environmental policy (Chapter 13)." In *Models, numbers & cases: Methods for studying international relations,* edited by D.F. Sprinz and Y. Wolinsky-Nahmias. University of Michigan Press, 2004.

King, A.M. *An assessment of reservoir filling policies under a changing climate for Ethiopia's Grand Renaissance Dam.* Drexel: Drexel University, 2013.

Kirkpatrick, D.D., and B. Hubbard. *Morsi defies Egypt army's ultimatum to bend to protest.* July 2, 2013. http://www.nytimes.com/2013/07/03/world/middleeast/egypt-protests.html (accessed December 7, 2016).

Kitissou, M., M. Ndulo, M. Nagel, and M. Grieco. *The hydropolitics of Africa: A contemporary challenge.* Newcastle, UK: Cambridge Scholars Publishing, 2007.

Kliot, N. *Water resources and conflict in the Middle East.* New York NY, USA: Routledge, 1994.

Korn, D.A. *Ethiopia, the United States and the Soviet Union.* London, UK: Southern Illinois University Press, 1986.

Krasner, S.D., ed. *International regimes.* Ithaca NY, USA: Cornel University Press, 1983.

Krishna, R. "The legal regime of the Nile River Basin." In *The politics of scarcity: Water in the Middle East,* edited by J.R. Starr and D.C. Stoll, 23–41. Boulder CO, USA: Westview, 1988.

Kumaraswany, P.R. *A to Z of the Arab-Israeli Conflict.* Lanham MD, USA: Scarecrow Press, 2006.

Laessing, U. *Sudan launches major dam to boost agricultural production, investment.* January 1, 2013. http://www.reuters.com/article/us-sudan-dam-idUSBRE9000B320130101 (accessed November 12, 2016).

Laidi, Z. *The super-powers and Africa: The constraints of a rivalry, 1960–1990.* Chicago IL, USA: University of Chicago Press, 1990.

Lawford, R., J. Bogardi, S. Marx, S. Jain, C.P. Wostl, K. Knüppe, C. Ringler, F. Lansigan, and F. Meza. "Basin perspectives on the Water-Energy-Food security nexus." *Current Opinion in Environmental Sustainability* 5, no. 6 (2013): 607–616.

Leb, C. "Share case studies: The Tigris-Euphrates Joint Technical Committee – Deadlocked." *IUCN.* March 15, 2010. http://cmsdata.iucn.org/downloads/tigris_euphrates.pdf (accessed December 22, 2015).

Legum, C. *Africa contemporary record: Annual survey and documents – Volume 20.* Africa Research Ltd., 1989.

Leila, R. *Great expectations, middling results.* May 29, 2013a. http://weekly.ahram.org.eg/News/2787/17/Great-expectations,-middling-results.aspx (accessed December 5, 2016).

—. *Dampening disputes.* May 29, 2013b. http://weekly.ahram.org.eg/News/2792.aspx (accessed December 5, 2016).

Lobo, J. *A voyage to Abyssinia by Father Jerome Lobo, a Portuguese missionary, by M. Le Grand, translated from the French by S. Johnson.* London, UK: Printed for Elliot and Kay, available at Bayer Staatsbibliothek, 1789.

Low, D.A., and A. Smith. *History of East Africa (Volume 3).* Oxford, UK: Clarendon Press, 1976.

Lowi, M.R. *Water and power: The politics of a scarce resource in the Jordan River Basin.* New York NY, USA: Cambridge University Press, 1993.

Luce, R.D., and H. Raiffa. *Games and decisions.* New York, USA: Wiley, 1957.

Lunn, J. "History." In *The Middle East and North Africa 2004: 50th edition*, edited by L. Dean, 294–320. London, UK: Europa Publications, Taylor & Francis Group, 2004.
Lytle, E.E. *The Aswan High Dam.* Monticello IL, USA: Council of Planning Librarians, 1977.
M.A.S, interview by M.M. Samaan. *Telephone interview with a top official at the Egyptian Ministry of Irrigation and Water Resources* (September 22, 2016).
M.M., interview by M.M. Samaan. *Face-to-face interview with a top official at the Ethiopian Ministry of Water Resources, Irrigation and Electricity during the Stockholm World Water Week* (August 30, 2016).
M.M.A., interview by M.M. Samaan. *Face-to-face interview with a top official at the Sudanese Ministry of Water Resources and Electricity during the Stockholm World Water Week* (August 31, 2016).
Maasho, A. *Ethiopia ratifies Nile treaty in snub to Egypt.* June 13, 2013. http://www.reuters.com/article/us-ethiopia-egypt-nile-idUSBRE95C0K020130613 (accessed December 6, 2016).
—. *Insight – Paying for giant Nile dam itself, Ethiopia thwarts Egypt but takes risks.* April 23, 2014. http://www.reuters.com/article/2014/04/23/ethiopia-energy-idUSL6N0N91QM20140423 (accessed July 8, 2015).
—. *Ethiopia leader rejects call for World Bank arbitration in dam dispute.* January 21, 2018a. https://www.reuters.com/article/us-ethiopia-egypt-dam/ethiopia-leader-rejects-call-for-world-bank-arbitration-in-dam-dispute-idUSKBN1FA0WE (accessed July 9, 2018).
—. *Ethiopia to take stake in Port of Djibouti, its trade gateway -state media.* May 1, 2018b. https://www.reuters.com/article/ethiopia-djibouti/ethiopia-to-take-stake-in-port-of-djibouti-its-trade-gateway-state-media-idUSL8N1S81QA (accessed July 13, 2018).
Madani, K., D. Rheinheimer, L. Elimam, and C. Connell-Buck. "A game theory approach to understanding the Nile River Basin conflict." In *A water resource: Festschrift in honor of Prof. Lars Bengtsson*, by K.M. Person, 97–114. Lund: Division of Water Resources, Lund University, 2011.
Maged, A. *Rabaa Al-Adaweya Square: From sit-in to cenotaph.* December 17, 2015. http://weekly.ahram.org.eg/News/15036.aspx (accessed December 8, 2016).
Mahajan, S. *British Foreign Policy, 1874–1914: The role of India.* London, UK: Routledge, 2003.
Mahoney, J. "The logic of process tracing tests in the social sciences." *Sociological Methods & Research* 20, no. 10 (2012): 1–28.
—. "Process tracing and historical explanation." *Security Studies* 24, no. 2 (2015): 200–218.
Malone, B. *Ethiopian PM warns Egypt off Nile war.* November 23, 2010. https://www.reuters.com/article/us-ethiopia-egypt-meles/ethiopian-pm-warns-egypt-off-nile-war-idUSTRE6AM5V820101123 (accessed November 21, 2016).
—. *Ethiopia to defy Egypt and build huge Nile dam.* March 30, 2011a. http://uk.reuters.com/article/ethiopia-dam-idUKLDE72T21M20110330 (accessed December 1, 2016).
—. *Ethiopia keeping Egypt in dark on Nile dam.* April 18, 2011b. http://www.reuters.com/article/2011/04/18/us-ethiopia-nile-idUSTRE73H59220110418 (accessed July 8, 2015).
Mangu, A.M.B. "The Democratic Republic of Congo and the Nile Cooperative Framework Agreement." In *Cooperative diplomacy, regional stability and national interests: The Nile River and riparian states*, edited by K.G. Adar and N.A. Check, 23–38. Pretoria, South Africa: Africa Institute of South Africa, 2011.
Mannes, A. *Profiles in terror: The guide to Middle East terrorist organizations.* Lanham MD, USA: Rowman and Littlefield Publishers, 2004.
Martin, L.L. "Heterogeneity, linkage and commons problems." *Journal of Theoretical Politics*(6) 4 (1994): 473–493.
Masri, M. *El-Watan reveals documents: For these reasons did Toshka fail (in Arabic).* October 6, 2014. http://www.elwatannews.com/news/details/571265 (accessed October 31, 2016).
Massoud, M.F. "Lawyers and the disintegration of the legal complex in Sudan." In *Fates on political liberalism in the British post-colony: The politics of the legal complex*, edited by T.C. Halliday, L. Karpik and M.M. Feeley, 193–218. New York NY, USA: Cambridge University Press, 2012.
Mazzucato, M. *The entrepreneurial state: Debunking the public vs. private myth in risk and innovation.* London, UK: Anthem, 2013.

McCaffrey, S. "The UN Convention on the Law of the Non-navigational Uses of International Watercourses: Prospects and pitfalls." In *International watercourses: Enhancing cooperation and managing conflict – Technical paper no. 414*, edited by S.M.A. Salman and L.B. De Chazournes, 17–28. Washington DC, USA: World Bank, 1998.

McCann, J. "Ethiopia, Britain and negotiations for the Lake Tana Dam, 1922–1935." *The International Journal of African Historical Studies* 14, no. 4 (1981): 667–699.

McCartney, M., T. Alemayehu, A. Shiferaw, and S.B. Awulachew. *Evaluation of current and future water resources development in the Lake Tana Basin, Ethiopia – IWMI Research Report no. 134.* Colombo, Sri Lanka: International Water Management Institute, 2010.

McDonald, M. *Nile Control (2 volumes).* Cairo, Egypt: The Egyptian Ministry of Irrigation, 1920.

McIntire, S. *Speeches in world history.* New York NY, USA: Files On File, 2009.

McIntyre, O. "Gabcikovo-Nagymaros Project: A test case for international water law (Section 13.13)." In *Transboundary water management: Principles and practice*, edited by A. Earle, A. Jaegerskog and J. Oejendal, 228–230. London, UK: Earthscan Ltd., 2010.

McIntyre, O., interview by M.M. Samaan. *Interview with the expert on international water law, conducted in person at the Stockholm World Water Week* (September 1, 2016).

McKenna, A., ed. *The Britannica guide to the world's most influential people: The 100 most influential world leaders of all time.* New York NY, USA: Britannica Educational Publishing, 2010.

Mekonnen, D.Z. "The Nile Basin Cooperative Framework Agreement negotiations and the adoption of a "water security" paradigm: Flight into obscurity or a logical cul-de-sac?" *The European Journal of International Law* 21, no. 2 (2010): 421–440.

—. "Between the scylla of water security and charybdis of benefit sharing: The Nile Basin Cooperative Framework Agreement – Failed or just teetering on the brink?" *Goettingen Journal of International Law* 3, no. 1 (2011): 345–372.

Meles Zenawi Memorial. *Ethiopian PM Meles Zenawi speech on launching GERD.* April 2, 2013. http://www.meleszenawi.com/ethiopian-pm-meles-zenawi-speech-on-launching-gerd-text-and-videos/ (accessed July 8, 2015).

Menzel, Ulrich. "The Role of Social Sciences and Economics in Sustainable Water Management." *the Five-Year EXCEED Closing Symposium.* Braunschweig: Technical University of Braunschweig, 2014. 56–64.

Metaferia, G. *Ethiopia and the United States: History, diplomacy, and analysis.* New York NY, USA: Algora Publishing, 2009.

Metawie, A.F. "History of co-operation in the Nile Basin." *Water Resources Development* 20, no. 1 (2004): 47–63.

Milas, S. *Egypt: Threats over Nile waters backfire As promise of war is not credible.* June 18, 2013. http://allafrica.com/stories/201306190263.html (accessed December 6, 2016).

Mirumachi, N., M. Zeitoun, and J. Warner. *Transboundary water interactions and the UN Watercourses Convention: Allocating waters and implementing principles.* Environment, Politics and Development Working Paper Series, Paper no. 53, London: Department of Geography, King's College London, 2013.

MIT. "The Grand Ethiopian Renaissance Dam: An opportunity for collaboration and shared benefits in the Eastern Nile Basin – An amicus brief to the riparian nations of Ethiopia, Sudan and Egypt from the International, Non-partisan Eastern Nile Working Group." *Abdul Latif Jameel World Water and Food Security Lab, the Massachusetts Institute of Technology.* November 13–14, 2014. http://jwafs.mit.edu/sites/default/files/documents/GERD_2014_Full_Report.pdf (accessed December 16, 2016).

Mogire, E. *Victims as security threats: Refugee impact on host state security in Africa.* Burlington VT, USA: Ashgate, 2011.

Moorehead, A. *The Blue Nile (reprint).* 1962/1985.

Mosley, P. *The politics of poverty reduction.* Oxford, UK: Oxford University Press, 2012.

Mostert, E. *Conflict and cooperation in the management of international freshwater resources: A global review – IHP Technical Documents in Hydrology no. 19.* Paris, France: UNESCO International Hydrologic Programme, 2003.

Mouterde, P. *Opposition eyes revolt on Morsi's first anniversary.* June 29, 2013. http://www.france24.com/en/20130623-egypt-morsi-opposition-tamarod-revolt (accessed December 6, 2016).
Mulira, J. "Independent Uganda and the Nile: Hydroelectric projects and plans." In *The River Nile in the post-colonial age: Conflict and cooperation among the Nile Basin countries*, edited by T. Tvedt, 125–160. New York NY, USA: I.B. Tauris, 2010.
Murad, M. *Sudan plays wait and see.* December 23–29, 1999. http://weekly.ahram.org.eg/Archive/1999/461/fr1.htm (accessed October 24, 2016).
NBI. Transboundary benefit sharing framework training manual: Socio-Economic Development and Benefit Sharing Project - P075952. Nile Basin Initiative, 2009.
—. "State of the River Nile Basin report." *Nile Basin Initiative.* 2012a. http://nileis.nilebasin.org/content/state-river-nile-basin-report (accessed July 8, 2015).
—. *Joint Multipurpose Program.* August 29, 2012b. http://entro.nilebasin.org/index.php/joint-muliti-purpose-program (accessed November 15, 2016).
—. *The Nile Information System: Shared Vision Program.* 2013. http://nileis.nilebasin.org/content/shared-vision-program (accessed October 22, 2016).
—. *One river, one people, one vision: Our centers.* 2016. http://www.nilebasin.org/ (accessed October 22, 2016).
Niasse, M. "Share case studies: Integrated management in the Senegal River." *IUCN.* March 15, 2010. http://cmsdata.iucn.org/downloads/senegal_1.pdf (accessed January 2, 2016).
Nilblock, T. *"Pariah states" & sanctions in the Middle East: Iraq, Libya, Sudan.* Boulder CO, USA: Lynne Rienner Publishers, 2002.
Nour El-Din, M.M. "Climate change risk management in Egypt: Proposed climate change adaptation strategy for the Ministry of Water Resources and Irrigation in Egypt – Prepared for the UNESCO Cairo Office." *The Egyptian Environmental Affairs Agency.* 2013. http://www.eeaa.gov.eg/portals/0/eeaaReports/CCRMP/7.%20CC%20Water%20Strategy/CC%20Final%20Submitted%208-March%202013%20AdptStrtgy.pdf (accessed October 31, 2016).
Novokshshenov, V. "Laboratory studies of the stone masonry in the Old Aswan Dam." *Materials and Structures* 26 (1993): 103–110.
OECD/IEA. *Electricity production from hydroelectric sources (% of total) after the IEA Statistics.* 2014. http://data.worldbank.org/indicator/EG.ELC.HYRO.ZS?locations=EG (accessed October 12, 2016).
Ofcansky, T. "Sudan: Recent history." In *Africa South of the Sahara 2003 (32nd edition)*, edited by K. Murison, 1007–1021. London, UK: Europa Publications, Taylor & Francis Group, 2003.
Ojakangas, R. *Earth science.* Portland, ME: Walch Publishing, 1997.
ORSAM. "Egypt." *Middle East/South Asia Daily Bulletin.* May 18, 2010. http://www.orsam.org.tr/eski/tr/trUploads/OrtadoguBulteni/20100518_18.pdf (accessed November 8, 2016).
Ostrom, E. *Governing the commons: The evolution of institutions for collective action.* Cambridge, UK: Cambridge University Press, 1990.
Othman, A., Rabeh, S., Fayez, M., Monib, M., & Hegazi, N. "El-Salam canal is a potential project reusing the Nile Delta drainage water for Sinai desert agriculture: Microbial and chemical water quality." *Journal of Advanced Research* 3, no. 2 (2012): 99–108.
Ovodenko, A. "Regional water cooperation: Creating incentives for integrated management." *Journal of Conflict Resolution*, 2014: 1–28.
Parada, H.H. *Chile-Israel relations, 1973–1990: The hidden connection.* Santiago, Chile: RIL Editors, 2012.
Parks, P.J. *Aswan High Dam.* San Diego CA, USA: Blackbirch Press, 2003.
Pearson, D. *Reds Quicker in Ethiopia, p. 5.* June 8, 1964a. https://news.google.com/newspapers?nid=1798&dat=19640608&id=q_UeAAAAIBAJ&sjid=6osEAAAAIBAJ&pg=4867,890963&hl=en (accessed September 22, 2016).
—. *Selassie sees education as Ethiopian salvation, p. 6.* June 11, 1964b. https://www.newspapers.com/newspage/20821081/ (accessed September 22, 2016).
Peel, S.C. *The binding of the Nile and the new Sudan.* London, UK: E. Arnold, 1904.

Perry, T., and Y. Saleh. *Egypt army topples president, announces transition.* July 3, 2013. http://www.reuters.com/article/us-egypt-protests-idUSBRE95Q0NO20130703 (accessed December 7, 2016).

Petterson, D. *Inside Sudan: Political Islam, conflict and catastrophe.* Boulder CO, USA: Westview Press, 2003.

Pflanz, M. *Egypt: 'all options open' in Nile dam row with Ethiopia.* June 12, 2013. http://www.telegraph.co.uk/news/worldnews/africaandindianocean/egypt/10113407/Egypt-all-options-open-in-Nile-dam-row-with-Ethiopia.html (accessed December 6, 2016).

Phillips, D.J.H. *Equitable and reasonable allocations of international watercourses: Water rights and the optimization of economic benefits.* Bonn, February 23–25, 2005: International Conference of Integrated Assessment of Water Resources and Global Change: A North-South Analysis, 2005.

Phillips, D.J.H., J.A. Allan, M. Claassen, J. Granit, A. Jägerskog, E. Kistin, M. Patrick, and A. Turton. *The TWO analysis: Introducing a methodology for the Transboundary Waters Opportunity – Report no. 23.* Stockholm, Sweden: Stockholm International Water Institute (SIWI), 2008.

Phillips, D.J.H., S. Attili, and J.S. Murray. "The Jordan River Basin: 1. Clarification of the allocations in the Johnston Plan." *Water International* 32, no. 1 (2007): 16–38.

Phillips, D.J.H., M. Daoudy, S. McCaffrey, J. Oejendal, and A. Turton. *Transboundary water cooperation as a tool for conflict prevention and broader benefit sharing.* Stockholm: Ministry of Foreign Affairs, Sweeden, 2006.

Qaddumi, H. *Practical approaches to transboundary water benefit sharing – Working paper no. 292.* London: Overseas Development Institute (ODI), 2008.

Raslan, H. *Egypt, Israel and the Nile water.* July 29, 2016. http://english.ahram.org.eg/NewsContent/4/0/236135/Opinion/0/Egypt,-Israel-and-the-Nile-water.aspx (accessed December 21, 2016).

Raub, W., & Weesie, J. "Cooperation via hostages". *Analyse and Kritik* 22, no. 1 (2000): 19–43.

Reus-Smit. "Constructivism (Chapter 8)." In *Theories of International Relations*, edited by S. Burchill, et al., 188–212. New York NY, USA: Palgrave Macmillan, 2005.

Reuters. *Egypt's Mubarak dismisses Ethiopia rebel claim.* November 25, 2010. https://af.reuters.com/article/topNews/idAFJOE6AO0BZ20101125 (accessed November 21, 2016).

—. *AU ends Egypt, Guinea Bissau suspension after elections.* June 17, 2014. http://af.reuters.com/article/guineaBissauNews/idAFL5N0OY55720140617 (accessed December 11, 2016).

—. *Egypt says it does not want war as tension grows with Sudan.* January 15, 2018a. https://www.reuters.com/article/us-egypt-sudan-politics/egypt-says-it-does-not-want-war-as-tension-grows-with-sudan-idUSKBN1F423D (accessed July 10, 2018).

—. *Egypt, Ethiopia leaders say Nile dam must not ruin relations.* January 18, 2018b. https://www.reuters.com/article/us-egypt-politics-ethiopia/egypt-ethiopia-leaders-say-nile-dam-must-not-ruin-relations-idUSKBN1F71Z1 (accessed July 9, 2018).

—. *Egypt, Ethiopia agree to settle differences over Nile mega-dam.* June 10, 2018c. https://www.reuters.com/article/us-egypt-ethiopia/egypt-ethiopia-agree-to-settle-differences-over-nile-mega-dam-idUSKBN1J60YL (accessed July 13, 2018).

—. *UAE to give Ethiopia $3 billion in aid and investments.* June 15, 2018d. https://af.reuters.com/article/ethiopiaNews/idAFL8N1TH4GJ (accessed July 13, 2018).

Roberts, C. *The logic of historical explanation.* University Park PA, USA: Pennsylvania State University Press, 1996.

Roberts, W.D. *Irrigation projects in the Upper Nile and their effects on tribal and local interests.* Cairo, Egypt, 1928.

Rolandsen, O.H., and M.W. Daly. *A history of South Sudan: From slavery to independence.* Cambridge, UK: Cambridge University Press, 2016.

Rubin, L. *Islam in the balance: Ideational threats in Arab politics.* Stanford CA, USA: Stanford University Press, 2014.

Rubinstein, R.A. *Peacekeeping under fire: Culture and intervention.* New York NY, USA: Routledge, 2008.

Sadat, A. *Anwar Sadat on international affairs – The Britannica Book of the Year.* Edited by F. Gibney. 1981. https://www.britannica.com/topic/Anwar-Sadat-on-international-affairs-1960770 (accessed October 2, 2016).
Sadoff, C.W., and D. Grey. "Beyond the river: Benefits of cooperation on international rivers." *Water Policy* 4 (2002): 389–403.
Sadoff, C.W., and D. Grey. "Cooperation on international rivers: A continuum for securing and sharing benefits." *Water International* 30, no. 4 (2005): 420–427.
Said, R. *The River Nile: Geology, hydrology and utilization.* Oxford, UK: Pergamon Press, 1993.
—. *Learning from past mistakes.* May 11–17, 2006. http://weekly.ahram.org.eg/Archive/2006/794/sc111.htm (accessed December 1, 2016).
Saleh, Y., and T. Finn. *More than 200 dead after Egypt forces crush protest camps.* August 14, 2013. https://www.reuters.com/article/us-egypt-protests/more-than-200-dead-after-egypt-forces-crush-protest-camps-idUSBRE97C09A20130814 (accessed December 10, 2016).
Saleh, Y., and P. Taylor. *Special Report: Mursi's downfall.* July 5, 2013. www.reuters.com/article/us-egypt-protests-downfall-specialreport-idUSBRE9640SP20130705 (accessed December 8, 2016).
Salem, A.A. "The myth and reality of the North Africa versus Sub-Saharan split in the Nile Basin." In *Regional integration in Africa: Bridging the North-Sub-Saharan divide*, edited by H.A. Hassan, 1–15. Pretoria, South Africa: Africa Institute of South Africa, 2011.
Salini Costruttori. "Financial statement, Directors' report." *Salini Impregilo.* December 31, 2010. http://www.salini-impregilo.com/static/upload/sal/salini-costruttori-spa_financial-statements-2010.pdf (accessed November 8, 2016).
Salini Impregilo. *Grand Ethiopian Renaissance Dam Project.* 2014a. http://www.salini-impregilo.com/en/projects/in-progress/dams-hydroelectric-plants-hydraulic-works/grand-ethiopian-renaissance-dam-project.html (accessed December 1, 2016).
—. *Ethiopia: Gibe III Hydroelectric Plant & the Grand Ethiopian Renaissance Dam (GERD) Hydroelectric Power Plant.* 2014b. http://reports.salini-impregilo.com/en/2014-fy/directors-report/operating-performance-geographic-region/ethiopia (accessed December 1, 2016).
—. *Beles multipurpose project.* February 5, 2016. http://www.salini-impregilo.com/en/projects/completed/dams-hydroelectric-plants/beles-multipurpose-project.html (accessed November 8, 2016).
Salman, S. *Law, justice, and development series: The World Bank policy of projects on international waterways.* Washington DC, USA: The World Bank, 2009.
Salman, S.M.A. "The United Nations Watercourses Convention ten years later: Why has its entry into force proven difficult?" *Water International* 32, no. 1 (2007): 1–15.
—. *Law, justice, and development series: The World Bank policy of projects on international waterways.* Washington D.C., USA: The World Bank, 2009.
—. "Grand Ethiopian Renaissance Dam: Challenges and opportunities." *Center for Infrastructure Protection and Homeland Security* 10, no. 4 (October 2011): 21–23, 29.
—. "The Nile Basin Cooperative Framework Agreement: A peacefully unfolding African spring?" *Water International* 38, no. 1 (2013): 17–29.
—. *Analysis of the Fourth Tripartite meeting on the Renaissance Dam.* August 28, 2014. http://sudanow.info.sd/analysis-of-the-fourth-tripartite-meeting-on-the-renaissance-dam/ (accessed December 11, 2016).
—. "The Grand Ethiopian Renaissance Dam: The road to the Declaration of Principles and the Khartoum Document." *Water International* 41, no. 4 (2016): 512–527.
Samaan, M. M. "The win-win-win scenario in the Blue Nile's hydropolitical game: Application on the Grand Ethiopian Renaissance Dam." *Forschungsberichte aus dem Institut für Sozialwissenschaften, TU-Braunschweig, Nr.113.*, 2014
Sambu, K.A. *The Misiri legend explored: A linguistic inquiry into the Kalenjiin People's oral tradition of ancient Egyptian origin.* Nairobi, Kenya: University of Nairobi Press, 2011.
Sampsell, B.M. *The geology of Egypt: A traveler's handbook.* Cairo, Egypt: The American University in Cairo Press, 2014.

Schemm, P. *After 20-year military standoff, Ethiopia and Eritrea agree to normalize ties in historic breakthrough.* July 8, 2018. https://www.washingtonpost.com/world/africa/ethiopias-leader-arrives-in-eritrea-for-unprecedented-summit-between-former-enemies/2018/07/08/a1144f96-826b-11e8-b9a5-7e1c013f8c33_story.html?noredirect=on&utm_term=.6665f611891a (accessed July 13, 2018).

Scheumann, W., I. Dombrowsky, and O. Hensengerth. "Dams on shared rivers: The concept of benefit sharing (chapter 8)." In *The global water system in the Anthropocene: Challenges for science and governance*, edited by A. Bhaduri, J. Bogardi, J. Leentvaar and S. Marx. Heidelberg, Germany: Springer, 2014.

Schmeier, S. *Governing international watercourses: Perspectives from different disciplines, A comprehensive literature review – Working paper no. 53.* Berlin, Germany: Hertie School of Governance, 2010.

Schmeier, S. *Governing international watercourses: River Basin Organizations and the sustainable governance of internationally shared rivers and lakes.* London, UK: Routledge, 2013.

Schwartz, M. *Encyclopedia of coastal science.* Dordrecht, Netherlands: Springer, 2005.

Sela, A. "The 1973 Arab war coalition: Aims, coherence and gain-distribution." In *Revisiting the Yom Kippur War*, edited by P.R. Kumaraswamy, 36–69. London, UK: Frank Cass, 2013.

Shapland, G. *Rivers of discord: International water disputes in the Middle East.* London, UK: C. Hurst & Co. (Publishers) Ltd., 1997.

Sharawi, H. "Israeli policy in Africa." In *The Arabs and Africa*, edited by K. Haseeb, 283–343. London, UK: Routledge, 2012.

Sharman, F. *Linguistic ties between ancient Egyptian and Bantu: Uncovering symbiotic affinities and relationships in vocabulary.* Boca Raton FL, USA: Universal Publishers, 2014.

Sharp, J.M. *Egypt: Background and U.S. relations.* Washington DC, USA: Congressional Research Service, 2009.

Shibl, Y.A. *The Aswan High Dam Project.* Beirut, Lebanon: Arab Institute for Research and Publishing, 1971.

Shinn, D.H., and T.P. Ofcansky. *Historical dictionary of Ethiopia.* Plymouth, UK: Scarecrow Press, 2013.

Shukrallah, S., and Y. Gaber. *What was religion doing in the debate on Egypt's Constitutional amendments?.* March 22, 2011. http://english.ahram.org.eg/NewsContent/1/64/8267/Egypt/Politics-/What-was-religion-doing-in-the-debate-on-Egypts-Co.aspx (accessed November 30, 2016).

Shukry, M.F. *The Khedive Ismail and Slavery in the Sudan 1863–1879.* Cairo, Egypt: Librairie La Resistance, 1938.

Sidahmed, A. "Islamism & the state." In *The Sudan handbook*, edited by J. Ryle, J. Willis, S. Baldo and J.M. Jok, 94–107. Rochester NY, USA: James Currey, 2011.

Sidahmed, A. "Oil and politics in Sudan (Chapter 6)." In *Sudan divided: Continuing conflict in a contested state*, edited by G.M. Sorbo and A.M. Ahmed, 103–120. New York NY, USA: Palgrave MacMillan, 2013.

Simpson, W., and M. Jones. *Europe 1783–1914.* New York NY, USA: Routledge, 2009.

SIWI. "Independent evaluation of the Nile Basin Trust Fund (NBTF): Final report by the Stockholm International Water Institute." *World Bank.* June 28, 2013. http://www-wds.worldbank.org/external/default/WDSContentServer/WDSP/IB/2014/07/10/000469252_20140710152723/Rendered/PDF/892770WP0P09230on0of0NBTF0June02013.pdf (accessed July 8, 2015).

Sneddon, C. *Concrete revolution: Large dams, Cold War geopolitics, and the U.S. Bureau of Reclamation.* Chicago IL, USA: University of Chicago Press, 2015.

Snidal, D. "The politics of scope: Endogenous actors, heterogeneity and institutions." In *Local commons and global interdependence*, edited by R.O. Keohane and E. Ostrom, 47–70. London, UK: Sage Publications, 1995.

Sokhem, P. "Share case studies: Mekong Agreement lacks requirements for national rules." *IUCN.* March 15, 2010. https://cmsdata.iucn.org/downloads/mekong_2.pdf (accessed December 22, 2015).

Spector, L., and G. Gruen. "Waters of controversy: Implications for the Arab-Israel peace process (Draft)." *American Jewish Committee*, 1980: 1–11.
Starr, J. "Water wars." *Foreign Policy* 82 (1991): 17–36.
Starr, J.R., and D.C. Stoll. *The politics of scarcity: Water in the Middle East.* Boulder CO, USA: Westview Press, 1988.
Stiglitz, J. *Globalization and its discontents.* New York NY, USA: Norton, 2003.
Stokstad, E. "Power play on the Nile: Ethiopia stunned neighbors with a colossal dam on the Blue Nile – Could it spread prosperity, not turmoil?" *Science* 351, no. 6276 (2016): 904–907.
Sudan Tribune. *Museveni urges Egypt to agree to equitable use of Nile waters.* April 1, 2004. http://www.sudantribune.com/spip.php?article2304 (accessed November 23, 2016).
—. *Sudan, China sign $396mln contract to heighten Roseires Dam.* April 28, 2008. http://www.sudantribune.com/spip.php?article26912 (accessed November 12, 2016).
—. *Ethiopia inaugurates Africa's first ever biggest power dam.* November 15, 2009. http://www.sudantribune.com/spip.php?article33117 (accessed November 7, 2016).
—. *Sudan downplays negative impact of Ethiopian dam project.* May 30, 2013a. http://www.sudantribune.com/spip.php?article46754 (accessed December 5, 2016).
—. *Nile Basin nations call for stronger relations.* June 23, 2013b. http://www.sudantribune.com/spip.php?article47051 (accessed December 7, 2016).
—. *Sudan affirms the internal nature of events in Egypt.* July 6, 2013c. http://www.sudantribune.com/spip.php?article47196 (accessed December 8, 2016).
—. *Sudan's support to Ethiopian dam is economically motivated: Bashir.* December 5, 2013d. http://www.sudantribune.com/spip.php?article49070 (accessed July 8, 2015).
—. *Sudan FM Criticises Egypt over Ethiopian Dam Dispute.* February 19, 2014a. http://www.sudantribune.com/spip.php?article50013 (accessed March 17, 2014).
—. *Sudan Slams Egyptian Media's Provocation over Ethiopian Dam.* May 9, 2014b. http://www.sudantribune.com/spip.php?article50942 (accessed May 11, 2014).
—. *Preliminary agreement reached on Ethiopia's Renaissance dam.* March 6, 2015a. http://sudantribune.com/spip.php?article54203 (accessed December 12, 2016).
—. *Ethiopia, Egypt vow to resolve Nile water dispute.* March 25, 2015b. http://www.sudantribune.com/spip.php?article54389= (accessed December 12, 2016).
—. *Eritrean president to visit Sudan on Thursday: state media.* June 11, 2015c. http://www.sudantribune.com/spip.php?article55296 (accessed December 13, 2016).
—. "Saudi Arabia to provide 1.7 billion dollars for Sudan's dam projects." *Sudan Tribune.* November 6, 2015d. http://www.sudantribune.com/spip.php?article56960 (accessed December 17, 2016).
—. *Sudan, Egypt and Ethiopia reach agreement on Renaissance Dam.* December 30, 2015e. http://www.sudantribune.com/spip.php?article57526 (accessed December 15, 2016).
—. "Chinese companies to grow one million feddans of cotton in Sudan." *Sudan Tribune.* August 18, 2016a. http://dev.sudantribune.com/All/Article/Index/8-18-2016-Chinese-companies-to-grow-one-million-feddans-of-cotton-in-Sudan/59954 (accessed December 17, 2016).
—. *Sudan lodges new complaint over Halayeb triangle: minister.* October 24, 2016b. http://www.sudantribune.com/spip.php?article60642 (accessed December 15, 2016).
—. *Egypt denies demanding Sudan exclusion from talks with Ethiopia over dam project.* January 2, 2018a. http://www.sudantribune.com/spip.php?article64407 (accessed July 9, 2018).
—. *Sudan recalls ambassador to Egypt for consultations.* January 5, 2018b. http://www.sudantribune.com/spip.php?article64418 (accessed July 10, 2018).
—. *Sudan speaks about Egyptian and Eritrean threats to its security.* January 12, 2018c. http://www.sudantribune.com/spip.php?article64470 (accessed July 10, 2018).
Swain, A. "Ethiopia, the Sudan, and Egypt: The Nile River dispute." *Journal of African Studies* (Cambridge University Press) 35, no. 4 (1997): 675–694.
—. *Managing water conflict: Asia, Africa and the Middle East.* New York NY, USA: Routledge, 2004.

T.B., interview by M.M. Samaan. *Face-to-face interview with an adviser to the Ethiopian Ministry of the Water Resources, Irrigation and Electricity during the Stockholm World Water Week* (September 1, 2016).

Taha, F. "The history of the Nile waters in the Sudan." In *The River Nile in the post-colonial age: Conflict and cooperation among the Nile Basin countries*, edited by T. Tvedt, 179–216. New York NY, USA: I.B. Tauris, 2010.

Taha, R.M. *Tamarod members shortly detained in Zagazig.* May 15, 2013. http://www.dailynews-egypt.com/2013/05/15/tamarod-members-shortly-detained-in-zagazig/ (accessed December 6, 2016).

Tal, D., and S. Tucker. "Overview of the Arab-Israeli Conflict." In *The encyclopedia of the Arab-Israeli Conflict: A political, social, and military history – Volume I: A-F*, edited by S. Tucker, 5–13. Santa Barbara CA, USA: ABC-CLIO, 2008.

Tarek, S. *Islamists win 70% of Egypt People's Assembly party list seats.* January 21, 2012. http://english.ahram.org.eg/NewsContent/33/0/32287/Elections-/0/Islamists-win%2D%2Dof-Egypt-Peoples-Assembly-party-lis.aspx (accessed November 30, 2016).

Tawfik, R. *Revisiting hydro-hegemony from benefit-sharing perspective: The case of the Grand Ethiopian Renaissance Dam – Discussion paper.* Bonn, Germany: Deutschs Institut fuer Entwicklungspolitik, 2015.

Taylor, R. "Rethinking water scarcity: The role of storage." *Transactions of the American Geophysical Union* 90, no. 28 (2009): 237–238.

Tekle, T.A. *Sudan agrees to tripartite committee over Ethiopia's Nile dam.* October 26, 2011. http://www.sudantribune.com/Sudan-agrees-to-tripartite,40531 (accessed December 3, 2016).

—. *Sudan's Bashir supports Ethiopia's Nile dam project.* April 5, 2012. http://www.sudantribune.com/spip.php?article41839 (accessed December 5, 2016).

—. *Foreign consultancy firms selected to study Ethiopia's Nile dam.* April 11, 2015. http://www.sudantribune.com/spip.php?article54583 (accessed December 14, 2016).

TFDD. *International water events: The Nile Basin.* 2008. http://gis.nacse.org/tfdd/event.php?BCODE_Input=NILE& (accessed October 7, 2016).

The Annual Register. *The Annual Register: A review of public events at home and abroad for the year 1924.* Edited by M. Epstein. Volume 166. London, UK: Longmans, Green and Co., 1925.

The Guardian. *Egyptian elections: Opposition alleges fraud – Ruling party accused of manipulating vote to ensure a sweeping victory in parliamentary elections.* November 29, 2010. https://www.theguardian.com/world/2010/nov/29/egyptian-opposition-alleges-election-fraud (accessed November 29, 2016).

—. *Egypt protests – as they happened.* January 29, 2011. https://www.theguardian.com/news/blog/2011/jan/30/egypt-protests-live-updates (accessed November 29, 2016).

Thomson, M. *Nile restrictions anger Ethiopia.* February 3, 2005. http://news.bbc.co.uk/2/hi/africa/4232107.stm (accessed November 23, 2016).

Tignor, R.L. *Egypt: A short history.* Princeton NJ, USA: Princeton University Press, 2011.

Timberlake, L. *Africa in crisis: The causes, the cures of environmental bankruptcy.* New York NY, USA: Earthscan, 1985/2013.

Turton, A.R. *Post-Apartheid institutional development in South African international river basins.* Pretoria, South Africa: Center for International Political Studies, University of Pretoria, 2003.

Turton, A.R., R. Meissner, P.M. Mampane, and O. Seremo. "A hydropolitical history of South Africa's international river basins: Report to the Water Research Commission." *Water Research Commission.* 2004. http://www.wrc.org.za/Knowledge%20Hub%20Documents/Research%20Reports/1220-1-04.pdf (accessed December 26, 2015).

Tvedt, T. *The River Nile in the age of the British: Political ecology and the quest for economic power.* London, UK: I.B. Tauris, 2004.

U.S. Department of State. *Papers relating to the foreign relations of the United States, 1928.* Volume 2. Washington DC, USA: U.S. Department of State, 1943.

—. *International Organization and Conference – Part 2, Issue 1.* Washington DC, USA: U.S. Department of State, 1956.

U.S. Senate Committee on Foreign Relations. *The Middle East between war and peace, November-December 1973 – A staff report.* Washington DC, USA: The U.S. Senate Committee on Foreign Relations, 1974.

UN. *Security Council Resolutions – 1996.* 1996. http://www.un.org/Docs/scres/1996/scres96.htm.230802 (accessed October 23, 2016).

—. *Press release: Security Council lifts sanctions against Sudan.* September 28, 2001. http://www.un.org/press/en/2001/sc7157.doc.htm (accessed October 24, 2016).

—. *Chapter XXVII: Environment, Convention on the Law of the Non-Navigational Uses of International Watercourses, Signature Status, United Nations.* 2014. https://treaties.un.org/Pages/ViewDetails.aspx?src=TREATY&mtdsg_no=XXVII-12&chapter=27&lang=en (accessed February 17, 2014).

UNEP. *Hydropolitical vulnerability and resilience along international waters: Africa.* Sioux Falls SD, USA: The United Nations Environment Programme, 2005.

UN-Water. "National water development report: Uganda – UN-Water/WWAP/2006/9." *UNESCO.* 2006. http://unesdoc.unesco.org/images/0014/001467/146760e.pdf (accessed October 19, 2016).

—. "Transboundary waters: Sharing benefits, sharing responsibilities (UN thematic paper)." *United Nations – Water.* 2008. http://www.unwater.org/downloads/UNW_TRANSBOUNDARY.pdf (accessed July 8, 2015).

USBR. *Land and water resources of the Blue Nile Basin, Ethiopia – Main report.* Washington DC, USA: Bureau of Reclamation, U.S. Department of the Interior, 1964a.

—. *Land water resources of the Blue Nile Basin, Ethiopia – Appendix I: Plans and estimates.* Washington DC, USA: Bureau of Reclamation, U.S. Department of the Interior, 1964b.

van der Zaag, P., I.M. Seyam, and H.G. Savenije. "Towards measurable criteria for the equitable sharing of international water resources." *Water Policy* 4, no. 1 (2002): 19–32.

Van Evera, S. *Guide to methods for students of Political Science.* London, UK: Cornell University Press, 1997.

Vassiliev, A. "The 1973 War (Chapter 23)." In *King Faisal: Personality, faith and times.* London, UK: Saki Books, 2012.

Veilleux, J.C. "Human security dimensions of dam development: The Grand Ethiopian Renaissance Dam." *Global Dialogue* 15, no. 2 (2013).

Wahish, N. *Toshka turns millennial green.* August 27, 1998. http://weekly.ahram.org.eg/archive/1998/392/ec1.htm (accessed October 30, 2016).

Wakabi, M. *Uganda: Country to test-run Kiira power station.* January 2, 2007. http://allafrica.com/stories/200701020590.html (accessed October 19, 2016).

Waldyes, T. *Ethiopia: Project X – Turning the energy tide.* February 7, 2011. http://allafrica.com/stories/201102080902.html (accessed November 30, 2016).

Waltz, K.N. *Theory of International Politics.* Long Grove IL, USA: Waveland Press, 1979/2010.

Ward, H. "Rational Choice (Chapter 3)." In *Theory and methods in political science*, edited by David Marsh and Gerry Stoker, 65–89. Palgrave Macmillan, 2002.

Ward, W.E.F., and L.W. White. *East Africa: A century of change 1870–1970.* New York NY, USA: Africana Publishing Corporation, 1972.

Warner, J. "The Toshka mirage in he Egyptian desert: River diversion as political diversion." *Environmental Science and Policy* 30 (2013): 102–112.

Water Technology. *Toshka project – Mubarak pumping station, Egypt.* n.d. http://www.water-technology.net/projects/mubarak/ (accessed October 30, 2016).

Waterbury, J. *Hydropolitics of the Nile Valley.* New York NY, USA: Syracuse University Press, 1979.

—. *The Nile Basin: National determinants of collective action.* Ann Arbor, MI: Yale University, 2002.

Waterbury, J., and D. Whittington. "Playing chicken on the Nile? The implications of microdam development in the Ethiopian Highlands and Egypt's New Valley project." *Middle Eastern Natural Environments*, 1999: 150–167.

Watkins, K. *Human Development Report 2006 – Beyond scarcity: Power, poverty and the global water crisis*. New York: United Nations, Human Development Report Office, 2006.

WCD. *Dams and development: A new framework for decision-making*. World Commission on Dams, Earthscan, 2000.

WDI. *World Development Indicators*. 2016. http://data.worldbank.org/data-catalog/world-development-indicators (accessed October 26, 2016).

WEF. *Water security: The Water-Energy-Food-Climate nexus, The World Economic Forum Water Initiative*. London, UK: Island Press, 2011.

Weir, M. "External relations." In *Egypt: Internal challenges and regional stability*, edited by L.C. Harris, 77–96. London, UK: Routledge and Kegan Paul, and Taylor and Francis e-Library, 1988/2005.

Westing, A.H., ed. *Global resources and international conflict: Environmental factors in strategic policy and action*. New York NY, USA: Oxford University Press, 1986.

Whittington, D. "Visions of Nile Basin development." *Water Policy* 6 (2004): 1–24.

Whittington, D., J. Waterbury, and M. Jeuland. "The Grand Renaissance Dam and prospects for cooperation on the Eastern Nile." *Water Policy* 16, no. 4 (2014): 595–608.

Whittington, D., X. Wu, and C.W. Sadoff. "Water resources management in the Nile Basin: The economic value of cooperation." *Water Policy* 7 (2005): 227–252.

Wichelns, D., J. Barry, M. Mueller, M. Nakao, L.D. Philo, and A. Zitello. "Co-operation regarding water and other resources will enhance economic development in Egypt, Sudan, Ethiopia and Eriterea." 19, no. 4 (2003): 535–552.

Wight, M. *International Theory: The three traditions*. Edited by G. Wight and B. Porter. Leicester, UK: Holmes & Meier for the Royal Institute of International Affairs, 1992.

Willis, J., O. Egemi, and P. Winter. "Early states on the Nile." In *The Sudan handbook*, edited by J. Ryle, J. Willis, S. Baldo and J.M. Jok, 10–22. Rochester NY, USA: James Currey, 2011.

Winder, J. *Unpublished reports on the Jonglei scheme*. Durham NC, USA: SAD, n.d.

Wishart, D.H. "The breakdown of the Johnston negotiations over the Jordan waters." *Middle Eastern Studies* 26, no. 4 (1990): 536–546.

Withana, R. *Power, politics, law: International law and state behaviour during international crises*. Leiden, Netherlands: Martinus Nijhoff Publishers, 2008.

Wittfogel, K. A. *Oriental despotism: A comparative study of total power*. New Haven CT, USA: Yale University Press, 1957.

Woldetsadik, T.K. *International watercourses law in the Nile River Basin: Three states at a crossroads*. New York NY, USA: Routledge, 2013.

—. "The Nile Basin Initiative and the Cooperative Framework Agreement: Failing institutional enterprises?" *SSRN*. July 1, 2016. https://papers.ssrn.com/sol3/papers.cfm?abstract_id=2803306 (accessed November 22, 2016).

Wolf, A.T. *Hydropolitics along the Jordan River: Scarce water and its impact on the Arab-Israeli Conflict*. New York NY, USA: United Nations University Press, 1995.

—. Criteria for equitable allocations: The heart of international water conflict." *Natural Resources Forum, a United Nations Sustainable Development Journal* 23, no. 1 (1999): 3–30.

—. "'Hydrostrategic' territory in the Jordan Basin: Water, war and Arab-Israeli peace negotiations." In *Water in the Middle East: A geography of peace*, edited by A.T. Wolf and H. Amery. Austin, USA: University of Texas Press, 2000.

—. "Shared waters: Conflict and cooperation." *The Annual Review of Environmental Resources* 32 (2007): 3.1–3.29.

Wolf, A.T., and J.T. Newton. "Case studies of transboundary dispute resolution (Appendix C)." In *Managing and transforming water conflicts*, edited by J. Delli Priscoli and A.T. Wolf, 169–248. Cambridge, UK: Cambridge University Press, 2009.

Woodward, P. *The Horn of Africa: Politics and international relations.* New York NY, USA: I.B. Tauris, 2003.

—. "Sudan's fragile state, 1956–1989." In *The Sudan handbook*, edited by J. Ryle, J. Willis, S. Baldo and J.M. Jok, 85–93. Rochester NY, USA: James Currey, 2011.

—. *U.S. foreign policy and the Horn of Africa.* New York NY, USA: Routledge, 2016.

World Bank. "The Inspection Panel report and recommendation on request for inspection – Uganda: Third Power Project (Credit 2268-UG) and the proposed Bujagali hydropower project." *World Bank.* 2001. http://siteresources.worldbank.org/EXTINSPECTIONPANEL/Resources/UgandaEligibilty.pdf (accessed October 20, 2016).

—. "Project Information Document (PID): Appraisal stage – UG Power generation project: Bujagali." *World Bank.* 2007. http://documents.worldbank.org/curated/en/555991468110935737/pdf/BujaProject0In1aisal0Stage312612007.pdf (accessed October 20, 2016).

—. *What is governance?* 2013a. http://web.worldbank.org/WBSITE/EXTERNAL/COUNTRIES/MENAEXT/EXTMNAREGTOPGOVERNANCE/0,,contentMDK:20513159~pagePK:34004173~piPK:34003707~theSitePK:497024,00.html (accessed February 28, 2016).

—. *Collective water management in Western Africa.* March 29, 2013b. http://www.worldbank.org/en/results/2013/03/29/collective-water-management-in-western-africa (accessed January 2, 2016).

—. *International Bank for Reconstruction and Development.* September 30, 2014. http://www.worldbank.org/en/about/what-we-do/brief/ibrd (accessed September 9, 2016).

—. *Nile Basin Trust Fund.* January 14, 2016. http://www.worldbank.org/en/programs/cooperation-in-international-waters-in-africa/brief/nile-basin-trust-fund (accessed December 19, 2016).

World Bulletin. *Ethiopia dam to start yielding power by late 2015.* March 13, 2014. http://www.worldbulletin.net/world/130950/ethiopia-dam-to-start-yielding-power-by-late-2015 (accessed July 8, 2015).

Wright, P. *Conflict on the Nile: The Fashoda incident of 1898.* London, UK: Heinemann, 1972.

Wu, X., and D. Whittington. "Incentive compatibility and conflict resolution in international river basins: A case study of the Nile Basin." *Water Resources Research* 42 (2006): 1–15.

Yoffe, S., A.T. Wolf, and M. Giordano. "Conflict and cooperation over international freshwater resources: Indicators of Basins At Risk." Edited by http://www.transboundarywaters.orst.edu/publications/abst_docs/Yoffe_Wolf_Giordano.pdf. *Journal of the American Water Resources Associations* Paper no. 02036 (2003): 1109–1126.

Zagare, F.C., and D.M. Kilgour. "Asymmetric deterrence." *International Studies Quarterly* 37 (1993): 1–27.

Zawahri, N. "Stabilizing Iraq's water supply: What the Euphrates and Tigris Rivers can learn from the Indus." *Third World Quarterly* 27, no. 6 (2006): 1041–1058.

—. "International rivers and national security: The Euphrates, Ganges-Brahmaputra, Indus, Tigris, and Yarmouk rivers." *Natural Resources Forum* 32 (2008): 280–289.

Zeitoun, M., J.A.T. Allan, and Y. Mohieldeen. "Virtual water 'flows' of the Nile Basin, 1998–2004: A first approximation and implications of water security." *Global Environmental Change* 20 (2010): 229–242.

Zeitoun, M., and N. Mirumachi. "Transboundary water interaction I: Reconsidering conflict and cooperation." *International Environmental Agreements: Politics, Law and Economics* 8, no. 4 (2008): 297–316.

Zeitoun, M., and J. Warner. "Hydro-hegemony: A framework for analysis of transboundary water conflicts." *Water Policy* 8 (2006): 435–460.

Zeitoun, M., J. Warner, N. Mirumachi, N. Matthews, K. McLaughlin, M. Woodhouse, A.E. Cascao, and J.A. Allan. "Transboundary water justice: A combined reading of literature on critical transboundary water interaction and "justice", for analysis and diplomacy." *Water Policy* 16 (2014): 174–193.

Zunes, S., and S. Tucker. "Aswan High Dam." In *The encyclopedia of Middle East wars: The United States in the Persian Gulf, Afghanistan, and Iraq conflicts (Volume II)*, edited by S. Tucker, 166–168. Oxford, UK: ABC-CLIO, LLC, 2010.

MIX
Papier aus verantwortungsvollen Quellen
Paper from responsible sources
FSC® C105338

If you have any concerns about our products,
you can contact us on
ProductSafety@springernature.com

In case Publisher is established outside the EU,
the EU authorized representative is:
Springer Nature Customer Service Center GmbH
Europaplatz 3, 69115 Heidelberg, Germany

Printed by Libri Plureos GmbH
in Hamburg, Germany